Nanomanufacturing and Nanomaterials Design

Nanomanufacturing includes bottom-up or top-down techniques, each of which gives an advanced, reliable, scaled-up and economical method in the production of nanomaterials. The text discusses fundamental concepts, advanced topics and applications of nanomanufacturing in a comprehensive manner.

Features

- Discussion of the design and fabrication of nano- and micro-devices in a comprehensive manner.
- Covers nanofabrication techniques for photovoltaics applications.
- Lists constitutive modelling and simulation of multifunctional nanomaterials.
- Introduces nanomanufacturing of nanorobots and their industrial applications.
- Presents nanomanufacturing of a high-performance piezoelectric nano-generator for energy harvesting.

Important topics include nanomanufacturing of high-performance piezoelectric nanogenerators for energy harvesting, nanosensor, nanorobots, nanomedicine, nano diagnostic tools, 3D nano printing, additive nanomanufacturing of functional materials for human-integrated smart wearables and nanofabrication techniques. *Nanomanufacturing and Nanomaterials Design* covers the latest applications of nanomanufacturing for a better understanding of the concepts.

The text provides scientific and technological insights on novel routes of design and fabrication of few-layered nanostructures and their heterostructures based on a variety of advanced materials. It will be a valuable resource for senior undergraduate, graduate students and researchers in the fields of mechanical, manufacturing, industrial, production engineering and materials science.

Nanomanufacturing and Nanomaterials Design
Principles and Applications

Edited by
Subhash Singh, Sanjay K. Behura,
Ashwani Kumar, and Kartikey Verma

CRC Press
Taylor & Francis Group
Boca Raton London New York

CRC Press is an imprint of the
Taylor & Francis Group, an **informa** business

Front cover image: Phonlamai Photo/Shutterstock

First edition published 2023
by CRC Press
6000 Broken Sound Parkway NW, Suite 300, Boca Raton, FL 33487-2742

and by CRC Press
4 Park Square, Milton Park, Abingdon, Oxon, OX14 4RN

CRC Press is an imprint of Taylor & Francis Group, LLC

ISBN: 978-1-032-08168-7 (hbk)
ISBN: 978-1-032-11586-3 (pbk)
ISBN: 978-1-003-22060-2 (ebk)

DOI: 10.1201/9781003220602

Typeset in Times
by MPS Limited, Dehradun

Dedication

"This book is dedicated to all engineers, researchers and academicians.."

Contents

Preface...ix
Acknowledgements...xiii
About the Editors xv
List of Contributors xvii

Chapter 1   Introduction and Origin to Nanomanufacturing.................................. 1

                S.V. Satya Prasad, S.B. Prasad, and Subhash Singh

Chapter 2   Challenges and Opportunities in Nanomanufacturing 17

                Subhash Singh, Ashwani Kumar, Sanjay K. Behura,
                and Kartikey Verma

Chapter 3   Intense Classification of Nanomanufacturing................................... 31

                H.J. Amith Yadav, D.H. Manjunatha, D. Gayathri, V.S. Patil,
                and M.N. Kalasad

Chapter 4   Experimental Investigation and Multi-Response
                Optimization of End Milling Process Parameters for
                Surface Integrity on Al7075-B_4C-BN Nanocomposites 45

                N. Zeelanbasha, M. Selvakumar, T. Ramkumar,
                and V. Sivananth

Chapter 5   Design and Manufacturing of Nano Sensors: Perspective
                and Applications.. 59

                Souvik Bag

Chapter 6   3D Nano Printing: Current Status and Emerging Trends of a
                Novel Fabrication Technique and Its Industrial Applications
                in Biomedicines.. 73

                F. Ahmad, A. Nazeer, and S. Ahmad

Chapter 7   Nanomanufacturing of Biomedicines: Current Status
                and Future Challenges... 95

                F. Ahmad, A.K. Ogra, and S. Ahmad

Chapter 8   Experimental Investigation on Spark Behaviour of
                ECDM for Potential Application in Nanofabrication 137

                Girija Nandan Arka, Mamata Kumari, and Subhash Singh

Chapter 9   Frequency Sensitivity Performance Analysis of Single-Layer and Multi-Layer SAW-Based Sensor Using Finite Element Method 149

Baruna Kumar Turuk and Basudeba Behera

Chapter 10   Nanomanufacturing for Energy Conversion and Storage Devices...... 165

Shubham Srivastava, Deepti Verma, Shreya Thusoo, Ashwani Kumar, Varun Pratap Singh, and Rajesh Kumar

Chapter 11   Nanofabrication Techniques for Solar Photovoltaic Applications 175

Girija Nandan Arka, S.B. Prasad, and Subhash Singh

Chapter 12   Emerging Nanomanufacturing Techniques with 2D Materials 189

Mamta Kumari, Ashok Kumar Jha, and Subhash Singh

Chapter 13   Biodegradable and Biocompatible Polymeric Nanocomposites for Tissue Engineering Applications 205

Sahana S. Sringari, Gourhari Chakraborty, and Arbind Prasad

Chapter 14   Design and Manufacturing of Nanorobots and Their Industrial Applications 227

Abhishek Shrivastava and Vijay Kumar Dalla

Chapter 15   Nanomanufacturing and Design of High-Performance Piezoelectric Nanogenerator for Energy Harvesting 241

Varun Pratap Singh, Ayush Dwivedi, Ashish Karn, Ashwani Kumar, Subhash Singh, Shubham Srivastava, and Kashika Srivastava

Index 273

Preface

Application-centric nanomaterials are discovered chronologically with technological civilization across the globe. Numerous technologies are entertained to engineer qualitative nanomaterials for the productive application in a wide range of application. Prominently, numerous inspiring properties of nanomaterials are promisingly incorporated to produce nanocomposite, nanomedicine and many more. Therefore, advanced research and development work in nanoscience and nanotechnology has impacted our lives in many ways by generating a new fundamental understanding of physical phenomena and material behaviour at the nanoscale. This knowledge has represented a scientific breakthrough in designing and development of new materials, structures and devices. Fabrication processes that can reliably produce features and structures with nanoscale resolution are critically important for nanoscale manufacturing and many other emerging applications. A comprehensive representative set of nanoscale technologies in processes, materials, pharmaceuticals, electronics, instruments, aerospace, automobile, photonic devices, robotics, renewable energy, biosensors, biomedicine and biotechnology have reached or are currently introduced to mass production. On the other hand, nanomaterials are playing a significant role in nanomanufacturing of newly developed devices for various novel practical applications. This is of particular importance to enthusiasts who are in search of materials with unique properties and with unmatched performance under testing conditions. The research and development being carried out in this particular area has led to novel material designs with optimized properties and to distinct cutting-edge routes of nanomanufacturing for the development and processing of such materials as well as devices.

Adhering to the above objectives, the chapters in the proposed book are meticulously planned on fundamentals and advances in the field of nanomaterials. The latest manufacturing techniques properties and applications of nanomaterials have been presented. The nanomaterials produced at nano scales possess unique and superior properties compared to conventional materials. Therefore, their design scenario and the modern manufacturing approaches for their synthesis is a highly important area requiring interdisciplinary efforts. This book unfolds on the topics related to the origin and the introduction of nanomanufacturing. It also enlists the importance of nanomanufacturing in the present day. Techniques such as additive or subtractive manufacturing, mass or replication conservation methods, top-down and bottom-up approaches for nanomanufacturing have been addressed in Chapter 1 However, numerous inherent challenges are encountered to transform nanomanufacturing. Therefore this book Chapter 2 dedicatedly discussed the numerous challenges and opportunities in nanomanufacturing for eminent applications. Furthermore, these nanomanufacturing processes could be stacked under several criteria based on familiar confinement. Thus, a comprehensive study has been exercised to explore nanomanufacturing and classified the different nanomanufacturing processes to select the based alternative for the productive outcome. Chapter 3 has highlighted above mention comprehensive fruitful discussion. Once the nanomanufacturing is decided, practical manufacturing could be a crucial practice to transit theory to reality. Thus Chapter 4 magnanimously discussed numerous design criteria to establish nanomanufacturing and vibrantly addressed the prosperity of economically viable nanomanufacturing techniques. Since the nanomanufacturing process could be governed by application-based nanomaterial. Thus Chapter 5 encouraged to discuss different applications of nanosensors and support elaborated discussion on different nanomanufacturing technology for producing nanosensors. Further, the different techniques have been highlighted to elevate the response of the gas sensors in-depth. This has been noticed that robots play a vital role in industry to make an accurate transition in hazardous surroundings or exceptional nano movements to establish superior end effects. Thus, nanorobots could play a significant role in the industry. Therefore Chapter 6 vibrantly addressed numerous designing parameters to make productive nanorobots and enlarged discussion on the nanorobot application in industries. Similarly, nanomedicine is acknowledged worldwide for its productive outreach. However, numerous challenges are countered to fabricate nanomedicine. To comprehend nanomedicine confinement Chapter 7 discussed designing parameters to manufacture nanomedicine and highlighted numerous manufacturing processes. Moreover, numerous challenges and opportunities also have been incorporated for exploring the determinant research scope. Nanomanufacturing applications also move towards electrical applications such as nanoelectromechanical systems. Chapter 8 discussed designing parameters and considerations to make nanoelectromechanical systems and their manufacturing process. Practical machining by electrochemical machining performed to justify nanomanufacturing potential exercised was narrated in Chapter 9 Moreover, electrochemical machining has huge potential to address nanomanufacturing. Nanomaterial application is remarkably exercised in energy conversion and storage devices for promising performance. Thus, Chapter 10 is devoted to highlight nanomanufacturing process to concoct energy conversion devices and storage devices. The study revealed that nanomaterials civilize the solar cells' application to a greater extent by showing promising results relative to the conventional solar cell. Thus, prolific nanomanufacturing approaches have been studied to comprehend nanofilm generation, challenges and opportunities for the promising application in photovoltaic solar cells application described in Chapter 11 Since 2D materials are acknowledged across the globe for their exceptional amicable properties, thus have been promisingly incorporated in numerous fields of application. Therefore Chapter 12 is devoted to exploring emerging nanomanufacturing techniques to produce productive 2D materials.

   The intention behind writing this book is to make it suitable for anyone willing to study about nanomanufacturing and nanomaterials design confined to principles and applications. The fundamental addressees of this book would be fruitful for the undergraduate and postgraduate students in diverse fields of engineering like Mechanical, Metallurgy, Manufacturing and Material Science, Biomedical, research scholars and scientists working in various research centers. Professors and lecturers working in various universities will also find this book extremely useful. Experts working in various industries related to the field of manufacturing and materials along with editors, reviewers and reading enthusiasts of numerous journals would find this book informative and interesting. Hence, it is trusted that this book would motivate and enthuse its readers to research this particular field of engineering.

Editors:
Dr. Subhash Singh
Dr. Sanjay K. Behura
Dr. Ashwani Kumar
Dr. Kartikey Verma

Acknowledgements

We express our gratitude to **CRC Press (Taylor & Francis Group)** and the editorial team for their suggestions and support during the completion of this book. We are grateful to all contributors and reviewers for their illuminating views on each book chapter presented in the book *"Nanomanufacturing and Nanomaterials Design: Principles and Applications."*

About the Editors

Dr. Subhash Singh is currently working as Associate Professor in the Department of Mechanical and Automation Engineering, Indira Gandhi Delhi Technical University for Women, New Delhi. He earned his B. Tech from KNIT, Sultanpur, 2003 and M. Tech from NIT Kurukshetra, 2007. He has completed Ph.D from IIT Roorkee in 2017. Dr. Singh specializes in areas such as modification of nanomaterials, thin coating, fabrication of MMCs, synthesis of 2D Materials, FSP, machining of biodegradable materials. He has published more than thirty research papers in various prestigious journals. He has also authored one book and twenty book chapters. He has completed three research projects.

Dr. Sanjay K. Behura is currently working as Assistant Professor in the Department of Physics, San Diego State University San Diego, USA. He has more than fifteen years of research and academic experience. He has received funding for seven international projects. He has published more than thirty research papers in various prestigious journals. He has applied for two patents which are in process of granting.

Dr. Ashwani Kumar received Ph.D. (Mechanical Engineering) in the area of Mechanical Vibration and Design. He is currently working as Senior Lecturer, Mechanical Engineering (Gazetted Officer Group B) at Technical Education Department, Uttar Pradesh (Under Government of Uttar Pradesh) Kanpur, India since December 2013. He has worked as Assistant Professor in the Department of Mechanical Engineering, Graphic Era University Dehradun India from July 2010 to November 2013. He has twelve years of research and academic experience in mechanical and materials engineering. He is the Series Editor of the book series "Advances in Manufacturing, Design and Computational Intelligence Techniques" published by CRC Press, Taylor & Francis, USA. He is Associate Editor for the International Journal of Mathematical, Engineering and Management Sciences (IJMEMS) Indexed in ESCI/Scopus and DOAJ. He is an editorial board member of 4 international journals and acts as a review board member of 20 prestigious (Indexed in SCI/SCIE/Scopus) international journals with high impact factors i.e., Applied Acoustics, Measurement, JESTEC, AJSE, SV-JME and LAJSS. In addition, he has published 90 research articles in journals, book chapters and conferences. He has authored/co-authored cum edited 20 books of Mechanical and Materials Engineering. He is associated with International Conferences as Invited Speaker/Advisory Board/Review Board member. He has delivered many invited talks in webinars, FDP and Workshops. He has been awarded as Best Teacher for excellence in academics and research. He has successfully guided 12 B.Tech. M.Tech and Ph.D. thesis. In administration he is working as coordinator for AICTE, E.O.A., Nodal officer for PMKVY-TI Scheme (Government of India) and internal coordinator for CDTP scheme (Government of Uttar Pradesh). He is currently involved in the research area of AI & ML in Mechanical Engineering, Advanced Materials and Manufacturing Techniques, Renewable Energy Harvesting, Heavy Vehicle Dynamics and

Sustainable Transportation. His current Google Scholar h-index is 10, RG index is 25 and ORCID ID is 0000-0003-4099-935X.

Dr. Kartikey Verma is currently working as Assistant Professor in the Department of Chemical Engineering, IIT Kanpur India. He has completed his B.Sc. from Dr. R.M.L Avadh University Faizabad, 2007 and M. Sc. from University of Lucknow, 2009. He has completed his Ph.D from University of Lucknow in 2015. Dr. Verma specializes in areas of Thin Films, Polymer Nanocomposites and Energy Storage Materials. He has published more than twenty research papers in various prestigious journals. He has also authored one book and many book chapters. Dr. Verma Received Young Scientist Award under the fast track scheme of Department of Science and Technology, New Delhi in 2015.

Contributors

F. Ahmad
Iris Worldwide Gurugram
Haryana, India

S. Ahmad
Eros Garden Charmwood Village
Faridabad, Haryana, India

S. Ahmad
Hamdard University
N. Delhi, India

Girija Nandan Arka
National Institute of Technology
Jamshedpur
Jharkhand, India

Souvik Bag
Ulsan National Institute of Science and
Technology (UNIST)
Ulsan, South Korea

Basudeba Behera
National Institute of Technology
Jamshedpur, Jharkhand, India

Dr. Sanjay K. Behura
Department of Physics
San Diego State University
San Diego, USA

Gourhari Chakraborty
National Institute of Technology
Andhra Pradesh, India

Vijay Kumar Dalla
National Institute of Technology
Jamshedpur, Jharkhand, India

Ayush Dwivedi
University of Petroleum and Energy
Studies
Dehradun, Uttarakhand, India

D. Gayathri
Davangere University
Davangere, India

Ashok Kumar Jha
National Institute of Technology
Jamshedpur, Jharkhand, India

M.N. Kalasad
Davangere University
Davangere, India

Ashish Karn
University of Petroleum and Energy
Studies
Energy Acres
Dehradun, Uttarakhand, India

Dr. Ashwani Kumar
Technical Education Department
Uttar Pradesh, Kanpur, India

Rajesh Kumar
Indian Institute of Technology (BHU)
Varanasi, India

Mamta Kumari
National Institute of Technology
Jamshedpur
Jharkhand, India

D.H. Manjunatha
Davangere University
Davangere, India

A. Nazeer
Amunix Pharmaceuticals Inc.
South San Fransisco, USA

A.K. Ogra
VBSOFT India Ltd.
Ahmedabad, Gujarat, India

Arbind Prasad
Katihar Engineering College
Katihar, Bihar, India

S.B. Prasad
National Institute of Technology
Jamshedpur, Jharkhand, India

S.V. Satya Prasad
National Institute of Technology
    Jamshedpur
Jharkhand, India

T. Ramkumar
Dr. Mahalingam College of
    Engineering and Technology
Pollachi, India

M. Selvakumar
Dr. Mahalingam College of
    Engineering and Technology
Pollachi, India

Abhishek Shrivastava
National Institute of Technology
Jamshedpur, Jharkhand, India

Subhash Singh
India Gandhi Delhi Technical
    University for Women
New Delhi, India

Varun Pratap Singh
University of Petroleum and Energy
    Studies
Dehradun, Uttarakhand, India

V. Sivananth
University of Technology and
    Applied Sciences
Ibri, Oman

Sahana S. Sringari
National Institute of Technology
Andhra Pradesh, India

Kashika Srivastava
Babu Banarasi Das University
Lucknow, India

Shubham Srivastava
Indian Institute of Technology (BHU)
Varanasi, India

Shreya Thusoo
Tokyo Institute of Technology
Japan

Baruna Kumar Turuk
National Institute of Technology
    Jamshedpur
Jharkhand, India

Deepti Verma
Central Water Commission:
    Government of India
India

Dr. Kartikey Verma
Indian Institute of Technology
Kanpur, India

H.J. Amith Yadav
Davangere University
Davangere, India

N. Zeelanbasha
PA College of Engineering and
    Technology
Pollachi, India

1 Introduction and Origin to Nanomanufacturing

S.V. Satya Prasad, S.B. Prasad, and Subhash Singh

CONTENTS

1.1 Introduction and Origins 1
1.2 Nanomanufacturing Challenges 6
1.3 Manufacturing at the Nanoscale 7
    1.3.1 Top-Down vs. Bottom-Up Approaches 8
    1.3.2 Top-Down Method 9
    1.3.3 Bottom-Up Method 10
1.4 Nanomanufacturing Applications 10
1.5 Conclusions and Future Scope 13
References 14

1.1 INTRODUCTION AND ORIGINS

The field of nanotechnology is vast, requires interdisciplinary skills and has the ability to change the way materials or components are manufactured. Therefore its research has had tremendous growth and led to exceptional development in the recent past. The year 1974 was the inception to nanotechnology in which Taniguchi coined the word "nanotechnology" and familiarized its concept to the world through production engineering conference held in Tokyo [1]. This concept of nanotechnology was described as manufacturing technology, which demanded extraordinary precision (about 1 nm and lower) and exceptionally fine dimensions for components such as memory devices, integrated circuits in computers, bearings, pump spares, aspherical lenses and optic-electronic components [2]. It is the tolerances of 0.1 to 100 nm (from atoms size to light's wavelength) that are significant in nanotechnology as described by Franks in the year 1987 [3]. The interests, development and practical implementations in the field of nano-technology stand on three pillars of the three historical events that took place. Richard Feynman's historic speech, "At the bottom there is a great space," in the year 1959 revolutionized the thoughts about materials and implanted the idea of nanotechnology [4]. The invention of scanning tunneling microscope in the year 1981 which made it possible to image at atomic–molecular levels for the very first time in the history of science, provided a significant tool to the nanotechnology field [5]. Lastly, the successful performance of single atomic operation by IBM Almaden Research Center, for the first time in 1990 reinforced the ideas about nanotechnology [6].

DOI: 10.1201/9781003220602-1

Generally, to signify the relevance of nanotechnology in the present as well as the future, it may be defined as the field which controls components or materials at atomic as well as molecular levels [7]. Nanotechnology may be considered as a marvel in the field of science. The research contributions over many decades in the fields of physics, biology and chemistry at nano levels have resulted in great discoveries. These fields along with various disciplines contribute to nanomanufacturing in a way where the boundaries of the output characteristics of this integration do not possess any clear differentiation.

The invention of semiconductors in physics has brought down the scales to 5–10 nm for research-related components and 10–22 nm for commercially available components respectively with adherence to Moore's law [8]. Likewise, advancements in chemistry have forced the researchers to move on from basic molecules and atoms to vast and analyze complex proteins comprising many such atoms. Even in biology, the behavioural aspects of cell membranes, the RNAs and DNAs are much more significant at nanoscale. This is represented in Figure 1.1a where the flowchart shows the areas of physics, biology and chemistry where nanotechnology has much significance.

The advancements in the manufacturing field over the last two decades made it possible to produce novel materials like graphene, 2D materials and carbon nanotubes. Some significant and novel tools like systems-biology techniques, atomic force microscopes, molecular-beam epitaxy and cold-atom physics to analyze and modify nanoscopic world could be manufactured. Some other nanoscaled products such as robots, gears, pumps, joints and bearings could be produced. Moreover, the commonly used manufacturing methods such as drilling, milling and cutting could be studied from the perspectives of nanoscale contributing to the optimization of the method by drawing comparisons with their corresponding phenomena at the macro levels. Figure 1.1b specifically shows the applications in the fields of biology, physics, chemistry and manufacturing where nanotechnology is the foundation. Moreover, these fields are also interrelated and interdependent where research development in one field is significant and useful in the other field. The structures and components in nanotechnology with extremely small dimensions exhibit large functional variations in comparison to their larger counterparts [9]. The contributions of science (physics, biology, chemistry) and manufacturing are extremely significant to nanomanufacturing and it interests the researchers to a greater extent. Table 1.1 lists out certain contributions.

The progressive advancements in the fields of nanotechnology and nanomanufacturing could result in advanced material characteristics in conjunction with nanoscaled manufacturing and engineering.

The field of nanotechnology as discussed earlier focuses on nanostructures and nanomaterials (natural or manmade) which are studied and analyzed at nanoscales (1–100 nm) to bring out the best, improved and novel functionalities of the concerned materials by restructuring them if necessary either at molecular or atomic levels [10,11]. This development of nanotechnology occurred in the initial decade of the 21st century and stationed it on the top of manufacturing methods. The subsequent decade shall not witness a remarkable production leap at the industry level but shall occur at biofactories, self-assembling lines which are automated and at chemical reactors that are programmed. The manufacturing shall be integral to

(a)

(b)

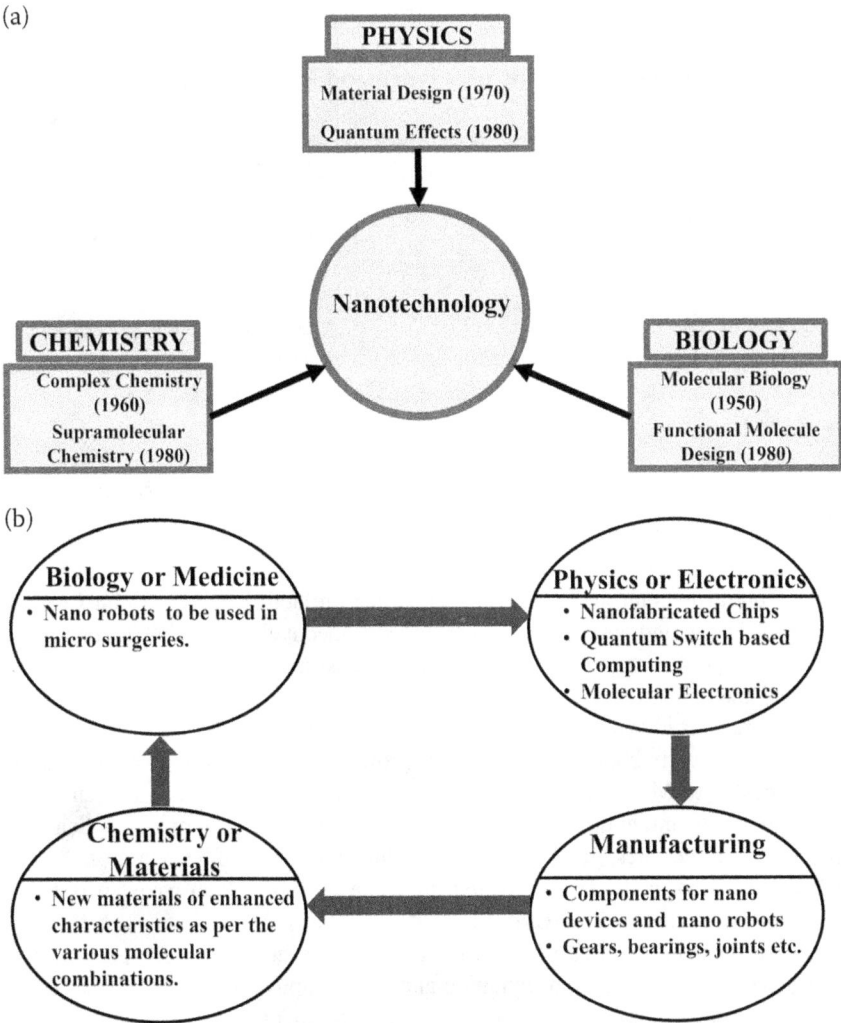

FIGURE 1.1 (a) Relevance of physics, biology and chemistry in nanotechnology. (b) Application of nanotechnology in fields of physics, biology, chemistry and manufacturing fields [9].

and customized as per the nature of application and dispersed in groups of technology. The focus of the researchers has been to control the physical and chemical forces and combine atoms to achieve a hierarchical self-assembly sequence that is predictable and also efficient.

It is to be understood and considered that developments in nanotechnology are long term when spoken in terms of commercial aspects. The majority of the commercially available products in the marketplace are 1st-generation products comprising nanostructures that are passive in nature and exhibit steadiness. Several organizations (large and small) have products of the 2nd and 3rd generations queued

TABLE 1.1

Contributions of Diverse Fields of Science and Manufacturing Towards Nanomanufacturing [9]

Area	Organic	Inorganic
Physics	Molecular Electronics	Mesoscopic Physics
		Scanning Electron Microscopy (SEM)
Biology	Biotechnology	–
	Medicine	
Chemistry	Physical Chemistry	Inorganic Chemistry
	Supramolecular Chemistry	Aerosol Science
		Computer Modelling
Manufacturing	–	Nanoscale Processes
		Material Science
		Precision Engineering

up, i.e., products comprising nanostructures that are active which exhibit dynamic behaviour in use and embryonic nanosystems, respectively. Concepts related to the products of 4th generation such as molecular nanosystems are in the research phase [12]. It is the industries of semiconductors that familiarized the consumers with sub 100 nm and paved way to the 1st-generation products exhibiting passive nanostructures in the early 2000s. The significant growth of nanotechnology resulted in the bulk nanostructured as well as electronics integrated materials such as nanowires, nanoparticles, quantum dots and nanocoatings of materials. They were applicable in sensors, optoelectronics, fluidics and biomedical devices. All of this was possible due to the exfoliation of the single-layered graphene in 2004 which paved way to the extensive research on 2D materials [13].

The subsequent years (2005–2010) comprised 2nd-generation components possessing nanostructures that are dynamic and with adaptive structures. They are actuators, sensors, amplifiers, transistors and targeted chemicals and drugs. The next few years from 2011 to 2015 saw the 3rd-generation products, which include 3D nanosystems with diverse synthesis methods and assembling methods. The examples are multi-scale architectures, bio assembling, nanoscale networking and robotics. The subsequent years gave way to the 4th-generation components of molecular nanosystems that are heterogeneous in nature, i.e., each molecule within the nanosystem possesses a definite structure and portrays a different role. Devices comprising molecules are used and this will pave way to novel functions with the fabricated architectures and structures. The above four generations will give rise to novel nanoscaled products with diverse scope for R&D which will allow products to be systematically manufactured at nanoscale [14]. It is most likely that every new generation product to include previous generation components at least to a certain extent if not completely.

With the research funds across the world from corporates and public sectors increasing, nanotechnology has a bright future. It has been found that annual investment of $5 billion for R&D on nanotechnology across the world has surpassed the $4 billion in funding provided by the government in developed countries like the United States in the year 2005. The estimates for the global market value of the components manufactured via nanotechnology is about $110 billion and has an increment rate of 25% per annum [15].

According to a survey conducted among 594 respondents, more than half the number have specified that their organizations are involved directly in the development of nanomanufacturing in the form of component suppliers, end-users or manufacturer integrators. The most surprising aspect is that a major proportion (approximately 19%) of educational institutions and academia is directly involved in developing nanomanufacturing technologies [16]. This is clearly evident in Figure 1.2 which represents some of the percentage contributions of the organizations towards the nanomanufacturing chain.

The major portion of the enhancements in products via nanomanufacturing has been estimated to be around $80 billion for enhanced materials and catalysts. This section has the domination of large organizations. The medium- and small-scale organizations are involved in developing newer products with a share of $30 billion across the world.

As the subsequent, major step, it is a challenge for nanotechnologists to commercialize the laboratory discovered products. That means components have to be manufactured at nanoscale in a cost-effective manner with practical scalable methods and feasible, controlled critical dimensions [17]. When it comes to reality, the economic constraints restrain the possibilities just like the laws of nature. The ultimate criterion is the cost of the manufacturing technique shouldn't be higher than that of the system or the component that is being manufactured. This is where

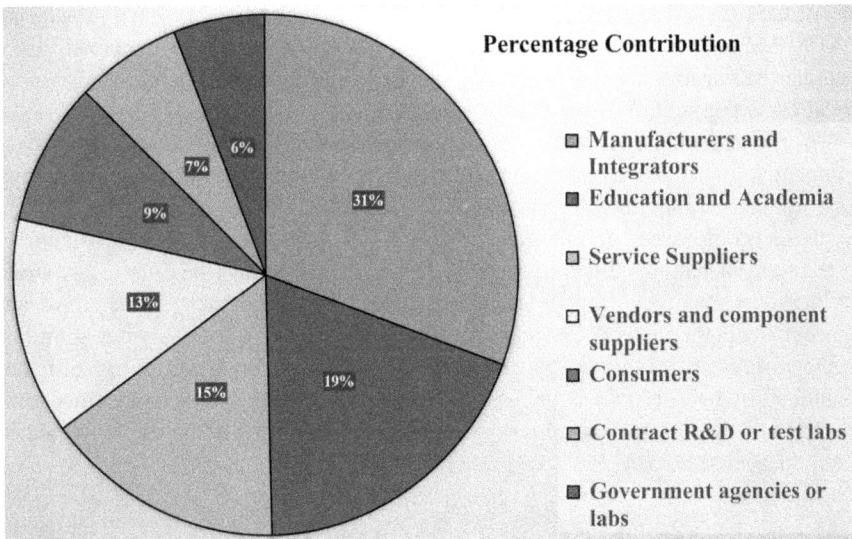

Percentage Contribution

- Manufacturers and Integrators
- Education and Academia
- Service Suppliers
- Vendors and component suppliers
- Consumers
- Contract R&D or test labs
- Government agencies or labs

FIGURE 1.2 The percentage contribution of the diverse organizations in the development of nanomanufacturing [16].

the systematic approach of the nanomanufacturing holds the key which has the ability to design, manufacture, change, control as well as assemble components of nanoscale by exploiting their characteristics at that scale in a most cost-effective manner and also enabling mass manufacturing [18,19]. The technique of nano-manufacturing involves multiple approaches such as the bottom-up directed as-sembling where blocks of nanostructure are built from the atoms, molecules and supramolecular scales. The top-down method (fragmentation and ultra-precision engineering) which is a high-resolution technique, the physic-chemical molecular, supramolecular engineering systems and hierarchical integration to systems of larger scale are also inclusive of nanomanufacturing. For such processes, it is es-sential to possess extreme control over the process to sense or actuate matter and capabilities at nanolevels. This facilitates scaling-up. The prime target is to lower the usage of materials as well as energy thereby reducing waste and impact over the environment. Therefore, economical and high rate of production will be feasible and this is ideal for implementation in the industrial sector [20].

Nanomanufacturing exploits the quantum as well as the surface effects in materials at nanoscale. Hence, most aspects of nanomanufacturing include multiple stages in diverse combinations so the tools like computer-aided design (CAD) and computer-aided manufacturing (CAM) will come in handy. The scalable nanomanufacturing is a mass production technique that produces nanoscaled structures as well as materials, assembles them into parts and subsystems, and combines these into systems of higher order. Nanomanufacturing acts as the bridge between the nanoscaled products pro-duced in the lab and the products manufactured for commercial purposes by elim-inating the scientific and technical constraints and endorses large-scale production. So research in this field of scalable nanomanufacturing helps in overcoming the issues of affordability, controllability, efficiency, quality, reliability, scalability and yield. In one of the previous studies, important attributes needed within a nanomanufacturing CAD/CAM system were listed and a prototype of the same was explained high-lighting the capabilities and future advancements needed for its comprehensive ap-plication in the field of nanomanufacturing [21].

To understand the significance of nanotechnologies, it is essential to have a complete knowledge on issues related to high-yield methods of nanomanufacturing for commercial manufacturing of nanoscale materials and structures. Moreover, an emphasis on the issues of nanoscale production should inspire further development in the field of nanomanufacturing [22]. The researchers must be able to elaborate problems of manufacturing in their respective work. This shall further have ad-vancements in the field of nanomanufacturing and will bring about diverse solutions to the various issues that are faced. Eventually, this will pave way to novel nano-manufacturing techniques at laboratories which will later on translate into com-mercial and large-scale techniques of nanomanufacturing and be implemented in the industrial sectors.

1.2   NANOMANUFACTURING CHALLENGES

The challenges of manufacturing can be classified into two sections. The outcomes desired and the suitable metrics.

The quality of the product, its performance, functionality and its durability, reliability and repeatability of a process, affordability, scalability, yield and efficiency of production are the outcomes desired. Whereas precision, resolution, size of the feature, complexity, density of the nanostructure, rate of formation and overlay registration are some metrics needed to be defined.

The above are the common challenges that one comes across during manufacturing but in the case of nanomanufacturing, control and measurements at nanoscales are extremely difficult and a very small range of errors could give way to failures that are large. Moreover, for large-scale production, processing or rate of formation, one has to compromise in consideration of the resolution and size of the feature. It is essential to validate every produced nanocomponent since each nanoscaled process has its own unique history of processing and might be untested [23]. So, for commercial applications, vast proven history, universal standards, targeted metrics and a supply chain that is reliable are extremely necessary. Moreover, standards for environment, toxicity, safety and health need to be followed and the market value for the nanoproduct to be produced should be determined.

Keeping the above factors in mind, the following areas of challenge have been identified [16]. They are:

- Design of nanostructured materials
- Nanoscale production
- Identification and conservation of radiological, biological and chemical explosives
- Metrology and instrumentation at nanoscale
- Photonics, magnetics and electronics that are nanoscaled
- Diagnostics, therapeutics and healthcare
- Research on nanosciences for energy demands
- Robotics and micro-crafts
- Nanoscaled methods to enhance the environment

Therefore, considering all the areas of challenge, the important challenges of nanomanufacturing can be summarized as follows.

- Control of the assembly process in 3D systems that are heterogeneous. This includes registration, alignment and interconnection of all the three dimensions having multiple functionalities.
- Processing and handling of nanoscaled structures in large-scale productions without foregoing the benefits of nanoscaled characteristics.
- Testing the reliability of nanoscaled components for long-term basis and also detection, removal or prevention of defects as well as contamination.

1.3 MANUFACTURING AT THE NANOSCALE

Nanomanufacturing is the process of producing nanoscaled materials or devices at the nanoscales including reliable, scaled up and cheaper methods. Research and development, integration of top-down and the bottom-up methods (self-assembly methods) are also integral to the nanomanufacturing. Simply it can be said that nanomanufacturing produces novel components with enhanced material characteristics. The top-down or the bottom-up approaches are the basic methods employed in the nanomanufacturing [24].

The top-down method reduces large-sized material up to nanoscales. This is analogous to a wooden block carving. Higher material quantities are needed for this method and there could be material wastage if excessive amounts of the material are discarded [25]. Contrastingly, the bottom-up method fabricates products by building them layer by layer right from the atomic or molecular stage. This process is extremely time-consuming but there is no material wastage in this process. The process is expensive in comparison to the top-down approach. The researchers are trying to come out with a concept such that the molecules in a component when placed in a certain order, instantaneously assemble on their own in the bottom up to form structures of uniform order. There are a number of novel techniques in the top-down as well as the bottom-up methods of nanomanufacturing which produce nano components of exceptional quality and simplify the process. Some of the processes are represented in Figure 1.3 [23].

1.3.1 TOP-DOWN VS. BOTTOM-UP APPROACHES

Prior to the understanding of the top-down and bottom-up methods, it is essential to understand the terms "deterministic" and "stochastic." Production methods which are deterministic in nature, exhibit extremely low distributions of the component performance mean as well as variation. But this doesn't rule away remarkable randomness at the nanoscales or atomic levels. The methods which are stochastic such as the bottom-up approach are statistical in nature with variations but can also possess

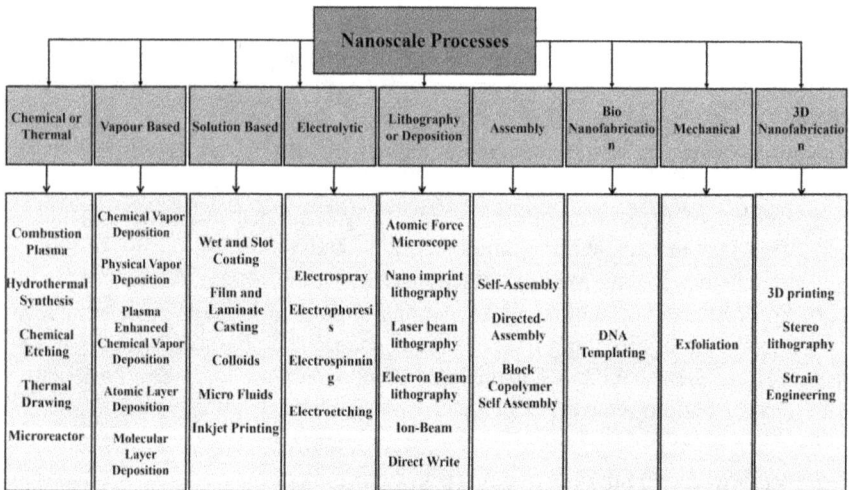

Chemical or Thermal	Vapour Based	Solution Based	Electrolytic	Lithography or Deposition	Assembly	Bio Nanofabrication	Mechanical	3D Nanofabrication
Combustion Plasma	Chemical Vapor Deposition	Wet and Slot Coating	Electrospray	Atomic Force Microscope	Self-Assembly	DNA Templating	Exfoliation	3D printing
Hydrothermal Synthesis	Physical Vapor Deposition	Film and Laminate Casting	Electrophoresis	Nano imprint lithography	Directed-Assembly			Stereo lithography
Chemical Etching	Plasma Enhanced Chemical Vapor Deposition	Colloids	Electrospinning	Laser beam lithography	Block Copolymer Self Assembly			Strain Engineering
Thermal Drawing	Atomic Layer Deposition	Micro Fluids	Electroetching	Electron Beam lithography				
Microreactor	Molecular Layer Deposition	Inkjet Printing		Ion-Beam				
				Direct Write				

(Top box: **Nanoscale Processes**)

FIGURE 1.3 Hierarchy of the various kinds of nanoscale top-down and bottom-up methods in nanoprocessing [23].

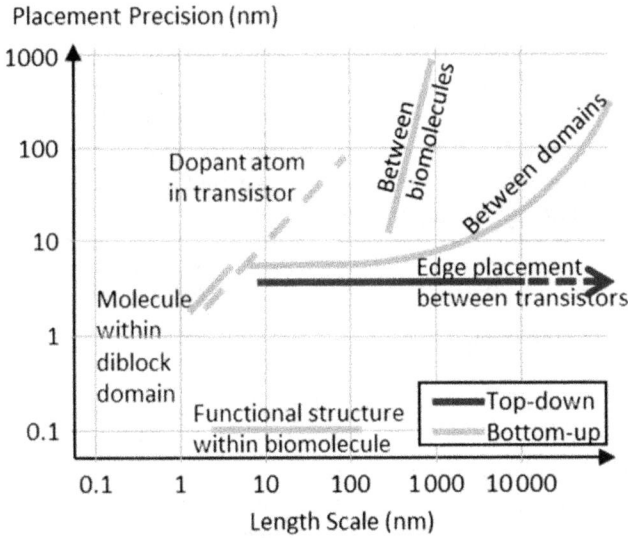

FIGURE 1.4   An example of how the stochastic processes control the structural precision in top-down approach, bottom-up approach and the damped-driven assembly [26].

distributions of the components' performance, variation and mean that are narrow when an average of large quantities of nanostructures are considered. An alternative method of analyzing this is by considering scale lengths where variations are seen (Figure 1.4) [26]. Certain stochastic methods can exhibit exceptional accuracy and precision over a short range, but tend to degrade very quickly at length scales longer than that of the individual unit. Synthesis of protein and folding can be taken as an illustration of this phenomenon. The deterministic models contrast the stochastic models in a way where disorder is seen at the nanoscale or the atomic levels but they possess excellent precision and accuracy over long ranges. The functional requirements of the nanoproduct determine the suitability of the considered approach.

The illustration in Figure 1.4, shows that dopants in a semiconductor will be effective along the dimensions of the device as per the present techniques of manufacturing. The atom lies only within the constraints of the created box. The edges may be placed in a portion of the size of the feature within the circuit's length scale. The diblock copolymers have no pattern in the self-assembly process, but they exhibit exceptional short-range orders, and as the distance varies exponentially, they decay. The individual molecules are placed inside the box's dimensions (domain). Highest control is seen in the biomolecules where the atoms can be placed specifically as per the secondary/tertiary structure's hierarchy. Precision placement decays quickly based on separation function in the non-bound biomolecules.

1.3.2   TOP-DOWN METHOD

The top-down approach is basically deterministic, in which the system's order is achieved by external forces applied. This signifies that the thermodynamic considerations of equilibrium do not influence the yield as in case of chemical reaction. Therefore, in this method, the yield can be customized as per the laws of physics. For example, IC production techniques have components that function with rate of error lower than 1 in 1012 and much lower defects of less than 0.1 defects per square centimetre. So the economic drivers are applicable until the limits of fundamental physics and there is an exceptional evolution in the top-down approach as per the conventional learning curves.

1.3.3   BOTTOM-UP METHOD

The kinetics and thermodynamics influence the yield in a bottom-up method to obtain the required structure. The bottom-up technique doesn't require any costly tools for creating structures at nanoscales. Moreover, the large scaling is also straightforward. By the method of chemical synthesis, carbon nanotubes, metallic nanowires, quantum dots, multifunctional particles for medical applications and plasmonically active particles have been manufactured as per desired quantities. The interaction among diverse components influencing the bottom-up self-assembly to fabricate complicated compounds and allowing them to evolve to their final stage is in the developmental stage [27].

The material characteristics and structures can be enhanced by the above-mentioned nanomanufacturing techniques. The produced nanomaterials shall possess exceptional characteristics like higher strength, lightweight, enhanced durability, resistance to scratches as well as ultraviolet- or infrared, anti-reflective and microbial, antifogging, self-cleaning, water-repellent, electrical conducting capabilities, etc. These exceptional characteristics shall widen the application of nanoscaled products in areas like sporting goods to fabricate baseball bats, tennis rackets, crude oil refining catalysts, applications where chemical and biological toxins can be identified, quicker, and highly powerful computer systems with enhanced efficiency in terms of energy due to the existence of nanoscale transistors, etc. The nanotechnology can also produce devices with enhanced capacity to store information like a small single chip [28]. Economical solar cells with enhanced efficiencies can also be produced in the field of energy by making use of nanotechnology.

1.4   NANOMANUFACTURING APPLICATIONS

The developments in the field of nanotechnology have resulted in the application over diverse sectors in industries which are capable of forming a completely new sector. Some of the major applications of the nano products are approximately in the percentages represented in Figure 1.5.

The other areas of application of the nanotechnology are in the following areas:

Manufacturing
- Materials that are light in weight, free of corrosion, greater strength and resistant to corrosion are produced.
- Coated panes which repel dirt and rain drops. Non-toxic dirt glass having decorative layers without toxicity.

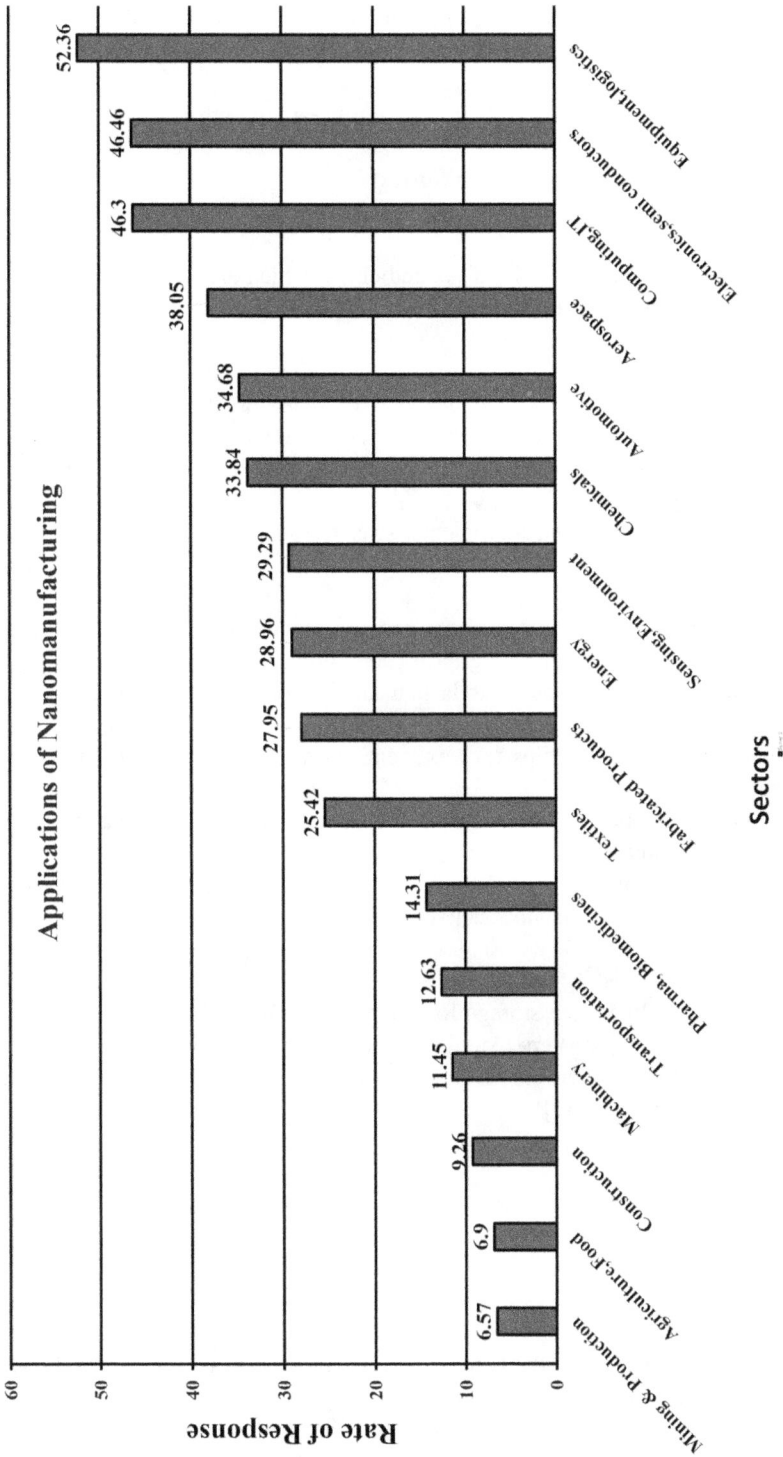

FIGURE 1.5 Graph showing the increment of nanomanufacturing applications in the various industrial sectors [16].

- Windows of plastic which are light in weight having protective layers that are hard as well as transparent or bearings having no lubricants to be used in automobiles.
- Nanocomposites, fibres, cables possessing great strength and low weight used in packaging [29].
- Paper, filters, membranes, fabric for textiles.

Electronics
- Devices for data storage like disks and chips, which are very small, have stability for long durations and a storage capacity equivalent to a national library stock.
- High-powered PCs.
- Storage memory that is flexible, 3D devices, transistors made of thin films, integrated circuits (ICs) and electromagnetic (EM) shielding.
- Optoelectronics applications for display, imaging, lighting, waveguides and metamaterials.
- Chemical, biological and multiplexed sensors [30].

Health
- Sensors for controlling the environment and functions of the body.
- Oral pharmaceuticals with long-term capabilities.
- Human tissue substitutes or foils that are biocompatible to close inner wounds.
- Economical, reusable chips for a preventive and diagnostic survey in the medical field.
- Amenities for sanitary with coatings that are active and photocatalytic to avert the accumulation of bacteria.
- In the biomedical field as orthopaedic implants, scaffolds of tissues, therapeutics, probes and diagnostics [31].

Energy
- Cost-effective hydrogen storage for the economy of regenerative energy.
- Novel solar cells for energy storage as per photosynthesis.
- Energy, harvesting, conversion and storage, in batteries, photovoltaic cells and supercapacitors [32].

Environmental
- Purifying water, treating waste water and analytical Separation.

Chemical
- Storage of gas and catalysis [33].

This clarifies that advances in the field of nanomanufacturing will influence the application in diverse industries by bringing novel components, components with enhanced characteristics or with novel functionalities. Nanoscaled materials with novel characteristics, diverse structures, forms and functionalities act as raw materials which can be modified through the methods of nanomanufacturing giving rise to nanosystems [34]. A recent discovery on the inventory of various methods of nano-manufacturing suggests a mind-boggling count of over 150 which can be categorized as top-down or bottom-up approaches. This suggests that an exceptionally wide range of nanomaterials and methods exists which can be employed to construct nanos-tructures and further be integrated to produce systems of large scale [35]. It has to be researched to ideally combine the materials and methods to obtain the desired products with enhanced features.

1.5 CONCLUSIONS AND FUTURE SCOPE

Manufacturing nanoscaled devices, structures and system require exceptional control of the process for matter actuation and sensing at nano levels. It is essential to obtain the desired product performance through integration in a hierarchy and convert the nanoscaled structures to products of micro and macro scales. The method can be either top-down or bottom-up of exceptional speed or resolution but the manufacturing process should possess high efficiency, minimize wastage of materials and energy, should have no environmental impact, must be cost-effective, repeatable and should be ideal for implementing in the industries. Considering these factors would produce new components and positively influence the economy and the society with positive influence of nanotechnology. The nanomanufacturing should primarily fulfil the following roles:

- It must give rise to novel materials, methods and services which isn't achievable through other techniques.
- The problems pertaining to the interests of the public like enhanced efficiency of work, health and safety, enhancement and conservation of the environment must be looked at.
- Provides support to industries by establishing novel and unconventional manufacturing schemes as well as their market.
- The growth of the economy must be revived by offering multiple opportunities of employment, dispersing the resources and activities and widening participation.
- The contemporary, advanced methods of manufacturing should be established across the world.

Revolutionary advancements in the fields of oncological diagnostics, information technology, and biotechnology and semiconductor electronics have created a need to develop nanoscale manufacturing to endorse the research growth in the manufacturing field. This has integrated the various disciplines of engineering to train budding researchers and professional workforce and make them competent in nano-technology so as make them a contributing factor in industries across the world.

Apart from the educational and research institutes, investments in nanomanu-facturing infrastructure give way to advanced sectors in the industries to design, fabricate, integrate manufacture and employ nano products. This advancement shall endorse value addition to the products making use of the advanced manufacturing methodologies thereby benefitting economies internationally. Nanomanufacturing is opening up new markets and reviving the economy across the world. Nanomanufacturing is the future of manufacturing and therefore it is finding its applications across diverse fields of science like physics, biology and chemistry. Some of the important applications are in the energy sector for storage of energy and electronics for manufacturing miniature components with exceptionally high data storage capacity. Nanomanufacturing is finding applications health and pharmaceutical sectors for manufacturing biomedical implants, tissue replacements and orally consumable drugs. The applications of nanomanufacturing are also positively impacting the environment where cleaning and purification of water is being done through nanomanufacturing. All of these make nanomanufacturing the heart of the industrial sector and also this indicates bright future for the field of nanotechnology through nanomanufacturing.

The research in the future on nanomanufacturing should be further narrowed to smaller scales such that manufacturing is carried out at atomic as well as close-to-atomic scales. This will significantly change the manufacturing field so more effort is needed in this to create a blueprint. In order to meet the growing demands of industries and society, nanomanufacturing should have high efficiency with lower follow-up methods. This should be of top priority to eliminate the side effects of manufacturing, encapsulation as well as alignment. The following aspects should be focused in future research:

- The potential of nano cutting is extremely high for advancements in the future. Currently, it is taking baby steps so efforts have to be put to develop a theory on nano cutting.
- High-efficiency methods of nanomanufacturing with lower costs must be introduced and must focus on large-scale production for much harder materials.
- Multi-scaled or multi-layered nanostructures perform exceptionally and are being used in majority of the areas. Hence study must be done to create an optimal design of multi-functional process.
- Metrology of the topography is extremely essential. Focus is needed on high-performance measurement should assess the quality of manufacturing for both research and application needs.
- The probable demands of quantum computers, health care as well as molecular circuits may be considered as it is worthy enough.

REFERENCES

1. Taretia R. Nanotechnology–A Review. *Journal of Orofacial & Health Sciences.* 2017; 8(1 to 3):5–8.
2. Shore P, Morantz P. Ultra-precision: Enabling our future. *Philosophical Transactions of the Royal Society A: Mathematical, Physical and Engineering Sciences.* 2012 Aug 28; 370(1973):3993–4014.
3. Franks J, Williams RF. Braids and the Jones polynomial. *Transactions of the American Mathematical Society.* 1987; 303(1):97–108.

4. Subramani K, Ahmed W, editors. *Emerging nanotechnologies in dentistry.* William Andrew Publishing; 2018 Jan.

5. Bayda S, Adeel M, Tuccinardi T, Cordani M, Rizzolio F. The history of nanoscience and nanotechnology: From chemical–physical applications to nanomedicine. *Molecules.* 2020 Jan; 25(1):112.

6. Mansoori GA, Soelaiman TF. Nanotechnology—An introduction for the standards community. *Journal of ASTM International.* 2005 Jun 1; 2(6):1–22.

7. Sahoo SK, Parveen S, Panda JJ. The present and future of nanotechnology in human health care. *Nanomedicine: Nanotechnology, Biology and Medicine.* 2007 Mar 1; 3(1):20–31.

8. Brock DC, Moore GE, editors. *Understanding Moore's law: Four decades of innovation.* Chemical Heritage Foundation; 2006.

9. Chryssolouris G, Stavropoulos P, Tsoukantas G, Salonitis K, Stournaras A. Nanomanufacturing processes: A critical review. *International Journal of Materials and Product Technology.* 2004 Jan 1; 21(4):331–348.

10. Thomas S, Pasquini D, Leu SY, Gopakumar DA, editors. *Nanoscale materials in water purification.* Elsevier; 2018 Nov 14.

11. Poole Jr CP, Owens FJ. *Introduction to nanotechnology.* John Wiley & Sons; 2003 May 30.

12. Aithal PS, Aithal S. Nanotechnology innovations and commercialization–opportunities, challenges & reasons for delay. *International Journal of Engineering and Manufacturing (IJEM).* 2016 Nov 8; 6(6):15–25.

13. Yang S, Zhang P, Nia AS, Feng X. Emerging 2D materials produced via electrochemistry. *Advanced Materials.* 2020 Mar; 32(10):1907857.

14. Chen H. *Mapping nanotechnology innovations and knowledge: Global and longitudinal patent and literature analysis.* Springer Science & Business Media; 2008 Dec 10.

15. Roco MC . The long view of nanotechnology development: The national nanotechnology initiative at 10 years. In *Nanotechnology research directions for societal needs in 2020.* Dordrecht: Springer; 2011 (pp. 1–28).

16. Busnaina A, Mehta M. Busnaina A . Introduction to nanoscale manufacturing and the state of the nanomanufacturing industry in the United States. In *Nanomanufacturing handbook.* CRC Press; 2017 Dec 19 (pp. 1–28).

17. Lu W, Lieber CM. Nanoelectronics from the bottom up. *Nanoscience and Technology: A Collection of Reviews from Nature Journals.* 2010:137–146.

18. Gibson I, Rosen DW, Stucker B, Khorasani M. *Additive manufacturing technologies.* Cham, Switzerland: Springer; 2021.

19. Kumar N, Kumbhat S. *Essentials in nanoscience and nanotechnology.* John Wiley & Sons; 2016 Apr 11.

20. Iqbal P, Preece JA, Mendes PM. Nanotechnology: The "top-down" and "bottom-up" approaches. *Supramolecular Chemistry: From Molecules to Nanomaterials.* 2012 Mar 15. Page 1–14.

21. Yoon HS, Lee HT, Jang KH, Kim CS, Park H, Kim DW, Lee K, Min S, Ahn SH. CAD/CAM for scalable nanomanufacturing: A network-based system for hybrid 3D printing. *Microsystems & Nano Engineering.* 2017 Sep 25; 3(1):–11.

22. Yeatman EM, Gramling HM, Wang EN. Introduction to the special topic on nanomanufacturing. *Microsystems & Nanoengineering.* 2017 Sep 25; 3(1):1–2.

23. Cooper K. Scalable nanomanufacturing—A review. *Micromachines.* 2017 Jan; 8(1):20.

24. Arole VM, Munde SV. Fabrication of nanomaterials by top-down and bottom-up approaches-an overview. *J. Mater. Sci.* 2014; 1:89–93.

25. Imboden M, Bishop D. Nanomanufacturing. *Physics Today.* 2014; 67(12):45.

26. Liddle JA, Gallatin GM. Nanomanufacturing: a perspective. *ACS nano.* 2016 Mar 22; 10(3):2995–3014.

27. Fang FZ, Zhang XD, Gao W, Guo YB, Byrne G, Hansen HN. Nanomanufacturing—Perspective and applications. *CIRP Annals*. 2017 Jan 1; 66(2):683–705.
28. Doumanidis CC. Nanomanufacturing: Technical advances, research challenges, and future directions. *Proceedings of the Institution of Mechanical Engineers, Part N: Journal of Nanoengineering and Nanosystems*. 2004 Dec 1; 218(2):51–70.
29. Hassan T, Salam A, Khan A, Khan SU, Khanzada H, Wasim M, Khan MQ, Kim IS. Functional nanocomposites and their potential applications: A review. *Journal of Polymer Research*. 2021 Feb; 28(2):1–22.
30. Wu W, Qiu G, Wang Y, Wang R, Ye P. Tellurene: its physical properties, scalable nanomanufacturing, and device applications. *Chemical Society Reviews*. 2018; 47(19):7203–7212.
31. Yi AY, Lu W, Farson DF, Lee LJ. Overview of polymer micro/nanomanufacturing for biomedical applications. *Advances in Polymer Technology: Journal of the Polymer Processing Institute*. 2008 Dec; 27(4):188–198.
32. Dou S, Xu J, Cui X, Liu W, Zhang Z, Deng Y, Hu W, Chen Y. High-temperature shock enabled nanomanufacturing for energy-related applications. *Advanced Energy Materials*. 2020 Sep; 10(33):2001331.
33. Oakes L, Hanken T, Carter R, Yates W, Pint CL. Roll-to-roll nanomanufacturing of hybrid nanostructures for energy storage device design. *ACS Applied Materials & Interfaces*. 2015 Jul 8; 7(26):14201–14210.
34. Rangasamy M. Nano technology: A review. *Journal of Applied Pharmaceutical Science*. 2011 Apr; 1(2):8–16.
35. Ramsden J. *Nanotechnology: An introduction*. William Andrew; 2016 May 11.

2 Challenges and Opportunities in Nanomanufacturing

Subhash Singh, Ashwani Kumar,
Sanjay K. Behura, and Kartikey Verma

CONTENTS

2.1 Introduction...17
2.2 Opportunities in Nanomanufacturing...21
    2.2.1 Integrating Functions..21
    2.2.2 Technology Convergence..22
2.3 Nanomanufacturing Processes 24
2.4 Challenges in Nanomanufacturing..25
2.5 Recommendations for Nanomanufacturing Research 26
2.6 Conclusion 28
References..29

2.1 INTRODUCTION

The field of nanotechnology has come a long way in the last few decades in terms of the investment made and also the amount of progress that has taken place. The initial stages of investment over basic investigation were completely justified as it paved way to a substantial development in the field of nanomaterials. Many novel production techniques with capabilities of producing two- and three-dimensional nanoscaled structures as well as novel device concepts came into existence which are the result of this extensive research carried out over the span of these 20 years. Nevertheless, many of the nanoscale components or products like the semi-conductors are produced via the top-down methods. When the financial aspects are considered, in the present-day scenario, fabricating nanoscale semiconductors with all of the features costs up to $7–10 billion in order to be constructed [1]. The economy of the equipment required for processing semiconductors has a range varying from hundred thousand dollars on the lower side to few million dollars among which, the equipment of lithography go way beyond the value of $50 million. With such high construction costs involved, the scope is cut off for small as well as medium-scale industries. Moreover, the existence of these huge facilities also huge impact on the environment. One such example is the amount of water needed in one day. This consumption of water amounts to over 4 million gallons each day by these industries. But if such barriers are slashed down and overcome, then it will pave way for novel

DOI: 10.1201/9781003220602-2

concepts and technological innovations that will result in a completely novel set of industries. Such innovation will allow any organization of any size to be able to produce devices or systems at the nanoscale at an extremely small fraction of the amount that is being invested today, let's say about 100th of the cost that incurs in the present-day scenario. This will unravel a creative wave which will make the manufacturing processes at nanoscale much more affordable and accessible to diverse kinds of industries like it had happened in the case of computing industry, where the invention of PC made technology accessible and affordable across all classes, breaking all kinds of socio and economic barriers. Such innovation will re-vitalize the process of manufacturing.

The method of manufacturing involves addition or subtraction of materials at the macro, micro as well as nanoscales. In the modern-day scenario, when we speak of micro or nanoscale production, methods such as physical vapour deposition (PVD), chemical vapour deposition (CVD) which include thin film deposition, wire bonding, packaging, etching, assembly as well as polishing are involved [2]. In order for the process of nanomanufacturing to be commercialized and be successful in producing nanoscale products, the entire system of nanomanufacturing needs to be multi-scale, robust and additive for it to be utilized in different ranges of in-dustrial applications. Already there seem to be indications of the nano production of diverse components and devices from the conventional methods involving vacuum. One such instance can be the commercial use of the inkjet printing methods for circuit patterns as well as the employing of screen-printing by the photovoltaic producers. There has been tremendous growth in additive manufacturing methods, which includes 3D (three-dimensional) electronic printing in recent times. But speaking in terms of the commercial aspects, the 3D printing methods can only reproduce at microscale and also the processes are slow [3]. For achieving nanoscale 3D printing, the process is extremely slow and also low resolutions are possible. Moreover, the slow nature of the additive manufacturing process makes the process extremely tedious and takes an unbearably long time for complex and larger sizes. The need for higher resolution further increases the production time and the cost. When the nature of 3D printing technique is top-down, bringing down the scale to nano level to fab-ricate nanostructures is extremely complex or not possible in certain cases of mate-rials. In certain cases, especially in case of fabrication of semiconductor-based electronics, the method of 3D nanoscale printing, irrespective of its limitations such as longer fabrication methods and lower printing resolutions, has proven to be sig-nificantly cost-effective in comparison to the conventional production practices [4]. One aspect to be understood here is that success isn't only dependable only on the technology there are other implications pertaining to the legal, societal and ethical issues which ought to be considered in the fabrication of nanoscale products. It is essential to understand and overcome the inadvertent implications of nanomanu-facturing upon the environment as well as on the health of humans. This is only possible through thorough observation and analysis of the implications when nano-particles are liberated during the entire lifecycle of the nanoproduct.

Nanomanufacturing involves manipulation, strong control of matter as well as the physical phenomena that occur at the nanoscale, about 1–100 nm length scales. The primary motto of nanomanufacturing is to achieve mechanical, physical, chemical as well as the biological characteristics which are either impossible to achieve in case of the bulk state materials or characteristics that are in complete contrast to their macroscale counterparts. The novelty and uniqueness of nanomanufacturing method lies in its integration of nanoscale top-down methods as well as the bottom-up processes that are complex and help build components which can have intricate geometries. Some of the top-down methods include PVD, CVD, pulsed layer deposition (PLD), atomic layer deposition (ALD), etc. The bottom-up approaches on the other hand are directed in self-assembly and spontaneous methods [5,6]. The factors such as cost-effectiveness, reliability and scalability must be looked into by the research and development (R&D) of the nanomanufacturing in order to differentiate this method from techniques such as nanofabrication or nanoprocessing. It is essential to integrate metrology, measurement and instrumentation into the R&D of nanomanufacturing. The producers or the manufacturing sectors are on the lookout of the technical methods that exhibit performance and exhibit features, such as repeatability and reproducibility, in the components. Methods with higher productivity such as parallel processing, patterning and printing should be included in the nanomanufacturing processes. Apart from large-scale production, there is also a great demand for large-scale customization, as in manufacturing large quantities of goods that can be customized individually. Certain other needs of the modern-day production are automation, "user-friendly" operation and flexibility of the design as well as production. In case of certain applications such as the military, where the demands are extremely high and requirements very stern, the nanosystems should be cost-effective, reliable as well as robust since the nanosystems should have the capability to perform in trying and extreme conditions where factors such as pressure, temperature, radiation, humidity, corrosion and electromagnetism are unpredictable and have greater range as well as fluctuation. In such cases, there should be no compromise in the performance of the product. The primary goal of nanomanufacturing is to ensure that novel nanocomponents and nanotechnologies are made available to the common public and also be useful commercially by lowering the verification time as well as the cost of verification of nanostructures [7]. Thus, the success in the field of nanomanufacturing will eventually lead to the growth and development of many diverse industries, pave way for many employment opportunities and thereby significantly influence the economy of the nation.

In the field of nanomanufacturing, the novel techniques can produce nanoproducts comprising multi-material mix in order to obtain function and length scale integration (FLSI). Irrespective of the challenges in terms of achieving such technology convergence, nanomanufacturing also is a good prospect for promising R&D in terms of innovation as well as value addition for the following demands [8].

- Nanoscale production via integration of innovation, developing production concepts as well as knowledge-based technologies in processing of non-silicon materials.
- The ability to predict the performance of a process or a product and avert or lower the risk at the time of nanoproduct development or production, thereby lowering the marketing time for the subsequent generation of products fabricated via nanomanufacturing.

- Ability of the product platforms in meeting the demands of the next-generation nanoproducts, spreading across all sections of the industries with stricter regulations along with environmental legislation.
- Effective and large-scale production to ensure efficient nanoproduct transfer and ideas pertaining to the technology from the laboratory into commercial, continuous production.

The field of nanomanufacturing is surely multi-disciplinary in nature. There is involvement of various aspects such as fabrication, processing and synthesis. Stages such as component design, modelling as well as simulation are included [9]. The nanoscale blocks of building, nanosystems, nanostructures, nanomaterials and nanodevices are integral to the process of nanomanufacturing. Diverse platforms, machines as well as tools are included in nanomanufacturing along with evaluation, testing and characterization of the nanoproducts. Moreover, dimensionality, visualization, resolution, sensing, registration and control are also integral to nanomanufacturing. Therefore, it can be said with utmost confidence that nanomanufacturing has impact on all of the stages within a value chain as represented in Figure 2.1 [8].

There are multiple challenges as well as opportunities in the field of nanomanufacturing during the R&D process which one comes across in all stages of the value chain and need to be addressed. Some areas of focus are as follows:

- Design of the nanoproduct and its fabrication across diverse scales such as the hierarchical nanostructures.
- Early predictions from the progress as well as the kinetics of the processes during nanosynthesis and nanoassembly.
- Need to develop advanced tools and instruments in the creation of nanoarchitectures to measure nanoscale characteristics.
- Ability to comprehend and monitor nucleation as well as nanomaterial or nanostructure growth in order to evaluate how the catalysts, thermodynamics,

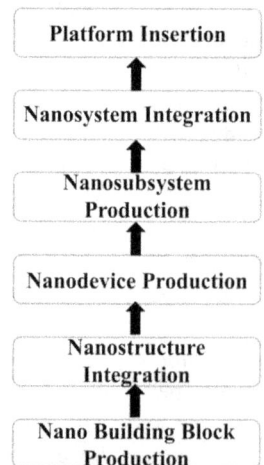

Platform Insertion

↑

Nanosystem Integration

↑

Nanosubsystem Production

↑

Nanodevice Production

↑

Nanostructure Integration

↑

Nano Building Block Production

FIGURE 2.1   Value chain of nanomanufacturing [8].

chemistry, environment and orientations of the crystal influence the surface morphologies and rate of growth.

• Establishment of the material or molecular behaviour away from the equilibrium. For instance, the synthesis as well as the nature of reactions are influenced by small differences in the energy along the geometry of the component.

2.2 OPPORTUNITIES IN NANOMANUFACTURING

The technology of nanomanufacturing reinforces the culmination of diverse functionalities within novel, advanced components. It is also essential that novel methods are needed to successfully combine products having multi-scale functions. The present section elaborates on the various opportunities provided by novel components along with novel concepts of designing a process. In particular, the focus is on the generated chances when novel concepts of design combine with diverse functions within a single nanocomponent.

2.2.1 INTEGRATING FUNCTIONS

When the present-day-based products obtained via novel and advanced technologies are observed, it can be seen that multiple functions are culminated together in extremely miniaturist enclosures. These miniaturist systems such as nanoelectronics, nano-actuators, nanosensors and nanofluidic devices embrace diversion functions in need of diverse length scale features which are packed within a small, solitary container [10]. With the use of the diverse benefits provided by embracing this nanomanufacturing, many aspects such as cost effectiveness, lower wastage of power, material and smaller sizes can be obtained through the focused endeavours of manufacturing experts and designers. The nanomanufacturing method therefore, strengthens culmination of diverse functionalities within novel components and their merger leads to the following opportunities.

• Enhanced-force nano-actuation: Certain goods produced via nanomanufacturing don't possess the capability to withstand the forces equivalent to their macro counterparts. In such cases, novel techniques are needed for manufacturing the components. Shape memory materials, magnetic and piezoelectric materials are some that come under this category which must be provided with elevated forces as well as duration of interaction [11].

• Environment resistance: Nanodevices utilized for applications pertaining to optics, biomedical, chemical and applications at extreme temperatures demand novel characteristics in comparison to their conventional counterparts which the conventional materials do not exhibit [12]. Hence, novel materials as well as nanomanufacturing techniques need to be employed to widen their usage in the corresponding fields. This shall widen the applications of nanosystems.

• High precision: To obtain the desired compatibility among structures manufactured via conventional techniques and also the nanocomponents fabricated made by non-silicon, an alteration is needed by the capable technology comprising multi-material manufacturing methods.

- Standardization and unification: The assembly of nanosystems to their systems and interfacing them to operating domains is an extremely essential stage of production and comprises 80% of the total cost of the system and it demands the usage of multi-material production techniques. In order to establish a technology that is stable and repeatable, standardization or unification of the components is needed for the multi-material packaging system for it to be produced.

The challenge of FLSI can therefore be defined as culmination of functional features with diverse scales of length within a solitary component. This definition arises considering the fundamental relation of machined structures as well as functional features. Thus, it can be concluded that integrating diverse functions within a solitary system is tough considering diverse features which need to be manufactured in a reliable and cost-effective manner at nanoscales. It should also be considered that such nanocomponents should be manufactured via advanced and complex technologies with unique combinations which could further lead to a new set of manufacturing challenges which might be having their own set of constraints and could be difficult to address. There are multiple upcoming ideas of components dependent on the concepts of FLSI which are the base of multi-disciplinary programs associated with R&D [13]. Some product examples can be considered in this case of design and production, some of which are shown in Figure 2.2.

One such example is the polymer platform comprising lab-on-chip to discover protein in applications pertaining to point-of-care testing created in the European FP6 project entitled SEMOFS, which combines a plane of plasmon resonance sensors and microfluidic actuators [13]. Another example is lab-on-a-chip possessing the biological laboratory functionality over a solitary substrate via sensors, channels or reservoirs to attain elevated sensitivity, lower consumption of samples, measurement automation as well as standardization along with a speedy analysis [14]. Production of contact lenses embracing nano interconnects of metal within a polymer that is biocompatible as well as light-emitting diodes is yet another instance of such technology [15]. This on the whole broadens the spectrum of novel components needed by consumers like see-through displays powered remotely as well as through wireless means which can have applications in training, gaming including manufacturing.

2.2.2 TECHNOLOGY CONVERGENCE

In the last 20 years, there has been tremendous research pertaining to high precision and nanomanufacturing which primarily worked on optimization of machining

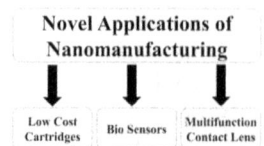

Novel Applications of Nanomanufacturing		
Low Cost Cartridges	Bio Sensors	Multifunction Contact Lens

FIGURE 2.2 Novel applications of nanomanufacturing [8].

processes in order to obtain exceptional resolution of features as well as good surface integrity. The major portions of these research studies removed material characteristics and microstructure from process variables at the time of nanostructured ultra-precision machining to attain the optimum response in machining. Likewise, other studies indicated growth in integrated processing chains which are positively influenced by concurrent optimization of material processing techniques in a cost-efficient way in adherence to the present needs of multiple existing as well as novel applications of engineering. There are many such studies in support of this novel trend and one major supporting study has come up in Japan, which is the incorporation of MEMS technology with nanophase material to produce third generation of MEMS. One hurdle in the exploration of the potential of such conjugated chains of process to achieve high accuracy and ultra-precision is the lack of information pertaining to material interactions at the nanoscale [16]. This further hampers the structuring and usage of nanomanufacturing for diverse spectrum of applications. This concept of combining the process of ultra-precision machining along with methods for depositing metallic nanophases as well as amorphous depositions has given rise to novel nanomanufacturing methods of the top-down and the bottom-up approaches. These novel methods paved way for unique platforms for manufacturing at nanoscales or the miniatured components with extremely high levels of precision.

The novel theory of top-down and the bottom-up approach generates new paths to make use of the modern advancements. This lays a foundation for exploiting refinement in depositions of materials, master making, replication technologies and direct structuring. This transforms the present capabilities of functional-scale incorporation, 3D processing as well as the ability to obtain extremely good surface integrity.

To achieve this kind of manufacturing ability, research needs to be carried out in multiple disciplines of engineering to give way to the novel advanced methods that are way different and higher in standard than the conventional technologies that are existing now. The optimization of the process parameters in such advanced manufacturing techniques will further bring down the production costs of the nanocomponents. Therefore, the pre-requisites needed for the growth of high-value-added nanocomponents shall be achieved and there shall also be advancements across diverse fields of the industry such as opto-electronics, biotechnology, automotive, ultra-precision engineering, medical, consumer goods and printed electronics [17].

The novel nanomanufacturing methods can help overcome the current challenges posed by nanomanufacturing. The nanomanufacturing methods such as directed assembly as well as transfer approaches involve the process of selective addition of materials which doesn't require removal of the material which can lower the wastage as well as the number of processes needed. The nanocomponents such as polymer nanostructures, nanoparticles and carbon nanotubes use the directed assembly technique with exceptional rate, directed and assembly that is precise in order to produce nanosystems. The techniques used shall employ the individualistic as well as the synergistic characteristics of nanoscale materials. These novel nanomanufacturing methods are carried out at ambient temperatures as well as pressures which significantly lowers the expenses pertaining to nanomanufacturing tools and machinery. This also ensures sustainability over a long period of time as wastage of the resources like consumables, energy along with costs are in control. The lower costs are due to the operation of the processes at ambient pressures as well as temperature which doesn't include extreme temperatures of vacuum. Thus, maintenance-related costs, equipment costs as well as energy used are in complete control. Also, the methods associated with directed assembly are simple like dip coating or spinning-based which lower the operational and tool-associated expenditure [18]. Moreover, lower quantities of processes, usage of energy, material as well as water shall greatly come down. Also, the techniques of transfer and directed assembly have higher rates and can be scaled. The beginning stages while designing the product comprise the opportunities which can lead to generation of environmentally friendly products which is possible when nanomanufacturing methods are assessed. This product analysis and assessment with respect to the economy and the environment, their usage as well as their disposal at the end while production, is essential during the developmental stage rather than facing the consequences later on when they rise.

2.3   NANOMANUFACTURING PROCESSES

Many nanomanufacturing methods for mass production of nanoproducts have been introduced. The two basic, diverse methods in which nanomaterials can be fabricated are the bottom-up and top-down approaches [19]. The primary difference between the two methods is that the bottom-up approach has a scientific research source of the nanoscience whereas the top-down approach doesn't have so. The bottom-up technique involves producing nanocomponents in same nanosphere which basically comprises the molecules, atoms as well as clusters. As a result of which this cluster or group can be easily managed leading to enhanced functionality of those nanostructures. The method of top-down gets its roots from the microelectronics which believes in reduction or tearing down of the bulk systems into nanosystems which being the current state thereby resulting in a highly effective technology in the existence. Therefore, the reduced scale of the components or devices thus obtained will be nanoscale. Figure 2.3 shows the basic principle involved in both the top-down as well as the bottom-up nanomanufacturing techniques in producing nanoscale structures [20].

Either of the manufacturing techniques produces components which have similar sizes. Moreover, there is a convergence in the aspect of size range in both the techniques when the structures are compared. In case of the bottom-up approach, the control at nanoscale, customization in terms of designs, nature of materials to be used are extremely high but in top-down technique, material acquisition is the only primary criteria and there isn't a good control as in the case of the former technique. At the nanoscale, top-down approach is advantageous for large orders and it facilitates macroscopic device connection [21]. The bottom-up method is ideal for arranging and producing short-range orders. Therefore, combining the two techniques would result in an extremely good integration needed for nanomanufacturing.

The top-down method is ideal for microscale components. The bottom-up method is best suited for nanoscale components as it combines clusters of atoms to form

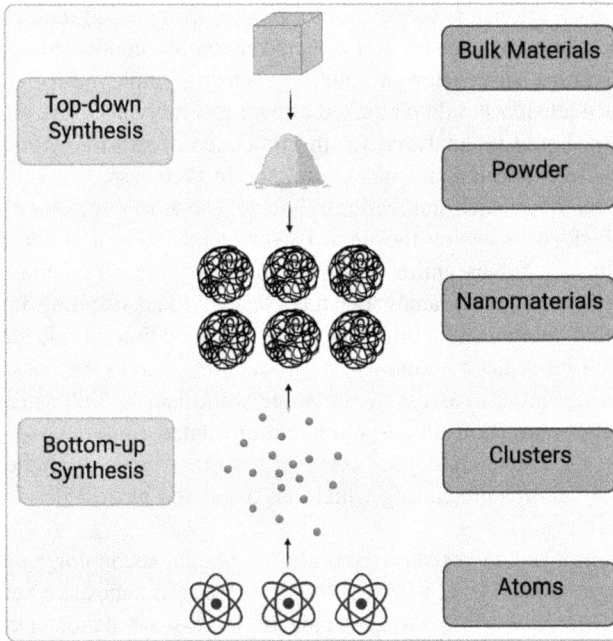

FIGURE 2.3   Top-down and bottom-up manufacturing methods [20].

nanoscale components. Therefore, integrating the two techniques shall broaden the spectrum of nanomanufacturing. Physical as well as chemical techniques are employed in bottom-up method for the production of nanoscale components by combining them to gradually produce larger structures [22]. Therefore, when contemplating the integration of both methods, when reduced to nanoscale, this technique forms the right counterpart for the top-down technique. The applications related to the biomedical field are the ideal motivation for this integration where natural forces bind with their chemical counterparts in production of nanostructures.

2.4   CHALLENGES IN NANOMANUFACTURING

Among the multiple challenges faced by nanomanufacturing technology, primarily the challenges can be cultural as well as technical. One such illustration is expenditure of billions of dollars by the electronics industry over manufacturing facilities and not changing its stance on upgradation unless there is significant enhancement performance-wise over the existing one apart from lower cost. Majority of the challenges w.r.t technology are associated with scalability without compromising on the characteristics of nanoscale [23,24]. Macroscopic systems with nanomaterials not having the nanoscale characteristics is the challenge when length scales are macroscopic. For this to be possible, large surface areas need to be controlled with forces being at nanoscale, resulting in a heterogeneous, multi-scale nanomanufacturing which can be fast, scaled and has repeatability along with high yield. Therefore, it is essential to address the aspects of flexibility and compatibility in nanomaterial-developed processes. For a process comprising multiple layers, for the several processes of transfer and directed assembly conducted in a sequence, influence of solvent, suspension or solution viscosity employed for single or next process, surface tension should be looked at over assembly needs. A process having many transfers should be analyzed for the influence of substrate compliance over assembly, transfer, adhesion and area of contact. In such cases, the substrate should be flexible. Moreover, such innovations shall give rise to components with nano-materials embedded, in which the impact over health as well as the environment must be established for its entire life cycle. It is essential to responsibly develop nanoproducts by effectively analyzing the life cycle and deciding instead of en-dorsing scalable nanomanufacturing or commercialization of the product since nanomanufacturing scales to commercial production. Due to the possibility of ha-zards, it is also essential to assess the probable workplace as well as environmental exposures at the fabrication time to eliminate unwanted consequences.

Apart from the above challenges some of the other types of challenges can be seen in case of nanomanufacturing which are discussed next [25].

- Tools employed in research laboratories of nanotechnology possess re-solution at atomic level to characterize surfaces at nanoscale but they are insufficient in meeting the demands of large-scale nanomanufacturing. Atomic force microscopy (AFM) can be considered as an illustration where the resolution is obtained at atomic level but at a rate that is ex-tremely slow and it won't be possible to make use of AFM for large-scale commercial operations.
- The methods of nanomanufacturing producing large-scale nanodevices are in their infancy, and meeting the market demands isn't possible with the industrial capacity currently available.
- The mechanical components that move in the nanotechnologies come across novel failure modes such as cracks, fatigue and stiction.
- It is a complex job to observe, characterize nanomaterials or nanomanu-facturing in which mechanisms are multi-scale.
- Nanoprocesses and nanosystems possess many sensing issues like diffi-culty in accessibility of signal source, inability to sense in-situ, weak and short nature of the signals, lower signal-to-noise ratio and difficulty in transduction of signals when quantized.
- Unclear information on the nature of nanomaterials leads to an inability to judge their toxic nature.
- There is no clear picture of the market scenario of nanoproducts irrespective of consumers' interests. Moreover, nanotechnology is driven primarily by interested scientists and not the public.

2.5   RECOMMENDATIONS FOR NANOMANUFACTURING RESEARCH

The following are some of the recommendations for research to be carried out on the nanomanufacturing process [25].

- More realistic statistical models based on physics must be developed keeping the hierarchy and nonlinearity in mind. These models will then make it possible to monitor, diagnosis, prognosis and analyze reliability.
- In order to evaluate how reliable nanodevices are, many physical domains such as barometric domains, fluidic, electrical, mechanical and optical must be studied.
- Probability distributions are to be established of critical variables in nanoprocesses failures as well as response variables of quality.
- Analysis of methods for evaluating accelerated degradation in the presence of many stresses and efficient test plans needs to be carried out.
- It is essential to analyze the capacitance as well as the sub-wavelength in case of optics for dimensional sensing.
- To find out about sensing of chemical species it is important to study the surface-enhanced spectroscopy methods such as Raman, fluorescence, electrical impedance and Fourier transform infrared should be analyzed.
- Development of a lexicon should be done for nanomanufacturing sensors.
- Methods must be introduced to study the chaos, non-linear dynamics, to detect rare events and to look at state variable spatio-temporal distribution.
- Methodologies to enhance precise placements of nanocomponents in places of desire in self-assembly are needed for large dimensional systems.
- Development in the principles of design for manufacturing, assembly and others is needed to facilitate scalable nanomanufacturing techniques.
- Since nanomanufacturing requires novel production methods, safety as well as precautions against health, an enhanced premium is needed for flexibility in manufacturing.
- Strict specifications are needed for nanomaterial characteristics, needs for nanointermediates special processing and precise control of critical parameters of performance parameters to facilitate efficiency and cost-effectiveness in nanomanufacturing supply chain for the long run.
- Languages specific to processes that exist for CVD as well as the molecular beam epitaxy are needed for nano EDM, dispersion, block patterning and direct-write e-beam nanomanufacturing methods.
- Research on nanoinformatics just like bioinformatics should be done to reveal the discoveries based on data.
- Tools of analysis as well as instrumentation for complete nanomaterial characterization are needed with online control of the process to be made to define specific characteristics for certain applications.
- Applications for enhancing or replacing the current goods or providing novel devices need to be identified. Such applications commercial success must be analyzed and costs must be decided.
- Nanoproducts of the consumers should be made to acquire wider acceptance which can enhance functional performance thereby reducing costs of the life cycle.
- Incentives of the government need to be looked into for a significant jump in the capabilities of the nanoprocesses.

- The issues related to control and planning of the overall life cycle should be investigated since there can be issues of end-of-life disposal in case of nanoproducts.
- Investigations on hazards of occupation, nanoproducts and safety of the workplace must be investigated thoroughly.
- The needs of nanomaterials as per Occupational Safety and Health Administration (OSHA) should be defined along with nanomaterial handling procedure defined.
- Studies on the influence of natural nanoparticles over environment as well as the health of the humans need to be done and safe limits of exposure to these materials need to be determined.
- Creation of opportunities must be carried out for the partnership developments of various academic communities to instigate research in collaboration.
- Course work on nanofabrication as well as nanomanufacturing must be inducted into the higher education system.
- Encouragement is needed for research on multi-discipline, collaboration, partnerships and transfer of information, discussions over issues from diverse perspectives such as materials engineer, IE, physicists, etc.
- Establishment of collaborative innovation is needed to overcome the issues of technology and science that are related to nanomanufacturing.
- The nanomanufacturing needs to be scaled up through the defining of diverse models of the products along with various standards such as ASME for tolerances, ASTM and ISO for engineering, IJES and STEP for geometry, specifications for biological as well as physical characteristics.
- A certain type of methodologies is needed for quality engineering to be determined to give the following:
  i. Guidelines in designing and analyzing experiments for optimization of the settings of the nanoprocesses.
  ii. Methods of online monitoring as well as diagnosis for reduction of nanoprocess variation as well as its downtime at the time of production.
  iii. Plans for continuous enhancement in quality as well as yield.

2.6   CONCLUSION

Multiple applications such as electronics, medical devices, sensors, functional structures and energy storage or harvesting can be produced via the methods of transfer process platform as well as directed assembly through integration of diverse nanocomponents comprising a functionality that is explicit. It is essential that engineers or scientists focus on this aspect of nanomanufacturing and understand how capable it is. This shifting paradigm in the field of nanomanufacturing shall pave way to novel, creative and sustainable methods which will help in the development of diverse sectors in other fields. Apart from providing exceptional potential and advancement in terms of technology, nanomanufacturing benefits the environment, by reducing the wastage of resources, reducing energy wastage, providing cost efficiency and improving the economy. Implementing a systematic method of nanomanufacturing shall effectively commercialize emerging industries. The advancements in the field of nanomanufacturing shall overcome the present barriers and this will pave way to innovation and bring diverse novel industries. This will eventually democratize the field of nanomanufacturing through play levelling. As multiple novel techniques come into existence for multi-material nanomanufacturing, the researchers should put their foot forward and try to analyze the complexity of these novel technologies and must try to understand and assess the issues related to them. This will further result in development and also help the experts to adopt and commercialize these novel nanomanufacturing techniques by overcoming the challenges that come their way and also implementing these methods in various opportunities across fields.

REFERENCES

1. Busnaina AA, Mead J, Isaacs J, Somu S. Nanomanufacturing and sustainability: Opportunities and challenges. *Nanotechnology for Sustainable Development*. 2013: 331–336.
2. Li L, Long R, Prezhdo OV. Why chemical vapor deposition grown MoS2 samples outperform physical vapor deposition samples: Time-domain ab initio analysis. *Nano Letters*. 2018 May 18; 18(6):4008–4014.
3. Keating S. Beyond 3D printing: The new Dimensions of additive fabrication. *Designing for Emerging Technologies: UX for Genomics, Robotics, and the Internet of Things*. 2014 Nov 7; 379–406.
4. Cantatore E . Applications of organic and printed electronics. In *Atechnology-enabled revolution*2013. New York: Springer. Pg 1–26.
5. Cooper KP, Wachter RF. Challenges and opportunities in nanomanufacturing. In *Instrumentation, Metrology, and Standards for Nanomanufacturing, Optics, and Semiconductors V*. 2011 Sep 20; 8105:810503. International Society for Optics and Photonics.
6. Thiruvengadathan R, Korampally V, Ghosh A, Chanda N, Gangopadhyay K, Gangopadhyay S. Nanomaterial processing using self-assembly-bottom-up chemical and biological approaches. *Reports on Progress in Physics*. 2013 May 30; 76(6):066501.
7. Rodrigues SM, Demokritou P, Dokoozlian N, Hendren CO, Karn B, Mauter MS, Sadik OA, Safarpour M, Unrine JM, Viers J, Welle P. Nanotechnology for sustainable food production: promising opportunities and scientific challenges. *Environmental Science: Nano*. 2017; 4(4):767–781.
8. Dimov S, Brousseau EB, Minev R, Bigot S. Micro-and nano-manufacturing: Challenges and opportunities. *Proceedings of the Institution of Mechanical Engineers, Part C: Journal of Mechanical Engineering Science*. 2012 Jan; 226(1):3–15.
9. Bartos P, Hughes JJ, Zhu W, Trtik P, editors. *Nanotechnology in construction*. Royal Society of Chemistry; 2004.
10. Evoy S, DiLello N, Deshpande V, Narayanan A, Liu H, Riegelman M, Martin BR, Hailer B, Bradley JC, Weiss W, Mayer TS. Dielectrophoretic assembly and integration of nanowire devices with functional CMOS operating circuitry. *Microelectronic Engineering*. 2004 Jul 1; 75(1):31–42.
11. Leng J, Lan X, Liu Y, Du S. Shape-memory polymers and their composites: Stimulus methods and applications. *Progress in Materials Science*. 2011 Sep 1; 56(7):1077–1135.
12. Tiquia-Arashiro S, Rodrigues DF. *Extremophiles: Applications in nanotechnology*. New York: Springer International Publishing; 2016 Sep 29.
13. Nestler J, Morschhauser A, Hiller K, Otto T, Bigot S, Auerswald J, Knapp HF, Gavillet J, Gessner T. Polymer lab-on-chip systems with integrated electrochemical pumps suitable for large-scale fabrication. *The International Journal of Advanced Manufacturing Technology*. 2010 Mar; 47(1):137–145.

14. Vazquez RM, Osellame R, Nolli D, Dongre C, van den Vlekkert H, Ramponi R, Pollnau M, Cerullo G. Integration of femtosecond laser written optical waveguides in a lab-on-chip. *Lab on a Chip*. 2009; 9(1):91–96.

15. Ho H, Saeedi E, Kim SS, Shen TT, Parviz BA. Contact lens with integrated inorganic semiconductor devices. In 2008 IEEE 21st International Conference on Micro Electro Mechanical Systems 2008 Jan 13 (pp. 403–406). IEEE.

16. Gaddi R, Van Kampen R, Unamuno A, Joshi V, Lacey D, Renault M, Smith C, Knipe R, Yost D. MEMS technology integrated in the CMOS back end. *Microelectronics Reliability*. 2010 Sep 1; 50(9–11):1593–1598.

17. Kumar K, Zindani D, Kumari N, Davim D. *Micro and nano machining of engineering materials*. Cham: Springer; 2019.

18. Maaskant E, de Wit P, Benes NE. Direct interfacial polymerization onto thin ceramic hollow fibers. *Journal of Membrane Science*. 2018 Mar 15; 550:296–301.

19. Biswas A, Bayer IS, Biris AS, Wang T, Dervishi E, Faupel F. Advances in top–down and bottom–up surface nanofabrication: Techniques, applications & future prospects. *Advances in Colloid And Interface Science*. 2012 Jan 15; 170(1–2):2–7.

20. Devatha CP, Thalla AK. Bhagyaraj Sneha, Oluwafemi O Samuel Kalarikkal Nandakumar, Thomas Sabu . Green synthesis of nanomaterials. In *Synthesis of inorganic nanomaterials* 2018 Jan 1 (pp. 169–184). Woodhead Publishing. New York

21. Smith KH, Tejeda-Montes E, Poch M, Mata A. Integrating top-down and self-assembly in the fabrication of peptide and protein-based biomedical materials. *Chemical Society Reviews*. 2011; 40(9):4563–4577.

22. Fang FZ, Zhang XD, Gao W, Guo YB, Byrne G, Hansen HN. Nanomanufacturing—perspective and applications. *CIRP Annals*. 2017 Jan 1; 66(2):683–705.

23. Imboden M, Bishop D. Nanomanufacturing. *Physics Today*. 2014; 67(12):45.

24. Chryssolouris G, Stavropoulos P, Tsoukantas G, Salonitis K, Stournaras A. Nanomanufacturing processes: A critical review. *International Journal of Materials and Product Technology*. 2004 Jan 1; 21(4):331–348.

25. Bukkapatnam S, Kamarthi S, Huang Q, Zeid A, Komanduri R. Nanomanufacturing systems: Opportunities for industrial engineers. *IIE Transactions*. 2012 Jul 1; 44(7): 492–495.

3 Intense Classification of Nanomanufacturing

*H.J. Amith Yadav, D.H. Manjunatha, D. Gayathri,
V.S. Patil, and M.N. Kalasad*

CONTENTS

3.1   Introduction...32
3.2   Semiconductor Nanocrystals...33
    3.2.1   Quantum Dots..33
    3.2.2   Quantum Rods...34
3.3   The Top-Down Approach...35
    3.3.1   Nanoimprint Lithography..35
    3.3.2   Ball Milling Technique...35
    3.3.3   Pulsed Laser Synthesis..35
    3.3.4   Radio Frequency Sputtering Method...35
3.4   The Bottom-Up Approach..36
    3.4.1   Chemical Vapour Deposition..36
    3.4.2   Dip Pen Lithography..36
    3.4.3   Self-Assembly...36
    3.4.4   Solution Combustion Synthesis..36
    3.4.5   Hydrothermal Synthesis..36
    3.4.6   Ultrasonochemical Synthesis..37
    3.4.7   Sol-Gel..37
    3.4.8   Co-precipitation Method...37
3.5   Nanomanufacturing Technologies...38
    3.5.1   Nanomechanical Machining..38
    3.5.2   Nanolithography..38
    3.5.3   Energy Beam Machining...39
    3.5.4   Deposition and Etching...39
    3.5.5   Nano Printing..39
    3.5.6   Nanoassembly...40
    3.5.7   Nanoreplication...40
3.6   Conclusion...40
References..41

DOI: 10.1201/9781003220602-3

3.1   INTRODUCTION

Significant progress in the developments in the field of nanoscale materials leads to the better understanding, of materials in the range of 1–100 nm have different properties and applications [1–3].

In the last three decades, nanomaterials' research gained the importance due to their distinctive properties anddue to their size and shape dependent properties. At the nanometre scale, the atomic and molecular interactions may strongly influence the macroscopic properties of materials, resulting in changes in mechanical, electrical and optical properties [4–6]. There are two important parameters which make these materials more attractive from a scientific and technological viewpoint, one is the high density of defects on the surface and second is the size of particles that is in quantum regime. Here, the range of nanoparticles is modest or tantamount to the de Broglie wavelength of electrons/holes. systems. Nanocrystals have unique properties compared to bulk solids. The number of atoms on the crystal's surface, for instance, is a significant fraction of the total number of atoms, and therefore will have a large influence on the overall properties of the crystal [7].

Materials in the nanometre scale have two distinctive effects, one is the size effect – in bulk materials energy levels are continuous, but when size of the particle is extremely small, energy levels become discrete, this effect is called the quantum confinement effect. This is shown in Figure 3.1; this effect modifies the physical properties of the materials [8–12]. Another size effect is the variation of density of states as a function of nanoparticles dimension is also shown in Figure 3.2.

The second effect is the surface or interface-induced effect as size of the particle becomes extremely small number of particles located at the surface goes on increasing, i.e., surface volume ratio increases enormously compare to that of bulk

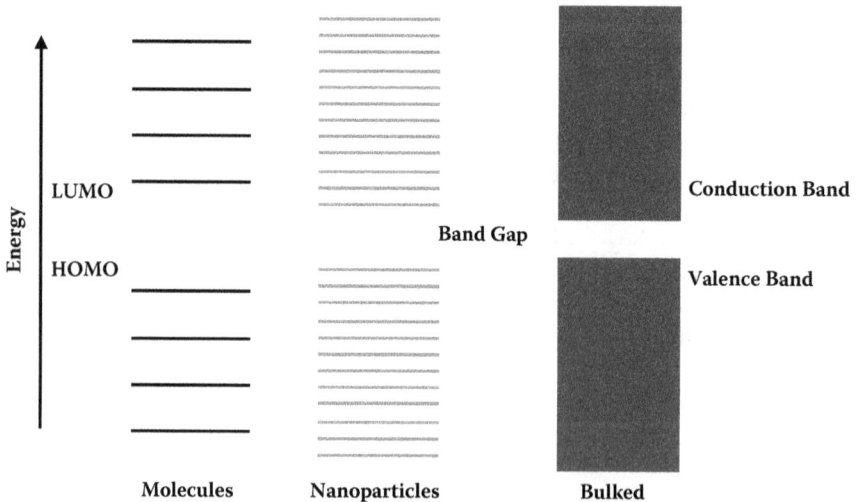

FIGURE 3.1   The quantum size effect: Available energy levels with different states of matter, such as molecular clusters, nanoparticles and bulk (periodic lattice).

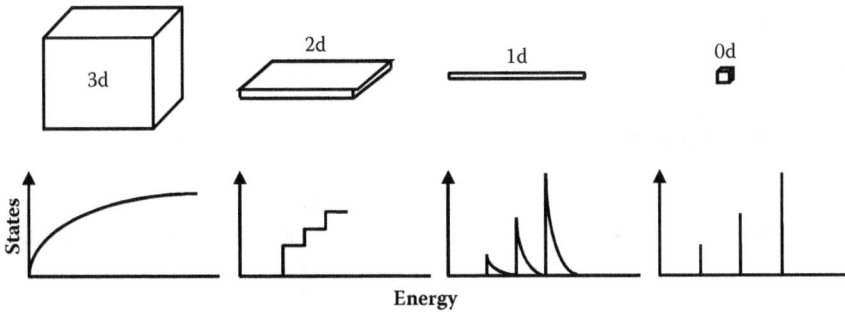

FIGURE 3.2 Evolution of the density of states (DOS) vs. dimensionality; confinement potential in Bulk, QWs, QWRs and QDs.

materials, which creates a different chemical environment, and this effect of nanoparticles also responsible for the changes in chemical, optical and mechanical properties. So we can say size of the particles also decides the thermo dynamical behaviour such as specific heat, melting point, thermal conductivity and vapour pressure of metals and semiconductors nanoparticles [13].

3.2 SEMICONDUCTOR NANOCRYSTALS

3.2.1 QUANTUM DOTS

Among these nanomaterials, semiconductor nanoparticles play a key role in the development of technological applications. Quantum dots (QDs) are tiny particles of a semiconducting material with diameters in the range of 2–10 nm. They are also known as "artificial atoms" because their quantized energy states are closer to atoms than bulk materials [7,14–16]. The electrons in QDs are restricted in small spaces and its radii are smaller than Bohr radius because of their small size. QDs exhibit highly size-dependent optical, electronic and fluorescence properties. The energy gap of QDs increases due to decreases in their size. The excellent photostability and bright emission make QDs powerful substitutes for organic fluorophores (fluorescent dyes) for variety of biological applications [17]. Further, these particles possess some special characteristics such as size-tunable colour, broad absorption, narrow emission, longer lifetimes and high extinction coefficient. QDs had many applications such as LEDs, photovoltaic, transistors and quantum computing [18–22]. Among the semiconductor quantum dots, metal chalcogenides class has been a promising candidate for various applications. Lot of work on the synthesis of metal chalcogenides has been reported, which includes single molecule precursor route, solvothermal route, sonochemical route, microwave irradiation, hydrothermal, the organo-metallic precursor route, etc. In most of the solution-based synthesis approaches, it is generally observed that the precursors and solvents used are not only expensive but also hazardous to nature. Hence, a robust, feasible approach resulting in superior quality quantum dots is quite essential for implementation in various applications [5,6,23].

3.2.2 QUANTUM RODS

Basically, quantum rods (QRs) are the single nanoparticles with elongated shapes that possess good advantages as compared to QDs like polarized emission. QRs show better photoluminescence properties [24,25]. Quantum rods are used for biomedical applications.

QRs have faster radioactive decay rate, larger absorption cross-section, bigger Stokes shift and multiple binding moieties. Generally, a single quantum dot exhibits plane polarized light but a single quantum rod will show linearly polarized emission. But in the case of QRs, the emission is reversibly switched on-off by external electric fields [26–28] and better charge separation/transport [29]. QRs bring new possibilities in the field of biological imaging. The transmission images of CdS nanorods are shown in Figure 3.3.

The concrete fruit of the nanotechnology revolution is nanomanufacturing, which is the financially adaptable and monetarily economical large-scale manufacturing of nanoscale materials. Nanomanufacturing procedures, unlike those employed in nanofabrication for research, must meet extra expense and time-to-showcase limitations.

Nanomanufacturing entails the production of nanomaterials at scaled-up, dependable and cost-effective levels [30]. Nanomanufacturing is the foundation of nanotechnology and nanoscience and consists of many techniques for controlling material structures and production at the range of 1–100 nm for 1–3 dimensions. Manufacturing of nanomaterials at accuracy and in the range of 1–100 nm in general is included in nanomanufacturing. The parts made by nanomanufacturing can be divided into two types:

1. Parts with nano- and microstructures having nanometre-scale. Components are either microscopic or macroscopic in size.
2. Parts having nanometric precision in geometrical correctness. On a macroscopic scale, the gadgets are used.

(a)   (b)

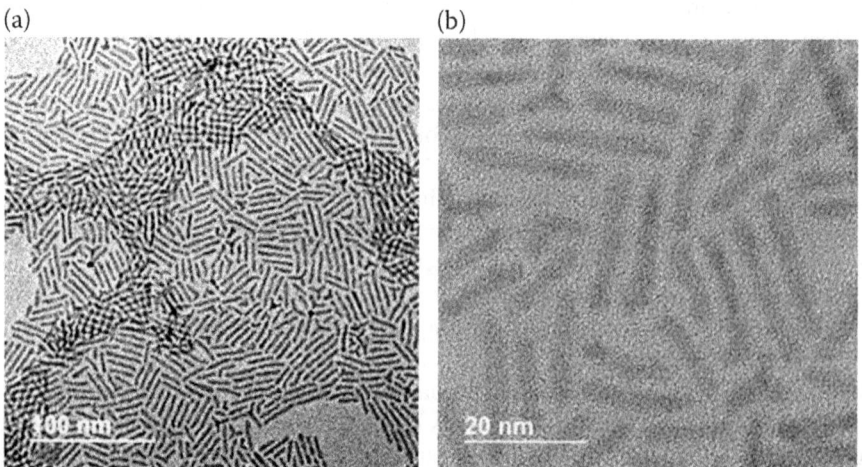

FIGURE 3.3   (a) TEM and (b) HRTEM images of CdS nanorods.

Nanomanufacturing can be approached in two ways. The first is the top-down technique, which entails lowering big materials to the size of nanoparticles and the next one is the bottom-up strategy, which entails developing things by assembling them from atomic- and molecular-scale components. Although the terms nanofabrication and nanomanufacturing are frequently interchanged, they are two distinct concepts with distinct economic implications. Nanofabrication is the study and testing of the feasibility of constructing nanoscale materials and processes, primarily in the laboratory, whereas nanomanufacturing is the production of nanotechnology-based things on an industrial scale, with a focus on low cost and reliability.

3.3   THE TOP-DOWN APPROACH

3.3.1   NANOIMPRINT LITHOGRAPHY

Nanoimprint Lithography is a technique for "stampeding" or "printing" tiny features onto a surface. Canon Nanotechnologies, the market and mechanical pioneer for high-resolution, minimal expense of-proprietorship nanoimprint lithography frameworks and answers for the semiconductor business, is an excellent example of such a method. Their ground-breaking *Jet and Flash Imprint Lithography* technology makes the tiny elements expected in the present best-in-class semiconductor memory gadgets.

3.3.2   BALL MILLING TECHNIQUE

Ball milling is a mechanical technique widely used to grind powders into fine particles and blend materials. Commercially available metal oxide powder is used as a precursor. Metal oxide powder was milled in a mechanically milling machine at some rpm for a duration of 15–25 h in stainless steel via atmospheric pressure and temperature. The ball to metal oxide ratio is taken into account for this method. Very fine nanoparticles are obtained [31].

3.3.3   PULSED LASER SYNTHESIS

This technique become more popular in 1980 because it was used to grow superconducting films. This technique is used for the synthesis of nanomaterials, quantum dots and nanotubes. Pulsed laser synthesis consists of laser beam of exact wavelength moved to transparent window of the reactor. At point when it encroaches outer layer of a strong objective, the light collaborates with matter and produces nanoparticles. It is feasible to dope the integrated material by picking a forerunner containing the dopant component as a reactant during laser irradiation [32].

3.3.4   RADIO FREQUENCY SPUTTERING METHOD

From this method, we get some desired applications of some metal oxide. Applying RF power in the sputtering chamber generates the plasma. The base pressure in the chamber was some torr and Argon gas was introduced. Obtained final product varies in thickness in terms of nanometres [33].

This process consumes a lot of energy, employs chemicals (some of which are quite dangerous) and generates trash. The outcomes are frequently one-of-a-kind and difficult to duplicate. Therefore bottom-up approaches are becoming more popular.

3.4   THE BOTTOM-UP APPROACH

3.4.1   CHEMICAL VAPOUR DEPOSITION

Chemical reaction that produces ultra-pure, high-performance films. Grolltrex is a single-layer graphene sheet manufacturer based in the United States that uses a unique transfer and processing chemical vapour deposition technology. This approach enables Grolltrex to produce high-performance graphene products while also lowering prices.

3.4.2   DIP PEN LITHOGRAPHY

Atomic force microscope's tip is "dipped" in a chemical solution and afterward used to "mark" on a surface, similar to an older style ink pen on paper.

3.4.3   SELF-ASSEMBLY

A method of assembling a group of components into an ordered structure without the use of outside guidance. This notion, which has proven particularly relevant in the field of nanotechnology, is still being researched by scientists. Indeed, as shrinking progresses to the nanoscale, traditional manufacturing approaches are failing because machinery capable of assembling nanoscale components into functioning devices has yet to be developed.

3.4.4   SOLUTION COMBUSTION SYNTHESIS

Prof. Patil was first to report the solution combustion synthesis. It is a useful, minimal expense technique for the creation of different nanoparticles [34]. Its reaction is controlled by monitoring the fuel-to-oxidizer ratio (F/O). Metal nitrides and fuel are used precursors required for synthesis. Metal nitrides are dissolved in distilled water. The reaction mixture is constantly stirred on a magnetic stirrer and then it will be heated in a muffle furnace until the final product is obtained. The advantage of this method is simple, fast, eco-friendly, efficient and scalable.

3.4.5   HYDROTHERMAL SYNTHESIS

Another name for hydrothermal synthesis is salvo thermal synthesis. In this method, morphology, pH, temperature and size of the nanomaterials can be altered. In this method autoclave is used, autoclave acts as the temperature gradient. The reaction mixture is stirred on a magnetic stirrer. Then the reaction mixture is transferred to an autoclave and it is placed in a hydrothermal oven for a certain temperature. The final product is obtained [35].

3.4.6 ULTRASONOCHEMICAL SYNTHESIS

Stoichiometric amount of metal nitrides was blended appropriately using a magnetic stirrer to get a clear solution. Ultrasound sonicator maintained at certain conditions irradiates the above solution [36]. The mixture is filtered and dried and the final product is obtained.

3.4.7 SOL-GEL

Stoichiometric amount of metal nitrides was mixed properly using a magnetic stirrer. stirrering is moveddone at a rate of some rpm maintained at a certain temperature after that solution is ultrasonicated the formed metal oxide in gel medium is filtered with Whatman filter paper. The separated metal oxide is washed with double distilled water to remove impurities and gel medium. Obtained product is dried in oven by keeping at certain temperature for 24 h. After drying, the final product is ready for characterization. Advantages of this method are simple, high purity, high productive efficiency and designing chemical composition [37].

3.4.8 CO-PRECIPITATION METHOD

The technique of coprecipitation becomes increasingly significant to produce nanomaterials because simple set up required. In this process, aqueous metallic salts are mixed at adequate temperatures with a base, which acts as a precipitant. The disadvantage of this method is persistent washing and calcination to accomplish an unadulterated period of phosphor [38].

Despite the enormous range of applications for nanomanufacturing (electronics, healthcare, energy, environmental difficulties and so on), substantial challenges explain why the move from lab demonstration to industrial-scale manufacturing has been gradual. The most significant challenges are as follows:

1. Developing economically feasible production procedures.
2. Controlling the accuracy with which nanostructures are assembled.
3. Establishing procedures for defect control and testing the reliability of the system. In the semiconductor business, defect control is now non-selective and consumes 20%–25% of overall manufacturing time. Defect removal for nanoscale systems is expected to take substantially longer due to the need for selective and thorough elimination of contaminants.
4. Maintaining nanoscale attributes and nanosystem quality in high-rate and high-volume production.
5. Evaluating the environmental and societal consequences of nanotechnology, which resulted in the pollution of trillions of microscopic plastic particles in the oceans, waterways, and even human bodies?

Nanomanufacturing takes more time and money to scale, and the health and environmental concerns associated with the manufacture of novel nanoelements must be addressed swiftly. If these difficulties are resolved, nanomanufacturing will play a vital role in many sectors' advances.

If nanomanufacturing was once mostly utilized for electronics, it is now finding uses in a variety of fields. Manufacturing flexible solar cell rolls has lowered installation costs in the solar energy sector, for example. Nanomaterials-based batteries can be recharged significantly better than traditional batteries. It also helps to improve air quality by enhancing the performance of catalysts that convert vapours escaping from autos and industrial plants into innocuous gases. Nanomanufacturing had many applications for the construction industry. It has provided strength to the buildings and has made maintenance easier. All these nanomanufacturing applications rely on a rising number of processes, as indicated above.

3.5   NANOMANUFACTURING TECHNOLOGIES

Nanomanufacturing aims to produce nanomaterials with accuracy. Nanomechanical machining, nanolithography, energy beam deposition/epitaxy and replication are only a few of the high-controllability nanoprocesses used in these technologies. Final product outcomes are crucial for performance control in the manufacturing industry, which is nanometrology's main responsibility.

3.5.1   NANOMECHANICAL MACHINING

Cutting, grinding and polishing are the most common mechanical machining operations. Material removal and machining accuracy at the nanometric level are both parts of nanomechanical machining. For producing high-performance functional surfaces, nanocutting and grinding are commonly utilized. Nanocutting can be utilized to achieve precise accuracy. Furthermore, it is capable of quickly roughing out massive amounts of material. The end-fly-cutting-servo approach, for example, has been used to machine multi-scale and biomimetic surfaces. [39]. Containing a mean height of 395 nm and a standard deviation of 15 nm, one type of ellipsoid micro-aspheric array with nanopyramids was manufactured [39]. However, as compared to the polishing process, the efficiency of these procedures is mostly influenced by machining accuracy [40].

3.5.2   NANOLITHOGRAPHY

The study and manufacturing of nanostructures having a range of 1–100 nm are known as nanolithography. Nanolithography was created to allow for mass manufacture of integrated circuits (ICs) and microelectromechanical systems in the industry. At the moment, its minimal feature size manufacturing capability is constantly improving, making it a very active field of study in academia [41,42] and nanoscience, including nanomedicine for diagnosis and treatment [43] and nanoelectronics for denser and faster computing. Nanoelectrode manufacturing that is repeatable is a good technique to get probes for nanoelectrochemistry research [44]. With a geometric surface area of Au nanoelectrodes, these nanoelectrodes can be made consistently and reproducibly and can be described using traditional electrochemical techniques [45]. On SiN insulating sheet, nanolithography allows the manufacture of Au nanoelectrodes with a geometrical surface area of 160 nm × 1 μm [46].

3.5.3 ENERGY BEAM MACHINING

A concentrated energy beam is utilized to melt or vaporise the undesirable material while also removing it from the parent material. Ion beams, laser beams and electron beams are examples of common energy beams.

For site-specific examination, deposition and ablation of materials, focused ion beam (FIB) is often utilized in the semiconductor industry, materials research and increasingly in the biological area [47]. FIB can create diamond-cutting tools with nanometric edge radii for ultra-precision machining [48,49]. However, because of the nanoscale lateral damage generated by FIB, the performance of such diamond tools and the lifespan of the cutting edge would be decreased [49]. The basic effects of nanocutting and the minimum thickness of the chip (MTC) are experimentally evaluated to reduce this damage, and the manufacturing process is adjusted by investigating the tool edge radius for nanocutting and MTC [50,51]. An FIB setup is a scientific tool that looks like a scanning electron microscope (SEM). High-energy FIB (30 kV) can view and quantify diamond tool edge radii of roughly 16 nm [52]. Low-energy FIB has been shown to be effective in reducing gallium ion implantation depth and in reducing the thickness of the FIB-induced amorphous layer. The thinnest chip thickness of roughly 9 nm was obtained in a nanocutting experiment utilizing a diamond tool edge radius of 22 nm that was created using low-energy FIB (5 kV) without a coolant. Only after optimizing the tool's performance did the ratio of minimum chip thickness to tool edge radius (MTC/r) drop from roughly 1.3 to 0.3–0.4. As a result, it was determined that reducing FIB-induced diamond tool damage and adding a coolant during nanocutting are advantageous for enhancing nanocutting research and achieving the minimum chip thickness.

3.5.4 DEPOSITION AND ETCHING

Any process for saving a slender film of material onto a substrate or onto recently formed layers is referred to as deposition, specifically thin film deposition. The control layer is only a few tens of nanometres thick. Chemical vapour deposition (CVD) is a chemical technique that yields high-quality, high-performance solids. In the semiconductor sector, CVD is frequently employed to create thin films. Physical vapour deposition (PVD), on the other hand, uses physical techniques to evaporate a material, subsequently deposited on the coated item. Deposition, also known as epitaxy, is the formation of a solid film on a crystalline substrate in which the growing film's atoms imitate the arrangement of the substrate atoms. Epitaxy is vital in both fundamental research on thin film growth methods and its application to create high-quality crystal layers from various materials to accomplish technically relevant functionalities [53].

3.5.5 NANO PRINTING

While classic methods with low accuracy, printing method is one specialized additive nanomanufacturing process. Droplets smaller than the nozzle may now be produced at room temperature and at high speeds, thanks to electro hydrodynamic jet printing. Furthermore, nanojets are capable of producing nanostructures [54]. Compared to previous technologies, Nano printing produces nanostructures. 3D Nano printing, particularly, builds an item or structure layer by layer, both vertically and horizontally.

3.5.6 NANOASSEMBLY

Assembly structure is used to make nanosystems and nanostructures with complicated 3D geometry. The spontaneous self-assembly process is driven by chemical or physical interactions between atoms and molecules such as Coulomb forces, Van der Waals' forces and hydrogen bonds, and is sensitive to molecular configuration and chemical and physical environments [55]. Depending on application, the size of the fundamental element might range from atomic to microscale.

Nanoelectronics, in which the circuit consists of nanodevices, nanowires, or even huge molecules, is a key application of nanoassembly. This technology can readily overcome the lithography process's poor resolution. High-performance computation, mass storage, and miniaturization might all be realized as a result [56]. Large DNA molecules, for example, can be utilized as a template to build an electric connection at the nanoscale [57]. The gold electrode array is produced first, then oligonucleotide monolayers are applied. After that, it's dipped into a DNA solution with pre-programmed sequences and sticky ends. As a result of self-assembly activities between complementary DNA sequences, the DNA network is produced. Finally, to accomplish electronic operations, certain metal grains are joined to the network.

3.5.7 NANOREPLICATION

Nanoscale features can be created on surfaces having functional qualities such as hydrodynamic, mechanical, biological, chemical or optical capabilities by replicating them. In a CIRP keynote paper [58], replication on micro- and nanoscale was thoroughly discussed. The transfer of a master geometry onto a substrate by copying the master geometry is referred to as replication. The transfer can be generated by heat, force, chemical activation or other energy input or activation, according to the discussion. A variety of impacts are often noticed. To assure the transfer of geometry, the master and substrate material are usually in physical contact. The so-called LIGA method, which combines X-ray lithography, electroforming and moulding, first appeared in the 1980s [59,60]. The master geometry is generated on a resist and then electroplated onto Ni. This shape is utilized to replicate polymers in the future.

3.6 CONCLUSION

Nanomanufacturing encompasses a wide range of materials, gadgets, products and processes. Numerous commercial and academic applications have been developed to aid nanomanufacturing fast development. Due to the UV wavelength, optical applications mostly focus on nanometric precision for big mirrors. Other applications rely heavily on the nanostructure's unique characteristics. To prevent the side effects of the manufacturing process, alignment and encapsulation, it is advised that highly efficient production with decreased follow-up operations should be the primary focus due to large demand from society and industry.

REFERENCES

1. W. H. Freeman and Company (1968) *Materials, A Scientific American Book*, San Francisco.
2. M. S. Whittingham (2004) *Chem Rev*, **104**, 4271.
3. J. M. Tarascon, and M. Armand (2001) *Nature*, **414**, 359.
4. A. P. Alivisatos, (1997) *Endeavour*, **21**, 56.
5. M. A. El-Syed (2004) *Acc. Chem. Res.*, **37**, 326.
6. A. Thiaville, and J. Miltat, (1999) *Science*, **284**, 1939.
7. G. Schmid (2004) *Nanoparticles: From theory to application*. Weinheim: WILEY-VCH Verlag GmbH & Co. KgaA.
8. R. Rossetti, J. L. Ellison, J. M. Gibson, and L. E. Brus (1984) *J. Chem. Phys.*, **80**, 4464.
9. A. L. Efros, and M. Rosen (2000) *Ann. Rev. Mater. Sci.*, **30**, 521.
10. A. P. Alivisatos (1996) *J. Phys. Chem.*, **100**, 13226.
11. H. Weller (1993) *Angew. Chem.*, **32**, 41.
12. U. Banin, Y. W. Cao, D. Katz, and O. Millo (1999) *Nature*, **400**, 542.
13. J. Jiang, K. Bosnick, M. Maillard, and L. E. Brus (2003) *J Phys Chem B*, **107**, 9964.
14. A. L. Rogach, A. Eychmuller, S. G. Hickey, and S. V. Kershaw (2007) *Small* **3**, 536.
15. O. I. Micic, C. J. Curtis, K. M. Jones, J. R. Sprague, and A. J. Nozik (1994) *J. Phys. Chem.*, **98**, 4966.
16. S. V. Kershaw, M. Burt, M. Harrison, A. L. Rogach, H. Weller, and A. Eychmuller, (1999) *Appl. Phys. Lett.*, **75**, 1694.
17. W. C. W. Chan, and S. M. Nie, (1998) *Science*, **281**, 2016.
18. C. R. Kagan, E. Lifshitz, E. H. Sargent, and D. V. Talapin (2016) Building devices from colloidal quantum dots. *Science*, **353**, 885.
19. M. Nirmal, and L. Brus (1999) *Accounts of Chemical Research*, **32**(5), 407.
20. E. H. Sargent (2012) *Nature Photonics*, **6**(3), 133.
21. Y. Zhao, and C. Burda (2012) *Energy & Environmental Science*, **5**(2), 5564.
22. I. L. Medintz, H. T. Uyeda, E. R. Goldman, and H. Mattoussi (2005) *Nature Materials*, **4**(6), 435.
23. S. V. Gaponenko (1998) *Optical properties of semiconductor nanocrystals*. Cambridge, UK: Univ. Press.
24. V. Wood, and V. Bulović (2010) *Nano Rev.*, **1**, 5202.
25. Y. Jiang, S.-Y. Cho, and M. Shim (2018) *J. Mater. Chem. C*, **6**, 2618.
26. H. Htoon, J. A. Hollingworth, A. V. Malko, R. Dickerson, and V. I. Klimov (2003) *Applied Physics Letters*, **82**(26), 4776–4778.
27. J. T. Hu, L. S. Li, W. D. Yang, L. Manna, L. W. Wang, and A. P. Alivisatos (2001) *Science*, **292**(5524), 2060–2063.
28. E. Rothenberg, M. Kazes, E. Shaviv, and U. Banin (2005) *Nano Lett.*, **5**(8), 1581–1586.
29. A. Salant, M. Shalom, Z. Tachan, S. Buhbut, A. Zaban, and U. Banin (2012) *Nano Lett.* **12**, 2095.
30. National Research Council. Committee to review the national nanotechnology initiative (2006) *A Matter of Size: Triennial Review of the National Nano-technology Initiative*.
31. N. Salah, et al. (2011) "High-energy ball milling technique for ZnO nanoparticles as antibacterial material." *International Journal of Nanomedicine*, **6**, 863.
32. S. Ibrahimkutty, P. Wagener, T. Rolo et al. (2015) "A hierarchical view on material formation during pulsed-laser synthesis of nanoparticles in liquid." *Sci Rep*, **5**, 16313.

33. L. S. Kibis et al. (2010) "The investigation of oxidized silver nanoparticles prepared by thermal evaporation and radio-frequency sputtering of metallic silver under oxygen." *Applied Surface Science*, **257**(2), 404–413.

34. S. T. Aruna, and A. S. Mukasyan (2008) "Combustion synthesis and nanomaterials." *Current Opinion in Solid State and Materials Science*, **12**, 44–50.

35. S. Mohan, M. Vellakkat, A. Aravind, and U. Reka. (2020) "Hydrothermal synthesis and characterization of Zinc Oxide nanoparticles of various shapes under different reaction conditions." *Nano Express*, **3**, 030028.

36. H. J. A. Yadav, et al. (2017) "Facile ultrasound route to prepare micro/nano super-structures for multifunctional applications." *ACS Sustainable Chemistry & Engineering*, **5**(3), 2061–2074.

37. D. Bokov et al. (2021) "Nanomaterial by sol-gel method: Synthesis and application." *Advances in Materials Science and Engineering*, **2021**, 1–21. 10.1155/2021/5102014

38. B. Jyothish, U. S. Geethu, and J. Jacob. (2022) "Influence of the Ag1+ and Co2+ doping on structural, optical and anti-cancer properties of $ZnFe_2O_4$ nanoparticles synthesized by co-precipitation method." *Materials Science and Engineering: B*, **276**, 115544.

39. S. To, and Z. Zhu (2015) "Novel end-fly-cutting-servo system for deterministic generation of hierarchical micro–nanostructures." *CIRP Annals—Manufacturing Technology*, **64**(1), 133–136.

40. F. Z. Fang, X. D. Zhang, A. Weckenmann, G. X. Zhang, and C. Evans (2013) "Manufacturing and measurement of freeform optics." *CIRP Annals—Manufacturing Technology*, **62**(2), 823–846.

41. M. Feldman (2013) *Nanolithography: The art of fabricating nanoelectronic and nanophotonic devices and systems*. WoodheadPub.

42. R. van den Berg (2005) "Extreme UV lithography preserves moore's law." *Optics & Laser Europe*, **129**, 29–31.

43. M. E. Davis, C. Zhuo, and M. S. Dong (2008) "Nanoparticle therapeutics: An emerging treatment modality for cancer." *Nature Reviews Drug Discovery*, **7**(9), 771.

44. G. Mészáros, S. Kronholz, S. Karthäuser, D. Mayer, and T. Wandlowski (2007) "Electrochemical fabrication and characterization of nanocontacts and nm-sized gaps." *Applied Physics A*, **87**(3), 569–575.

45. P. R. Gray, P. J. Hurst, S. H. Lewis, and R. G. Meyer (2008) "Analysis and design of analog integrated circuits." *5th Edition International Student Version. Best Practice & Research Clinical Gastroenterology*, **22**(4), 617–624.

46. S. E. F. Kleijn, A. I. Yanson, and M. T. M. Koper (2012) "Electrochemical char-acterization of nano-sized gold electrodes fabricated by nano-lithography." *Journal of Electroanalytical Chemistry*, **666**(3), 19–24.

47. A. A. Tseng (2004) "Recent developments in micromilling using focused ion beam technology." *Journal of Micromechanics & Microengineering*, **14**(14), 15–34.

48. J. Sun, X. Luo, W. Chang, J. M. Ritchie, J. Chien, and A. Lee (2012) "Fabrication of periodic nanostructures by single-point diamond turning with focused ion beam built tool tips." *Journal of Micromechanics & Microengineering*, **22**(22), 115014.

49. Z. W. Xu, F. Z. Fang, S. J. Zhang, X. D. Zhang, X. T. Hu, Y. Q. Fu, et al (2010) "Fabrication of micro doe using micro tools shaped with focused ion beam." *Optics Express*, **18**(8), 8025–8032.

50. N. Ikawa, S. Shimada, and H. Tanaka (1992) "Minimum thickness of cut in micro-machining." *Nanotechnology*, **3**(1), 6.

51. S. M. Son, S. L. Han, and J. H. Ahn (2005) "Effects of the friction coefficient on the minimum cutting thickness in micro cutting." *International Journal of Machine Tools & Manufacture*, **45**(4–5), 529–535.

52. W. Wu, Z. W. Xu, F. Z. Fang, B. Liu, Y. Z. Xiao, J. Chen, et al (2014) "Decrease of Fib-induced lateral damage for diamond tool used in nano cutting." *Nuclear Instruments & Methods in Physics Research*, **330**(4), 91–98.
53. F. Z. Fang, X. D. Zhang, W. W. Gao, Y. B. Guo, G. Byrne, and H. N. Hansen (2017) "Nanomanufacturing- perspective and applications." *C I R P Annals*, **66**(2), 683–705. 10.1016/j.cirp.2017.05.004
54. C. Ru, J. Luo, S. Xie, and Y. Sun (2014) "A review of non-contact micro- and nano-printing technologies." *Journal of Micromechanics & Microengineering*, **24**(5), 053001.
55. B. W. Ninham, and Nostro P. L. (2010) *Molecular forces and self assembly: In colloid, nano sciences and biology*. Cambridge University Press.
56. E. C. Kan, and Z. Liu (2000) "Directed self-assembly process for nano-electronic devices and interconnect." *Superlattices & Microstructures*, **27**(5–6), 473–479.
57. Y. Eichen, E. Braun, U. Sivan, and G. Ben-Yoseph (2010) "Self-assembly of nanoelectronic components and circuits using biological templates." *ActaPolymerica*, **49**(10–11), 663–670.
58. H. N. Hansen, R. J. Hocken, and G. Tosello (2011) "Replication of micro and nano surface geometries." *CIRP Annals—Manufacturing Technology*, **60**(2), 695–714.
59. L. Alting, F. Kimura, H. N. Hansen, and G. Bissacco (2003) "Micro engineering." *CIRP Annals—Manufacturing Technology*, **52**(2), 635–657.
60. J. A. McGeough, M. C. Leu, K. P. Rajurkar, A. K. M. De Silva, and Q. Liu (2001) "Electroforming process and application to micro/macro manufacturing." *CIRP Annals—Manufacturing Technology*, **50**(2), 499–514.

4 Experimental Investigation and Multi-Response Optimization of End Milling Process Parameters for Surface Integrity on Al7075-B₄C-BN Nanocomposites

N. Zeelanbasha, M. Selvakumar, T. Ramkumar, and V. Sivananth

CONTENTS

4.1 Introduction 46
4.2 Experimental Procedure 47
    4.2.1 Materials and Methods 47
    4.2.2 Experimental Setup for Measurement of Temperature Rise 47
4.3 Statistical Analyses 51
    4.3.1 ANOVA for Temperature Intensification 51
    4.3.2 ANOVA for Micro-Hardness 52
4.4 Results and Discussions 52
    4.4.1 Response Surface Methodology of Temperature Rise: Direct
           Consequence of Spindle Speed and Feed Rate on Temperature
           Rise 52
    4.4.2 Response Surface Methodology of Micro-Hardness 53
    4.4.3 Direct Consequence of Spindle Speed and Feed Rate on
           Micro-Hardness 54
    4.4.4 Direct Consequence of Spindle Speed and Radial Depth
           of Cut on Micro-Hardness 55
4.5 Conclusion 57
References 57

DOI: 10.1201/9781003220602-4

45

4.1 INTRODUCTION

For several decades, in aerospace and automotive industries, aluminium alloy is widely used as a structural material. Among the various aluminium series, AA7075 is one of the most utilized materials in automobile, spacecraft and missile applications due to its lightweight, low density, high strength and excellent hardness properties. Using a single ceramic reinforcement in AMCs are enhanced the microstructural and mechanical properties as compared to the pure aluminium alloy.

Machining difficulties were caused much more in alloys while machining when the heat was generated [1]. Chip morphology, cutting force and surface roughness in machining operations were greatly affected during the cutting temperature. Thus having a thorough understanding of the changes in the cutting temperature during the milling process was much essential [2,3]. High cutting temperature mainly causes tool wear and also affects machined surface quality. Further, thermal stress was led by the temperature difference between workpiece surface and inner layer. It was coupled with residual stress in the blank and causes distortion during machining [4]. Liu et al. [5] discussed that improving the performance of cutting processes and temperature distribution prediction of workpieces was of great significance. K-type fast response thermocouples were employed to determine the temperature of workpiece aspects, which results from the cutting course. Micro-hardness is characterized as an important factor in the study of surface integrity during the machining process. During the machining process, the milled surface undergoes both physical and metallurgical changes due to the combined effects of high local temperature, severe plastic deformation, and rapid quenching rate. The microstructure changes of the milled surface usually occurred. The surface modifications for the milled surface are analyzed in depth for the selection of appropriate cutting conditions. Adopting the recommended machining condition will lead to preventing the poor mechanical properties [6–8].

During the milling process, the effects of various conditions (spindle speed 1000, 2000, 3000 rpm, feed rate 0.2, 0.5, 0.8 μm/tooth and axial depth of cut 20,60,100 μm), on the surface integrity were investigated by [9,10]. Feed per tooth and spindle speed were found to have more impact on micro-hardness with the contributions of 30.21% and 69.25% respectively. The spindle speed of 10,000 rpm, axial cutting depth of 20 μm and feed rate of 0.2 μm/tooth resulted in the lowest hardness value.

Limited research has been conducted on the optimization techniques using multi-response optimization tools. Therefore, the current research is focused on multi-response optimization of the end milling process to improve the surface integrity of Al7075- B_4C-BN nanocomposites using integrated RSM and MOGA approach. The composites were characterized using SEM and EDAX analysis. The prediction of temperature rise becomes essential for the improvement of mechanical properties on milled surfaces using multi-objective optimization. Parameters such as feed rate (F), spindle speed (N), radial depth of cut, (Dr), axial depth of cut (Da) and radial rake angle (γ) are taken into consideration. RSM and MOGA were employed for the planning of experiments and the prediction of the optimal hardness and cutting temperature. Micro-hardness is characterized as an important factor in the study of surface integrity during the machining process. During the machining process, the milled surface undergoes both physical and metallurgical changes due to the combined effects of cutting temperature.

4.2 EXPERIMENTAL PROCEDURE

4.2.1 MATERIALS AND METHODS

The ceramic powders B_4C (2.52 g/cm^3) and BN (2.1 g/cm^3) 98% of purity and 1 μm particle size were purchased from Sigma Aldrich-Germany. Before fabrication of composites, the secondary powders were ball milled continuously and the particle sizes were reduced to less than 100 nm. Three different wt.% of BN particles (3, 6 and 9) were selected for preparing these hybrid nanocomposite materials and 3 wt.% of B_4C is kept constant for all composite materials. The conventional stir casting technique is used to fabricate the composites. The composition details of the hybrid nanocomposites are given in Table 4.1. The density of the prepared hybrid nano-composites was evaluated by Archimedes' principle. It is an average value of three measurements observed for each sample as shown in Table 4.2. The microstructure of the prepared hybrid composites was observed using SEM. Following that the che-mical composition of the fabricated composites is observed using EDAX spectrum and represented in Figure 4.1(a and b). The secondary particles such as BN and B_4C were homogeneously dispersed in the matrix medium. Vickers micro-hardness tester is used to estimate the hardness of the prepared composite materials. Using a circular base diamond intender with 5 kg of load and a dwell time period of 10 s, three tests were executed on specimens and the mean value was acquired through measured values. Based upon the significant improvement of hardness Al7075/9BN/3B₄C was selected to perform the end milling operation [11–13].

4.2.2 EXPERIMENTAL SETUP FOR MEASUREMENT OF TEMPERATURE RISE

Figure 4.2(a) and (b) illustrates the experimental setup used for estimating a tem-perature rise. A hole of 1 mm is drilled on workpiece at 2.5 mm lower than the

TABLE 4.1
Wt.% of BN and B₄C in the Composites

Sample code	AA7075	BN	B4C
AA7075/BN/B₄C@	94	3	3
AA7075/BN/B₄C	91	6	3
AA7075/BN/B₄C	88	9	3

TABLE 4.2
Density Values of Prepared Hybrid Composites

Sample code	Actual density (g/cm³)	Relative density	Hardness (HV)
AA7075/BN/B₄C	2.76	0.91	64
AA7075/BN/B₄C	2.72	0.88	75
AA7075/BN/B₄C	2.66	0.90	84.2

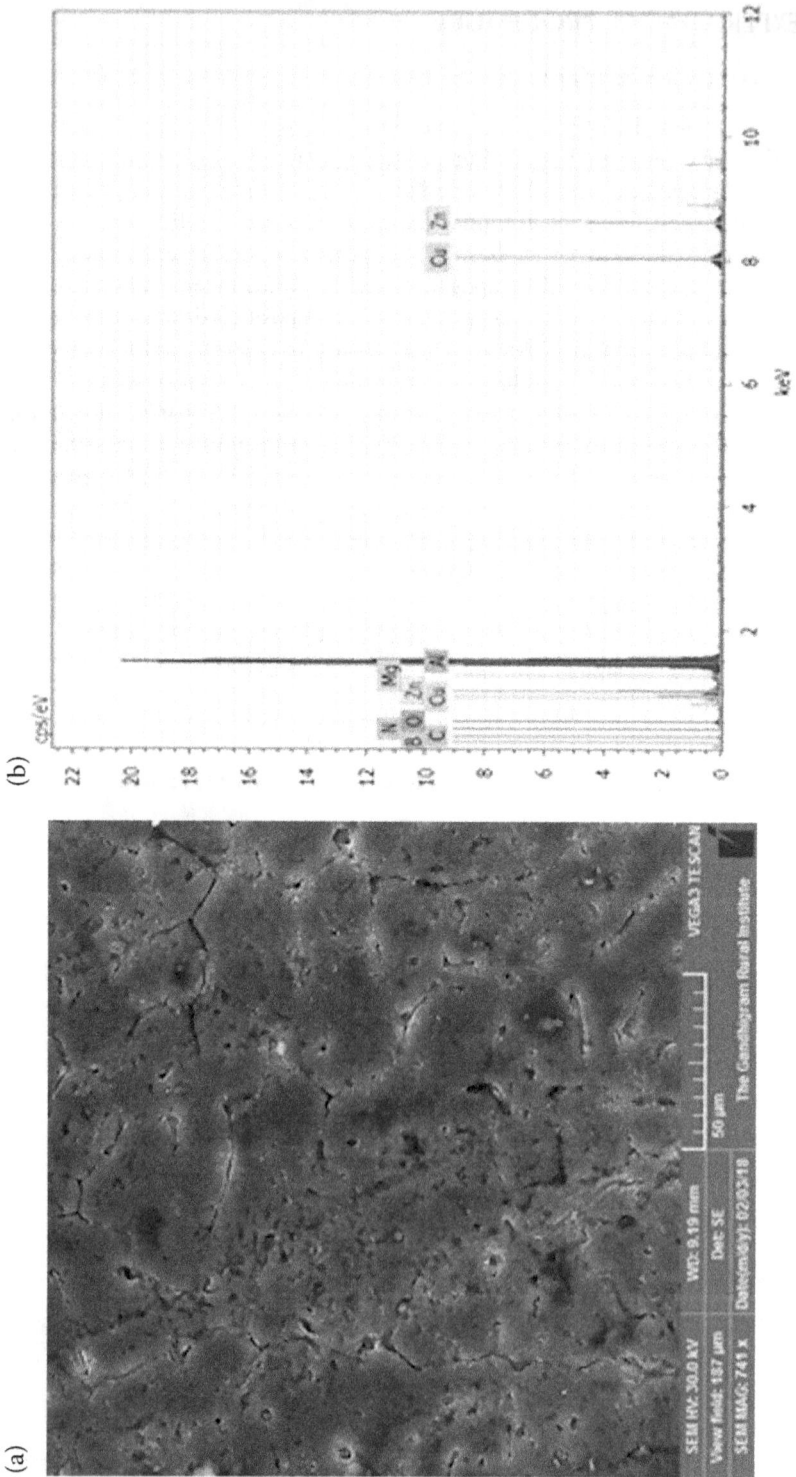

FIGURE 4.1 (a) SEM micrograph of AA7075/9BN/3B4C. (b) Corresponding EDAX.

(a)   (b)

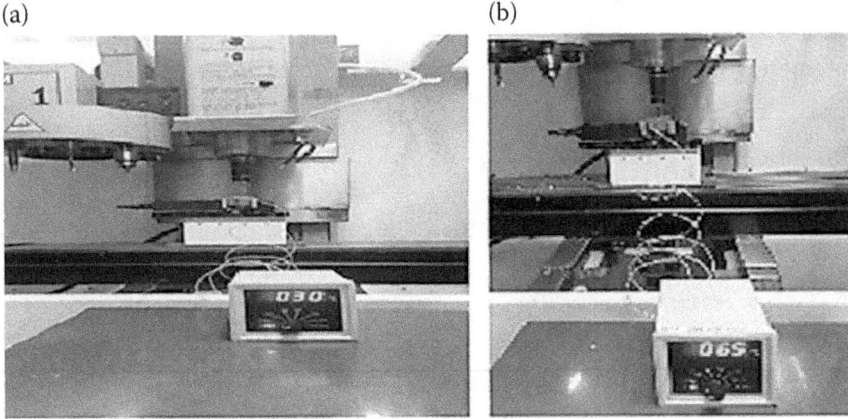

FIGURE 4.2 (a and b) Measurement of temperature rise.

FIGURE 4.3 Drilled workpiece.

machining surface (Figure 4.3). The initial temperature is measured by inserting the probe of K-type thermocouple in the drilled surface of workpiece at room temperature as depicted in Figure 4.2(a). The maximum temperature is measured by inserting the probe of K-type thermocouple in the drilled surface of workpiece during the end milling machining process as evidenced in Figure 4.2(b). The difference between the initial temperature and maximum temperature is therefore noted as the temperature rise. Figure 4.3 shows the photographic view of drilled workpiece samples. For an accuracy purpose, the experiments were conducted thrice and average value was tabulated for thirty experiments [14]. The parameters and their levels for machining the workpiece are shown in Table 4.3 and Table 4.4.

TABLE 4.3

Parameters for Machining and Levels

Sl. no.	Cutting parameters	Unit	levels		
			−1	0	1
1	Spindle speed (N)	rpm	1400	2500	3600
2	Feed rate (F)	(mm/rev)	0.04	0.08	0.12
3	Axial depth of cut (D_a)	(mm)	0.4	0.7	1
4	Radial depth of cut (D_r)	(mm)	0.4	0.7	1
5	Radial rake angle (γ)	(degree)	12	18	24

TABLE 4.4

Recorded Responses Temperature Rise and Micro-Hardness

Sl. no.	Spindle speed (N)	Feed rate (F)	Axial depth of cut (Da)	Radial depth of cut (Dr)	Rake angle (γ)	Temperature rise (Tr)	Micro-hardness (MH)
	rpm	mm/rev	mm	mm	°	°C	HV
1	3600	0.12	1	1	24	40	75.10
2	3600	0.04	0.40	1	24	39	75.30
3	1400	0.12	1	0.40	24	37	75.30
4	3600	0.12	0.40	0.40	24	37	75.30
5	3600	0.04	1	1	12	35	75.90
6	1400	0.12	0.40	1	24	32	76.10
7	3600	0.04	1	0.40	24	39	76.30
8	3600	0.12	0.40	1	12	39	76.50
9	1400	0.04	1	0.40	12	22	76.50
10	3600	0.12	1	0.40	12	36	76.50
11	1400	0.04	1	1	24	22	76.60
12	1400	0.04	0.40	0.40	24	20	76.90
13	3600	0.04	0.40	0.40	12	23	77.50
14	1400	0.04	0.40	1	12	20	77.60
15	2500	0.08	0.70	1	18	31	80.50
16	2500	0.08	1	0.70	18	30	82.20
17	2500	0.08	0.70	0.40	18	31	82.30
18	2500	0.08	0.70	0.70	24	31	82.70
19	2500	0.08	0.70	0.70	18	36	83
20	2500	0.04	0.70	0.70	18	25	83.30
21	2500	0.08	0.40	0.70	18	29	83.40
22	2500	0.08	0.70	0.70	18	31	84.10
23	1400	0.12	0.40	0.40	12	19	84.30
24	2500	0.08	0.70	0.70	18	28	84.30
25	2500	0.12	0.70	0.70	18	31	84.50

TABLE 4.4 (Continued)
Recorded Responses Temperature Rise and Micro-Hardness

Sl. no.	Spindle speed (N)	Feed rate (F)	Axial depth of cut (Da)	Radial depth of cut (Dr)	Rake angle (γ)	Temperature rise (Tr)	Micro-hardness (MH)
	rpm	mm/rev	mm	mm	°	°C	HV
26	2500	0.08	0.70	0.70	18	29	84.60
27	2500	0.08	0.70	0.70	18	31	84.70
28	3600	0.08	0.70	0.70	18	30	84.80
29	1400	0.08	0.70	0.70	18	28	84.80
30	2500	0.08	0.70	0.70	18	27	85.10
31	1400	0.12	1	1	12	38	85.70
32	2500	0.08	0.70	0.70	12	30	85.90

4.3 STATISTICAL ANALYSES

4.3.1 ANOVA FOR TEMPERATURE INTENSIFICATION

It is evident from Table 4.5 that the influence of spindle speed, feed rate, radial rake angle, axial depth of cut and radial depth of cut (Prob = 4.38E-06, Prob = 5.58E-05, Prob = 0.003107, Prob = 0.01532, Prob = 0.009052) on the temperature rise are statistically significant. This circumstance indicates that the radial rake angle, radial depth of cut, spindle speed, feed rate and axial depth of cut exhibit a substantial effect on the temperature variation in the end milling. The regression equation for temperature rise is shown in empirical Equation (4.1)

$$
\begin{aligned}
Tr = {} & -39.65236 + 0.006881 * N + 151.01010 * F + 38.91835 * Da + 35.28199 * Dr \\
& + 1.47622 * \gamma - 0.036932 * N * F - 0.003030 * N * Da + 0.000758 * N * Dr \\
& + 0.000095 * N * \gamma + 41.66667 * F * Da + 41.66667 * F * Dr - 1.56250 * F * \gamma \\
& - 20.83333 * Da * Dr - 0.694444 * Da * \gamma - 1.11111 * Dr * \gamma;
\end{aligned}
$$

$$(4.1)$$

TABLE 4.5
Model Adequacy for Temperature Rise

Source	Sum of squares	DF	Mean squares	F-cal	P-Value Prob> F	R2	Whether the model is adequate
Model	483.118	15	34.559	25.93832	4.38E-06	0.97	adequate
Residual	9.722	12	0.711094	–	–	–	–
Lack of	5.202	9	2.050	4.440064	0.009052	–	Inadequate

TABLE 4.6

Model Adequacy for Micro-Hardness

Source	Sum of squares	DF	Mean squares	F-cal	P-Value Prob> F	R2	Whether the model is adequate
Model	493.118	20	24.6559	30.39832	6.61E-07	0.98	adequate
Residual	8.922036	11	0.811094	–	–	–	–
Lack of	6.302036	6	1.050339	2.004464	0.231249	–	Inadequate

4.3.2   ANOVA FOR MICRO-HARDNESS

Table 4.6 illustrates the influence of spindle speed, radial rake angle and feed rate (Prob = 0.00022, Prob = 0.00491, Prob = 0.479005) on the micro-hardness are statistically significant and axial depth of cut, radial depth of cut (Prob = 0.683463, Prob = 2.23E-05) is statistically insignificant [15,16]. This circumstance indicates that the radial rake angle, feed rate and spindle speed influence the micro-hardness in the end milling machining process. The regression equation for micro-hardness is shown in empirical Equation (4.2)

$$
\begin{aligned}
MH = {}& 50.91489 - 0.000656 * N + 238.22183 * F + 22.06908 * Da + 47.85695 * Dr \\
& + 0.401747 * \gamma - 0.021875 * N * F + 0.00000000000000000583 * N * Da \\
& - 0.001098 * N * Dr + 0.000140 * N * \gamma + 12.50000 * F * Da + 19.79167 * F * Dr \\
& - 4.89583 * F * x(5) + 11.94444 * Da * Dr + 0.034722 * Da * \gamma - 0.055556 * Dr * x(5) \\
& - 0.0000000777057 * N * N - 621.26494 * F * F - 23.26693 * Da * Da \\
& - 38.82249 * Dr * Dr \\
& - 0.016501 * \gamma * \gamma
\end{aligned}
$$

$$(4.2)$$

4.4   RESULTS AND DISCUSSIONS

4.4.1   RESPONSE SURFACE METHODOLOGY OF TEMPERATURE RISE: DIRECT CONSEQUENCE OF SPINDLE SPEED AND FEED RATE ON TEMPERATURE RISE

Figure 4.4(a) and (b) reveals the values of spindle speed and feed rate had substantial result on temperature rise, whereas feed rate (0.06–0.12 mm), at higher spindle speed 3600 rpm resulted to induces the temperature rise. Generation of heat in the cutting zone increments, bringing about high cutting temperature during the end milling process [17]. Relatively, the lower spindle speed (1400–2000 rpm) and feed rate (0.04–0.05 mm/rev) reduce the heat in the cutting tool. This is due to reason; if the spindle speed is small, the heat generation due to friction is lower to be transferred to the tool and the cutting chip, subsequently, heat dissipation in the bottom of the chip, the extent of metal removal decreases consistently to unit time,

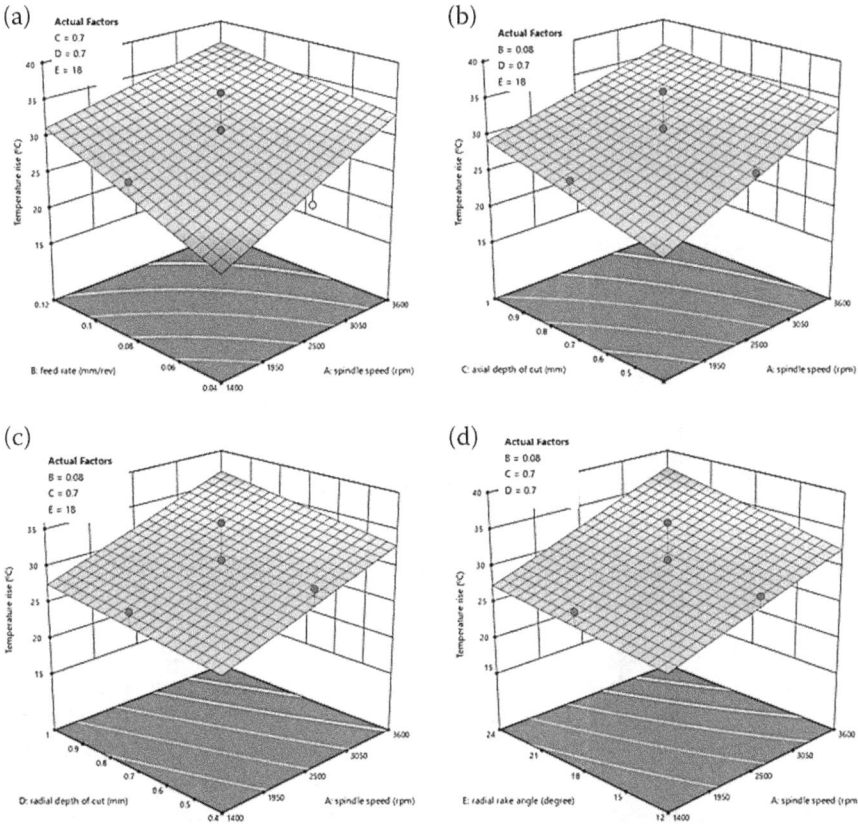

FIGURE 4.4 (a–d) Interactive effect of cutting parameters on temperature rise.

and the power consumption reduces; as an outcome the temperate rise decreases. Hence, it has been concluded that temperature rise is reduced at spindle speed (1400–2000 rpm) and feed rate (0.04–0.05 mm/rev).

4.4.2 Response Surface Methodology of Micro-Hardness

The optical micrograph of the base metal and a typical case of machined surface (trial No.2) is shown in Figure 4.5(a) and (b). The average grain size in the base metal has been determined by using the mean linear intercept method and found to be 35 µm. After machining, it is observed that there was a gradual decrease in the grain size; preferably more elongated grains were observed, as a result of severe plastic deformation induced during the machining process. The average grain size of the grains after the machined surface was 25 µm. During plastic deformation, the dislocation density of the grains is increased by the dislocation movements and dislocation generation within the crystal structure. The absence of grain mobility after severe deformation attributes to the micro-hardness enhancement in the machined surface.

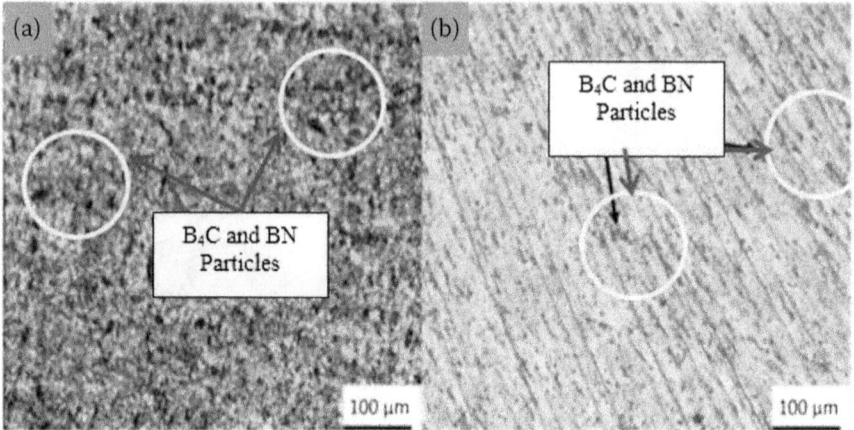

FIGURE 4.5   (a) Microstructure of the Al 7075/9BN/3B4C. (b) Machined surface of the same.

4.4.3   DIRECT CONSEQUENCE OF SPINDLE SPEED AND FEED RATE ON MICRO-HARDNESS

Figure 4.6(a) and Figure 4.7(a) and (b) illustrate the impact of spindle speed (N) and feed rate (F) on micro-hardness, it is observed that the process parameters show a notable effect on the micro-hardness. The micro-hardness is highly influenced by the effect of feed rate, where an increase in the feed rate results in higher micro-hardness. But, the spindle speed depicts an inverse relationship on the micro-hardness where as the micro-hardness decrease proportionally with the rise of spindle speed, and improved micro-hardness can be attained at lower spindle speeds. The micro-hardness (75–84.2 HV) of the machined workpiece is improved during the feed rate changes from 0.06 to 0.09 mm/rev at a lower spindle speed. The micro-hardness decreases proportionally with the feed rate and a spindle speed ranging between 3050 and 3600 rpm. This is due to the reason that the thermal softening effect of the material at this level to over-ageing at an elevated cutting temperature formed at the local surface, which leads to the reduction in the hardness on the machined surface and subsurface [18]. Hence, it has been concluded that the higher micro-hardness was owing to significant effect of work hardening at higher feed rate and lower spindle speeds in the milling process.

From Figure 4.6(b) and Figure 4.7(c), it is obvious that axial depth of cut displays a parabolic trend with an increase in the spindle speed and the micro-hardness of the machined surface decrease with an increase in the spindle speed. The increase in the micro-hardness was observed in the optimal range of axial depth of cut between 0.6 and 0.8 mm for the increase in the spindle speed. This was due to the effect of strain hardening and a rise in temperature during the milling. Strengthening of the subsurface is influenced by the increasing depth of cut through the compressive force of the tool, exactly below the machined surface [19].

At a lower depth of cut, the less interaction of the tool with the machined surface causes less metallurgical changes, owing to the low temperature rise and thus,

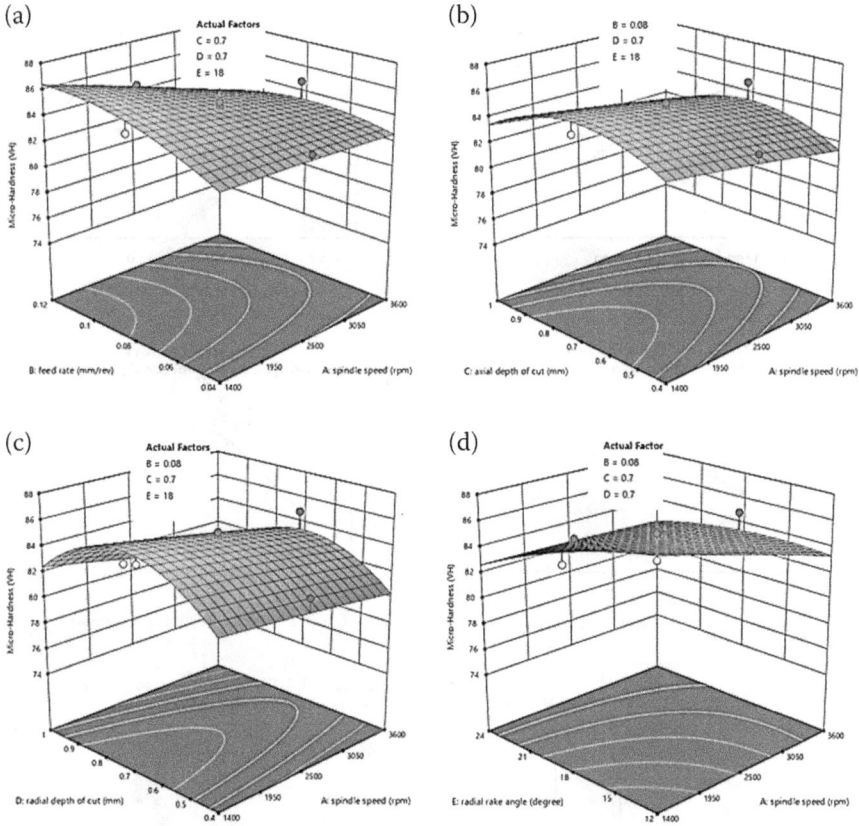

FIGURE 4.6 (a–d) Interaction between micro-hardness and cutting parameters.

resulting in lower micro-hardness. But, at higher depth of cut, the increase in the temperature rise would significantly induce metallurgical changes on the machined surface, owing to thermal softening, leading to decreasing hardness. Hence, it is clear that higher micro-hardness can be achieved for axial depth of cut between 0.6 and 0.8 mm for the case of lower spindle speed of 1400 rpm.

4.4.4 DIRECT CONSEQUENCE OF SPINDLE SPEED AND RADIAL DEPTH OF CUT ON MICRO-HARDNESS

From Figure 4.6(c) and Figure 4.7(d), the plot between spindle speed and depth of cut. It is perceived that the depth of cut depicts a parabolic trend of micro-hardness for the increase in the spindle speed. Micro-hardness achieved maximum between the optimal range of 0.6 and 0.8 mm of radial depth of cut. An increase in the depth of cut causes strengthening of the subsurface by the compressive force of the tool. At a lower depth of cut, the less interaction of the tool with the machined surface causes less metallurgical changes, owing to the low temperature rise and thus, resulting in lower micro-hardness. The probability of obtaining higher micro-hardness

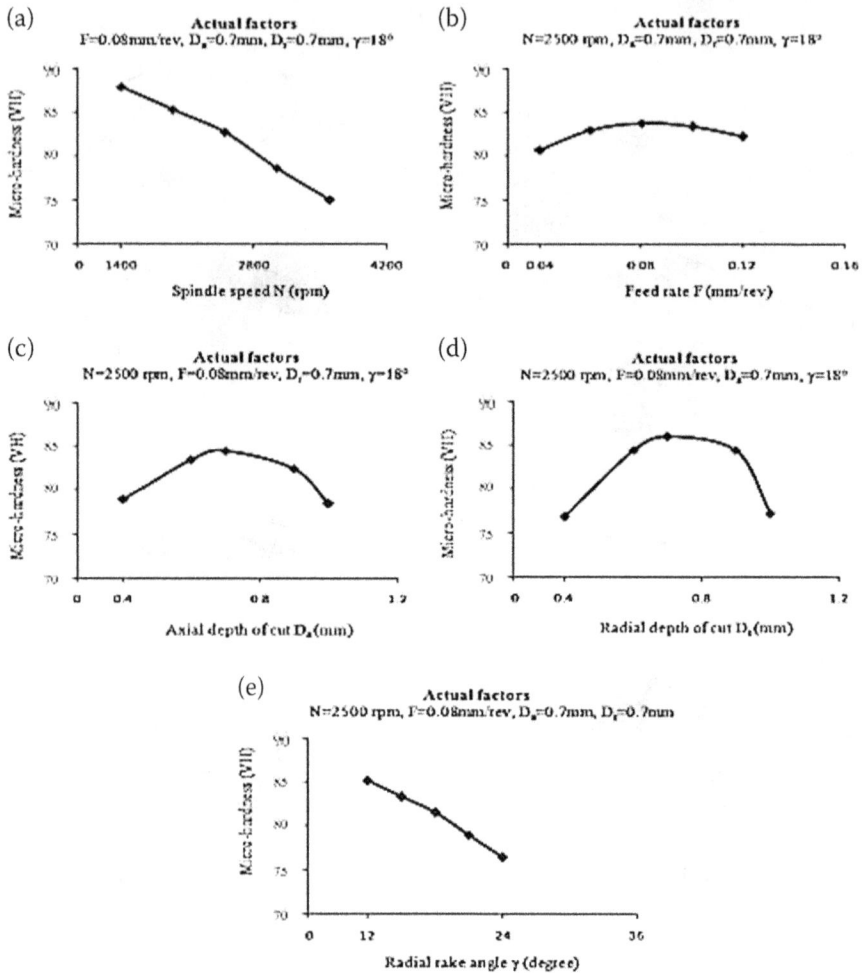

FIGURE 4.7 (a–e) Direct influence of parameters on micro-hardness.

is preferred for optimal radial depth of cut in between 0.6 and 0.8 mm for the case of relatively spindle speeds.

Figure 4.6(d) and Figure 4.7(e) illustrate the impact of radial rake angle and spindle speed on micro-hardness. The deviation of micro-hardness with radial rake angle is shown in Figure 4.7(e). The micro-hardness improved at a lower radial rake angle. From Figure 4.6(d), it is factual that micro-hardness is reduced as the radial rake angle is increased from 20° to 24° at spindle speed (3050–3600 rpm), due to the reason that while a rise in rake angle escalates the area of interaction of the tool and workpiece. Consequently, leading to higher friction and causes to produce a higher temperature rise. More cutting temperature is developed at the local surface causing a reduction in the hardness of the machined surface and subsurface [20]. Also higher rake angle (20°–24°), the tool becomes weak which accelerates to reduce the life of the cutting tool. Hence, it is clear that higher micro-hardness can be achieved for radial rake angle between 12° and 15° for the case of a lower spindle speed of 1400 rpm.

4.5 CONCLUSION

F-statistical test values for temperature rise reveal that spindle speed has a considerable effect on the increase in temperature compared to feed rate, radial rake angle, radial depth of cut and axial depth of cut.

F-statistical test values for micro-hardness indicate that rake angle has a substantial influence over micro-hardness when compared with spindle speed, feed rate, axial depth of cut and radial depth of cut.

The statistical values of R^2 (91%) for temperature rise and R^2 (97%) for micro-hardness represent that the average value would be more of a better predictor.

The interaction and direct surface plot analysis of temperature rise disclose that the temperature can be attained minimum at 1400 and 2000 rpm spindle speed, 0.04 and 0.05 mm/rev feed rate, 0.4 and 0.6 mm axial depth of cut, 0.4 and 0.6 mm depth of cut and 12° and 15° radial rake angle.

REFERENCES

1. S. Bharathi Raja, N. Baskar, Application of particle swarm optimization technique for achieving desired milled surface roughness in minimum machining time, *Expert Syst Appl.* 39, 5982–5989 (2012).
2. G. Campatelli, L. Lorenzini, A. Scippa, Optimization of process parameters using a response surface method for minimizing power consumption in the milling of carbon steel, *J. Clean. Prod.* 66, 309–316 (2014).
3. X. Cui, J. Guo, J. Zhao, Y. Yan, Chip temperature and its effects on chip morphology, cutting forces, and surface roughness in high-speed face milling of hardened steel, *Int J Adv Manuf Tech.* 77, 2209–2219 (2014).
4. E.l. Hakim, M.A. Shalaby, M.A. Veldhuis, S.C. Dosbaeva, Effect of secondary hardening on cutting forces, cutting temperature, and tool wear in hard turning of high alloy tool steels, *Meas.* 65, 233–238 (2015).
5. H. Hassanpour, M.H. Sadeghi, H. Rezaei, A. Rasti, Experimental study of cutting force, microhardness, surface roughness, and burr size on micromilling of Ti6Al4V in minimum quantity lubrication, *Mater Manuf Process.* 31, 1654–1662 (2016).
6. T. Kivak, Optimization of surface roughness and flank wear using the Taguchi method in milling of Hadfield steel with PVD and CVD coated inserts, *Meas.* 50, 19–28 (2014).
7. K. Abdullah, W.C. David, A.E. Smith, Multi-objective optimization using genetic algorithms: A tutorial, *Reliab Eng Syst Safe.* 91, 992–1007 (2006).
8. J. Yan, L. Li, Multi-objective optimization of milling parameters–The trade-offs between energy, production rate and cutting quality, *J. Clean. Prod.* 52, 462–471 (2013).
9. M. Santhanakrishnan, P.S. Sivasakthivel, R. Sudhakaran, Modeling of geometrical and machining parameters on temperature rise while machining Al 6351 using response surface methodology and genetic algorithm, *J Braz Soc Mech Sci.* 39, 487–496 (2015).
10. G. Mahesh, S. Muthu, S.R. Devadasan, Prediction of surface roughness of end milling operation using genetic algorithm, *Int J Mach Tool Manu.* 77, 369–381 (2015).

11. K.S.S. Rao, K.V. Allamraju, Effect on micro-hardness and residual stress in CNC turning of aluminium 7075 alloy, *Mater Today: Proc.* 4, 975–981 (2016).
12. M. Sayuti, A.D. Ahmed, M. Sarhan, An investigation of optimum SiO2 nanolubrication parameters in end milling of aerospace Al6061-T6 alloy, *Int J Adv Manuf Technol.* 67, 833–849 (2013).
13. S. Kouadri, K. Necib, S. Atlati, B. Haddag, M. Nouari, Quantification of the chip segmentation in metal machining: Application to machining the aeronautical aluminium alloy AA2024-T351 with cemented carbide tools WC-Co, *Int J Mach Tool Manu.* 64, 102–113 (2013).
14. Y. Yusoff, M.S. Ngadiman, A.M. Zain, 'Overview of NSGA-II for optimizing machining process parameters, *Procedia Eng.* 15, 3978–3983 (2011).
15. L. Maiyar, M. Ramanujam, R. Venkatesan, K. Jerald, Optimization of machining parameters for end milling of Inconel 718 super alloy using Taguchi based grey relational analysis, *Procedia Eng.* 64, 1276–1282 (2013).
16. J. Ratava, M. Lohtander, J. Varis, Modelling cutting instability in rough turning 34CrNiMo6 steel, *Int. J. Opt Res.* 25, 518–531 (2016).
17. G. Quintana, J. Ciurana, J. Ribatallada, Modelling power consumption in ball-end milling operations, *Mater Manuf Process.* 26, 746–756 (2011).
18. P. Palanisamy, I. Rajendran, S. Shanmugasundaram, R. Saravanan, Prediction of cutting force and temperature rise in the end-milling operation, *P I Mech Eng B-J Eng.* 220, 1577–1587 (2006).
19. Z. Peng, J. Li, P. Yan, S. Gao, C. Zhang, X. Wang, Experimental and simulation research on micro-milling temperature and cutting deformation of heat-resistance stainless steel, *Inte J of Ad Manu Tech.*95, 2495–2508 (2018).10.1007/s00170-017-1091-6
20. J.R. Santosh, M.C. Machado, A.R. Barrozo, M.A.S. Jackson, E.O. Ezugwu, Multi-objective optimization of cutting conditions when turning aluminum alloys using genetic algorithm, *Int J Adv Manuf Technol.* 76, 1123–1138 (2014).

5 Design and Manufacturing of Nano Sensors

Perspective and Applications

Souvik Bag

CONTENTS

5.1 Introduction...59
5.2 Chemiresistive Type Gas Sensor Based on Nanomaterials.........................62
    5.2.1 Pure Metal and Metal Oxide-Based Gas Sensor...............................62
    5.2.2 Transition Metal Dichalcogenides (TMDs) a Two Dimensional
        (2D)-Based Gas Sensor..63
5.3 Film Deposition Techniques 64
5.4 Gas Sensing Parameters..65
    5.4.1 Baseline..65
    5.4.2 Sensor Response..65
    5.4.3 Calibration Curve and Sensor Sensitivity 65
    5.4.4 Selectivity 66
    5.4.5 Sensor Response and Recovery Time 66
    5.4.6 Repeatability...66
    5.4.7 Drift...67
    5.4.8 Limit of Detection (LOD)..67
5.5 Method to Improve Gas Sensing Response 67
5.6 Conclusion 69
References..69

5.1 INTRODUCTION

There are various kinds of natural, chemical and artificial species available in our surrounding environment where some of them are very essential while others are more harmful or less. Figure 5.1 shows the typical gas concentration levels in global environment. The essential gases/species such as oxygen (O_2), nitrogen (N_2) and humidity should be maintained at a sufficient level in global atmosphere while excess emission of hazardous or toxic gases can harm living atmosphere. In this regard, burning of fossil fuels such as petroleum, coal and natural gas are prime concerns for global air pollution [1–4]. Moreover, this process produces, compounds containing

DOI: 10.1201/9781003220602-5

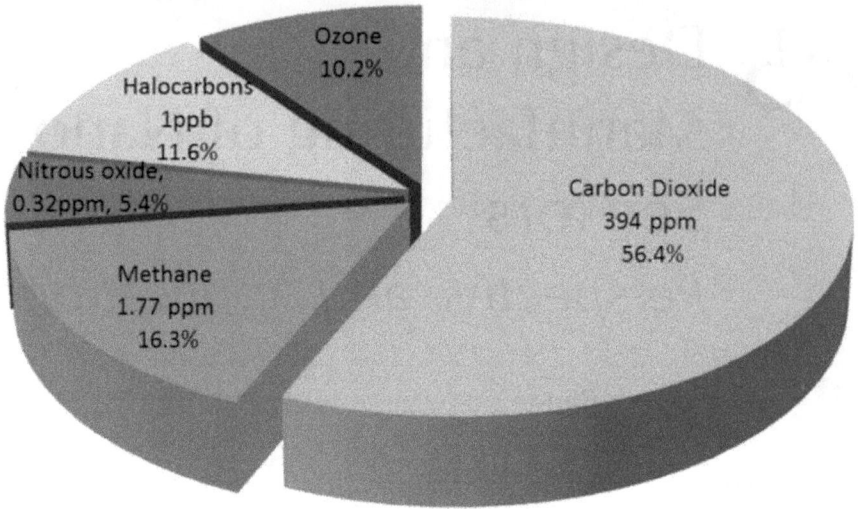

FIGURE 5.1 Typical gas concentration levels in global environment.

sulphur, carbon and nitrogen to generate gaseous oxides by reacting with air. These toxic gases are harmful to the living atmosphere in parts per million (ppm) or even parts per billion (ppb). In addition, the leakage due to the explosion of lower hydrocarbons and hydrogen (H_2) from fuels also promotes the air pollution; Where the alarming level of each gas is kept at 1/10 of the lower explosion limit (LEL) [5]. At the present scenario, the increase of hazardous or toxic gases in global environment is a major issue concerning the environment monitoring system where gas sensor shows the possible solution by detecting these toxic gases in a precise way. The permissible exposure limits (PELS) for volatile organic compounds (VOCs), various gases and other air pollutants have been legislated by either occupational safety concentrations or environmental protection agency regulations based on the strength of toxicity or offensiveness of each gas. For the last two decades, many hazardous gases or VOCs are predominantly considered for sensory detection, nevertheless, some of them are still yet to be confronted.

A sensor is a device which detects/receives a physical or chemical stimulus and responds with an analytically useful signal. In most of the cases, the sensor output is in the form of electrical or electronic. The sensor may be classified in two broad categories i.e., physical and chemical sensors, respectively. Out of which chemical sensors could be defined as a device which transforms chemical information of target species into useful analytical signal. On the other hand, physical sensor could be demarcated as a device which gives information about the physical property of the system to be investigated. Gas sensor is a family member of chemical sensor, which detects and/or quantifies the presence of gases as well as concentration of the gases in a fixed volume where dry air is generally used as a reference environment. Basically, chemical sensors or gas sensors are composed of two basic functional units named as (a) receptor part and (b) transducer part. In some cases, a separator is used (e.g., membrane). The function of the receptor part is to recognize the specific gas species and to transform this information in the form of energy. On the other hand, function of the transducer is to transform this energy into a useful analytical output signal. The output signal of the gas sensor may be resistive and capacitive type, optical type, electromotive force, resonant frequency, etc. The nature of the output signals depends on materials used as both receptor and transducer. Particularly, the selectivity of a gas sensor to a specific gas depends on materials used as receptors; and sensitivity depends on transducer function. Moreover, the selectivity can be improved by introducing a new kind of material or modifying the surface of the foreign additives in the receptor unit, while sensitivity can be increased by improving transducer function which depends on various external factors like device structure, micro or nano-structure of the materials, etc. A schematic of the typical chemical sensor as well as gas sensor is illustrated in Figure 5.2.

The gas sensor can be classified into different categories based on transduction principles. Some common examples include electrochemical (potentiometric and amperometric), calorimetric, gas chromatograph, mass-sensitive (surface acoustic wave (SAW), quartz crystal microbalance (QMB), and microcantilever), optical, capacitive, magnetic, conductometric (current or resistive type), etc. Among these gas sensors, conductometric gas sensor (particularly, resistive type), also called as chemiresistive type gas sensor is the furthermost auspicious process due to their high sensitivity, selectivity, low power consumption, portable in size, etc. So, we will mainly focus on chemiresistive type gas sensors, while other sensors will be introduced briefly.

Thus, the present research work is focused on toxic gas detection using carbon nanofillers based on thin sensing films. As described regarding the advantages of

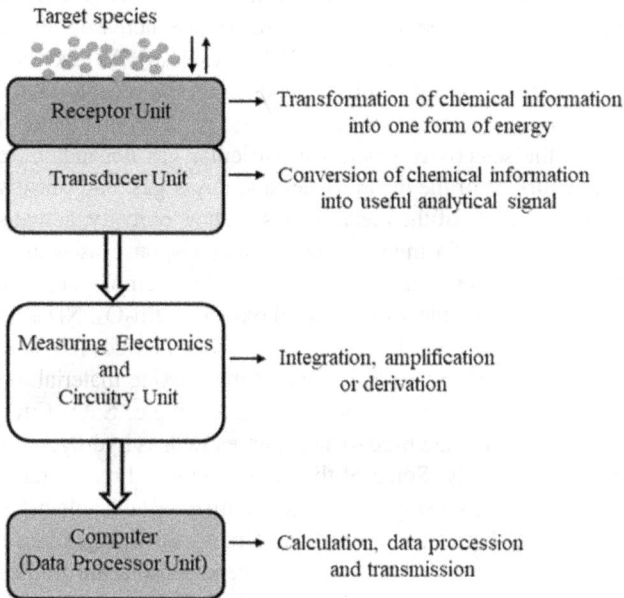

FIGURE 5.2   Representation of typical chemical sensor or gas sensor.

chemiresistive type sensor over other gas sensors, a detailed literature review has been exercised on the development of toxic gas sensors using various nanomaterials (especially carbon-based nanomaterials), carbon nanofiller like CB, CNTs, RGO, etc. reinforced polymer composite based sensor towards toxic gas detection like methanol and carbon dioxide. In addition, different useful parameters for chemiresistive gas sensing measurement, different sensing film deposition techniques and methods to improve gas sensing response have been studied.

5.2 CHEMIRESISTIVE TYPE GAS SENSOR BASED ON NANOMATERIALS

5.2.1 Pure Metal and Metal Oxide-Based Gas Sensor

Chemiresistive type gas sensors constructed on semiconductive metal oxide are one of the utmost promising gas sensors used in various applications. These sensors are also named as "oxide semiconductor gas sensors" primarily classified into two subgroups: surface sensitive and bulk sensitive [6]. Here, we are going to focus only on surface-sensitive gas sensors. The main advantages of using these metal oxide materials as sensing elements are easy fabrication, low cost and simple operation. In addition, these types of sensors are compact in size and have significant durability [7]. The basic operating principle is based on the change in sensor resistance in the occurrence or absence of a target gas. Semiconducting metal oxide can also be classified into two groups: n-type (SnO_2, ZnO, $AgVO_3$, TiO_2, In_2O_3, WO_x, MoO_3, CdO, etc.) where electrons are majority carriers and p-type (CuO, NiO, TeO_2, etc.) in which holes are majority carriers. As per the sensing principle, the sensor resistance decreases after adsorption of reducing gases (electron donors like H_2, H_2S, CH_2O, CO, etc.) on n-type metal oxide surface. On the other hand, sensor resistance increases after adsorption of oxidizing gases (electron acceptors like O_2, NO_2, etc.) [6,8]. As discussed earlier, the sensitivity and selectivity of a gas sensor depends on the receptor and transduction function of the sensor, respectively; nevertheless, in semiconducting metal oxide-based gas sensor, the selectivity towards a particular gas depends on chemical interaction between surface of the metal oxides and target gas, and sensitivity depends on microstructural property of the metal oxides such as porosity, active surface area, grain size, neck size, etc. To improve the sensing response as well as sensitivity towards specific gases, numerous approaches have been employed, including noble metals (Pt, Pd, Ag, etc.), complex/mixed metal oxides ($CdIn_2O_4$, $NiTa_2O_6$, $Bi_2Fe_4O_2$, etc.), etc. In addition, some other foreign metals or metal oxides like Mg, Ni, Cu, Zn, PdO, CuO, Cr_2O_3, etc. are also loaded to the primary oxide material to improve the gas selectivity along with sensing response. For example, SnO_2-CuO and In_2O_3-Fe_2O_3-based oxide materials are used to detect gases such as hydrogen sulphide (H_2S) and Ozone (O_3), respectively. Some of the pure metals and oxide materials used in various chemiresistive gas sensing claims are enumerated in Table 5.1 [6,7].

Though metal oxide-based gas sensors have many advantages like reversibility, rapid response, low cost, robustness, etc. still these sensors are limited by several disadvantages like high-temperature operation which leads to high power consumption, poor selectivity, inability to operate in a vacuum or inert atmospheres, long-term

TABLE 5.1

Examples of Semiconducting Metal Oxide-Based Materials Used as Chemiresistive Gas Sensor

Materials	Operating temperature (°C)	Additives	Target gas
In_2O_3	50–100	Pd, Pt, Li_2O, Er_2O_3, WO_3, CaO, Al_2O_3, SnO_2, SiO_2, etc.	O_2, CO, O_3, H_2, H_2S, CH_4, C_2H_5OH, NO, NH_3, etc.
SnO_2	100–500	Pd, Sb, Ag, Pt, Au, Rh, In_2O3, TiO_2, Fe_2O_3, CaO, Bi_2O_3, etc.	H_2, CO, O_3, C_2H_5OH, CCl_4, C_4H_{10}, $CHCl_3$, CH_4, CH_3, CN, NH_3, SO_2, H_2S, NO, PH_3, NO_2, HF, O_2, H_2O, CH_2SH, etc.
WO_3	200–500	Sb, Pt and Au.	O_2, O_3, H_2S, CH_2O and NO.
ZnO	200–600	Ni, Li_2O, Er_2O_3, SnO_2, CaO, In_2O_3, WO_3, Al_2O_3, etc.	O_3, CO, O_2, H_2, CH_4, NH_3, NO_2, C_2H_5OH, etc.
Fe_2O_3	200–500	Pd, SnO_2, Mg, TiO_2, MgO and SO_4.	CH_4, C_4H_{10}, H_2, CO, NH_3, F_2, C_2H_5OH, etc.
MoO_3	100–500	Pt and ZrO_2	H_2 and CO.
In_2O_3-SnO_2	327	-	NO
SnO_2-LaOCl	350–425	-	CO_2

instability due to aging, cross-sensitivity (most of the cases metal oxides are sensitive to humidity) [8,9]. Recent research works on metal oxide-based gas sensor have paid attention to the improvement of these inabilities by using various modifications of oxide materials and other sensing-related parameters [10].

5.2.2 Transition Metal Dichalcogenides (TMDs) a Two Dimensional (2D)-Based Gas Sensor

2D constructed TMDs materials are a group of materials having a chemical formula presented as MX_2, where M refers to transition metals (e.g., Nb, Zr, W, Hf, Re, Ti, Mo, Ta, etc.) and X refers to chalcogen (Se, S, Te, etc.) [11]. TMDs with 2D structures have fascinated much more devotion in gas sensing applications owing to excellent sensing ability at the molecular level and remarkable chemical and physical properties such as semiconducting property, high surface-to-volume ratio, etc. [12]. These materials could be engineered by several synthesizing routes, including chemical vapour deposition (CVD) [13], micromechanical exfoliation and liquid exfoliation [14]. Especially, liquid exfoliation is more suitable for gas sensing and other electronic applications. Some examples of these 2D-layered materials magnanimously incorporated as gas sensing elements are MoS_2, WS_2, ReS_2, $MoSe_2$, WSe_2, $ReSe_2$, etc. [15]. TMDs found gas sensors are able to be operated at room temperature, but poor recovery needs thermal assistance or UV rays which lead to high power consumption [16]. The limitations in selectivity, stability and poor recovery can be avoided by surface functionalization or designing heterostructures of TMDs nanomaterials [17].

5.3 FILM DEPOSITION TECHNIQUES

In gas sensing technology, the active sensing layer or sensing film plays a vital role in gas adsorption. Various film deposition techniques have been used by many researchers for making sensing films. Common deposition techniques (especially for the polymer-based film) are described below:

   a. *Electrochemical deposition:* It is one of the convenient film deposition techniques where film thickness could be controlled by governing the whole charge passing through the electrochemical cell. Generally, this technique is used to deposit the polymer-based material on patterned microelectrodes and can be used as chemiresistive sensor [18].
   b. *Dip-coating:* In this technique, a substrate is used to dip into a chemical solution in such a way that part of the solution is deposited on the substrate to make the sensing film. The sensing film thickness is frequently controlled by the time of dipping [19].
   c. *Spin-coating:* Spin-coating is a film deposition method where a prepared chemical solution is deposited on a spinning substrate [20]. A thin film is formed after the evaporation of that solution. The above process can be repeated to control the film thickness. Also, film thickness depends on the concentration of the chemical solution and the spinning speed of the substrate.
   d. *Langmuir-Blodgett (LB) technique:* It is an eminent technique for making a thin film, especially for polymer-based materials. Generally, the LB film deposition technique produces an ultrathin film which can also be thicker by repeating the deposition process [21].
   e. *Layer-by-layer self-assembly technique:* It is a very useful technique widely applied in various applications where multilayer thin film is required. The films are developed by the deposition of oppositely charged material layer by layer [22]. The cascading deposition is usually done by various common techniques, including spin-coating, dip-coating, etc. The thickness of the sensing film depends on the number of deposited opposite layers.
   f. *Thermal evaporation:* In this film deposition process, the sensing films are prepared by heating the chemical solution on a substrate in a vacuum environment where the solution evaporates gradually. The thickness of the film depends on evaporation time [23].
   g. *Screen printing:* It is a useful technique for the deposition of sensing film on various substrates like alumina or silicon [24]. A mask is used to deposit the paste solution onto the substrate. After that, it is dried at a suitable temperature for getting sensing film.
   h. *Drop-coating:* It is the simplest way to deposit sensing film on a substrate. In this method, the chemical solution is drop-cast onto the substrate in a controlled manner. After evaporation of the deposited material, the sensing film is formed [25]. The sensing film thickness depends on the number of

deposited drops, solution viscosity and density. Also, applied pressure during film deposition influences film thickness. Usually, this method is very popular for making the sensing film in many gas sensing applications. However, fabricated sensing film is usually not uniform.

5.4   GAS SENSING PARAMETERS

Various gas sensing parameters are available to accomplish the chemiresistive gas sensing measurements, which are described briefly below.

5.4.1   BASELINE

The baseline of a sensor can be defined as a stabilized sensor signal after equilibrium with its surrounding environment. Normally, the sensor shows large fluctuations in output signal before baseline stabilization. So, it is necessary to stabilize a sensor at its baseline before the commencement of gas sensing measurements. For a chemiresistive type gas sensor, the baseline of the sensor is determined in terms of resistance (R).

5.4.2   SENSOR RESPONSE

The sensor response of a chemiresistive sensor could be defined as the ratio of relative changes in resistance due to the target gas to the baseline resistance, which can be expressed as follows [26]:

$$\text{Sensor response} = (\Delta R/R_a)\% = \frac{R_g - R_a}{R_a} \times 100 \tag{5.1}$$

where R_g and R_a are the sensor resistance in target gas and dry air, respectively. The sensor response expressed in Equation (5.1) is applicable for n-type-based material exposed to oxidizing gases. But Equation (5.1) can also be applicable for reducing gases. In that case, the sensor response can be shown in a negative direction [27]. In another way, when n-type-based material is exposed to reducing gases, the sensor response can be expressed as Equation (5.2) [28]:

$$(\Delta R/R_a)\% = \frac{R_a - R_g}{R_a} \times 100 \tag{5.2}$$

In the case of p-type-based sensing material, when the sensor is exposed to reducing and oxidizing gases, the sensor response can be expressed as Equations (5.1) and (5.2), respectively.

5.4.3   CALIBRATION CURVE AND SENSOR SENSITIVITY

In any chemiresistive type gas sensor, the sensor output signal could be obtained in terms of resistance (R). Moreover, a calibration curve of a chemiresistive gas sensor is defined as a relationship between sensor response and executed gas concentrations (C), which can be expressed as $R = f(C)$ or it can be written as sensor response with the function of input gas concentrations [29]. The calibration graph can be plotted between $\Delta R/R_a$ and input gas concentrations.

The sensitivity of a gas sensor is defined as the ability to detect the gas concentration from a particular gas concentration or range of gas concentration and could determine from the slope of the calibration curve. The sensitivity (S) of a gas sensor can be expressed as follows [30]:

$$S(\%) = \frac{dy}{dx} \times 100 \tag{5.3}$$

where y denotes the sensor response for a given gas concentration range (x).

5.4.4 SELECTIVITY

Normally, including the target gas, a gas sensor responds to other gases existing in the surrounding environment. So, a selectivity test is required to know whether a sensor is prepared for a particular gas or not. For the selectivity test, the sensitivity of other interfering gases is also desired under the same operating condition. Selectivity of a gas sensor can be expressed as follows [30,31]:

$$\text{Selectivity} = \frac{(\text{Sensitivity of interfering gas})}{(\text{Sensitivity of target gas})} \tag{5.4}$$

5.4.5 SENSOR RESPONSE AND RECOVERY TIME

The response time (R_s) of a gas sensor could be defined as the time taken for sensor response to reach 90% of its saturation value after exposing the sensor to the target gas. Similarly, recovery time (R_e) of a gas sensor could be defined as the time taken for sensor response to drop 90% of its saturation value after switching off the target gas supply.

5.4.6 REPEATABILITY

Repeatability of a gas sensor is the ability to sense a specific target gas repeatedly for a given gas concentration under the same measuring condition [32]. The measuring condition includes:

a. Same measurement process as executed during that specific gas measurement for a given concentration or a specific range of concentration.
b. Same instrument used for gas measurement under the same condition.
c. Same observer who operated the gas sensing measurement.
d. Same place or laboratory where gas sensing measurements were executed.
e. Repetition of the specific gas concentration for a short period of time.

5.4.7 DRIFT

The sensor drift is defined as the random variations of the output sensing signal after exposure to the same concentration of target analytes under the same conditions [33]. The key factors for drift are fluctuation of environment parameters and measurement conditions, including target gas concentration, surrounding relative humidity (%RH) and temperature (°C), measuring time, flow rate variation of the target gas, instrumental error, thermo-mechanical degradation, poisoning, etc. The above-mentioned factors can change the baseline as well as the sensitivity of the sensor over a period of time. Besides, drift causes to produce poor sensor repeatability.

5.4.8 LIMIT OF DETECTION (LOD)

Limit of detection (LOD) is a minimum level of detection through which a sensor can detect a specific gas. It can be defined as three times the standard deviation of sensor response of the blank sample in dry air (without target gas) to the sensor sensitivity of the specific gas [34].

5.5 METHOD TO IMPROVE GAS SENSING RESPONSE

Several methods are available to improve the sensor response towards various gases. These methods include doping, surface modification or functionalization of the materials, controlling the thickness of sensor film and grain size, etc. Brief discussions on these methods are described in this section.

a. ***Doping:*** Doping is one of the significant methods to elevate the sensor selectivity as well as sensitivity towards a particular gas or vapour. Various impurities or dopants such as noble metals (Pt, Pd, Rh, Ag, Au, etc.), nonmetals (Se), alkaline earth metals (Mg, Ca, Sr, Ba, etc.), transition metals (Co, Fe, Cu, etc.), metalloids (Si, B, etc.), etc. are usually added to the semiconducting materials to improve the gas sensing property [35]. Some of these dopants act as "catalyst" or "accelerators," while other materials act as "inhibitors" in various sensing processes. In maximum cases, noble and transition metals are used as doping additives in metal oxide-based semiconducting materials. Three different doping methods are used to introduce dopants into semiconducting material, which include diffusion, ion implantation and neutron transformation or transmutation [9]. Basically, the dopant creates an ionic and electronic defect which changes Fermi-level position of the semiconducting material and enhances the gas sensing properties. Doping is also applicable for polymer-based gas sensors. After doping, an organic polymer having insulating or semiconducting behaviour shows a drastic change in electrical, magnetic, optical and structural properties which influence the gas sensing behaviour of that doped polymer [36]. As discussed earlier, the presence of delocalized π electrons in the doped polymer are primary reasons for the enhancement of conductivity as well as gas sensing properties towards a specific gas.

The doping process in the polymer can be done through different routes, including redox doping (ion doping), photo and charge-injection doping and non-redox doping [36]. Like, metal oxide semiconducting materials, all the conductive polymers belong to either n-type or p-type depending upon the dopant. By using the doping process, many research works on the gas sensor have been exercised so far. For example, Xu et al. conveyed various alcohol sensors based on zinc oxide (ZnO) nanomaterials loaded with different dopants like Ru, Mg, Pd, Y, La and Na. It has been suggested that ZnO nanomaterials loaded with an optimal amount of Ru can increase the sensor response towards ethanol (C_2H_5OH) [37]. Moon et al. stated Pd-doped TiO_2 nanofibers for NO_2 gas sensing where remarkable enhancement in sensor response has been achieved than that of pure TiO_2 at low NO_2 concentration level [38]. Sadek et al. reported doped and dedoped poly-aniline nanofibers for hydrogen (H_2) gas sensing where doped polyaniline showed higher sensitivity than that of dedoped polyaniline [39].

b. *Surface modification or functionalization of the materials:* Surface modification portrays a principal character in gas sensing technology. Chemical modification of sensor surface increases the selectivity as well as sensitivity towards a particular gas. Surface modification techniques can be divided into two categories (i) covalent modification and (ii) non-covalent modification. In the case of a covalent bond, the chemical species or functional groups are chemically bonded with the surface of the sensor material. In contrast, nanomaterials attached to sensor surface through non-covalent bonds are weakly bonded which can also be called as van der Waals interactions. Non-covalent surface modifications are basically su-pramolecular complexation through different adsorptive and wrapping forces which include weak van der Waals and π-stacking interactions. The nature of the functional groups depends on the materials to be functio-nalized or modified. Several functional groups are available for the mod-ification of material surfaces. For example, carboxyl group (-COOH), hydroxyl group (-OH) and amino group (-NH$_2$) are commonly used to functionalize CNTs [4]. It has been reported that CNTs attached with COOH group strongly respond to the various VOCs due to polar-polar interaction [40]. Pristine CNTs are unable to detect CO gas, but COOH group functionalized CNTs can absorb CO gas molecules through weak hydrogen bonding [41]. Tran et al. described SWCNTs based NO_2 gas sensor fabricated on 3-aminopropyltriethoxysilane (APTES) functiona-lized oxidized silicon wafer (SiO_2) [42]. The sensor response with APTES treated surface was found to be twice that of without APTES treated surface at the same concentration level of NO_2 gas. It has been suggested that the presence of amino groups (-NH$_2$) in APTES donates electrons to semiconducting SWCNTs and these electrons are transferred again to oxidizing NO_2 gas, which results in higher sensor response. Many gas sensing approaches through non-covalent surface modifications have been reported [43], where different nanoparticles (Pd, Au, Pt, Ru, etc.) were attached to CNTs wall through various reduction processes. Also, the wrapping of polymers on CNTs can influence the gas sensing properties [44]. However, surface modification through covalent bonding deteriorates the CNT's properties confined to electrical and mechanical But, the non-covalent surface modification does not affect the mechanical and electrical properties of CNTs [4]. The above-reported results reveal that surface modification portrays a vital role in gas sensing technology by increasing selectivity as well as sensitivity.

c. **Thickness and grain size of sensing film:** Thickness of sensing film is an important parameter for gas sensing applications. Sensitivity of the sensor strongly depends on the thickness of the sensing layer [45]. Generally, high sensitivity is expected when depletion width matches that of film thickness. Grain size is another vital parameter to advance the sensor response for semiconducting metal oxide-based sensors. The surface area of the sensing film increases when grain size decreases. As a result, more number of target gas molecules can be adsorbed at the grain boundary [46]. Also, defects such as vacancies and dangling bonds influence the electron transport properties of the sensor material.

5.6 CONCLUSION

This chapter describes the detailed literature study on chemiresistive sensor (conduct metric type) using various nanomaterials, different useful parameters for chemiresistive gas sensing measurement, different sensing film deposition techniques, methods to improve gas sensing response, various polymer composites using carbon-based nanofillers for methanol sensing and carbon dioxide gas sensing using polymer composites reinforced with different carbon nanofillers. Moreover, brief information about the need for gas sensors, development of gas sensors, momentum of nanotechnology in gas sensing applications and different types of gas sensors were discussed.

REFERENCES

1. Barbir, F., Veziroğlu, T.N. and Plass Jr., H.J., 1990. Environmental damage due to fossil fuels use. *International Journal of Hydrogen Energy*, 15(10), pp.739–749.
2. Lelieveld, J.O., Crutzen, P.J., Ramanathan, V., Andreae, M.O., Brenninkmeijer, C.A.M., Campos, T., Cass, G.R., Dickerson, R.R., Fischer, H., De Gouw, J.A. and Hansel, A., 2001. The indian ocean experiment: Widespread air pollution from South and Southeast Asia. *Science*, 291(5506), pp.1031–1036.
3. Bose, B.K., 2010. Global warming: Energy, environmental pollution, and the impact of power electronics. *IEEE Industrial Electronics Magazine*, 4(1), pp.6–17.
4. Mittal, M. and Kumar, A., 2014. Carbon nanotube (CNT) gas sensors for emissions from fossil fuel burning. *Sensors and Actuators B: Chemical*, 203, pp.349–362.
5. Yamazoe, N., 2005. Toward innovations of gas sensor technology. *Sensors and Actuators B: Chemical*, 108(1–2), pp.2–14.
6. Jaaniso, R. and Tan, O.K. eds., 2013. *Semiconductor gas sensors*. Elsevier.
7. Korotčenkov, G.S., 2013. Handbook of gas sensor materials: Properties, advantages and shortcomings for applications. *Conventional approaches*. Springer.
8. Kim, H.J. and Lee, J.H., 2014. Highly sensitive and selective gas sensors using p-type oxide semiconductors: Overview. *Sensors and Actuators B: Chemical*, 192, pp.607–627.

9. Taylor, R.F. and Schultz, J.S., 1996. *Handbook of chemical and biological sensors.* CRC Press.

10. Vlachos, D.S., Papadopoulos, C.A. and Avaritsiotis, J.N., 1997. Characterisation of the catalyst-semiconductor interaction mechanism in metal-oxide gas sensors. *Sensors and Actuators B: Chemical*, 44(1–3), pp.458–461.

11. Tan, T.L., Ng, M.F. and Eda, G., 2016. Stable monolayer transition metal dichalcogenide ordered alloys with tunable electronic properties. *The Journal of Physical Chemistry C*, 120(5), pp.2501–2508.

12. Bhimanapati, G.R., Lin, Z., Meunier, V., Jung, Y., Cha, J., Das, S., Xiao, D., Son, Y., Strano, M.S., Cooper, V.R. and Liang, L., 2015. Recent advances in two-dimensional materials beyond graphene. *ACS Nano*, 9(12), pp.11509–11539.

13. Bergeron, H., Sangwan, V.K., McMorrow, J.J., Campbell, G.P., Balla, I., Liu, X., Bedzyk, M.J., Marks, T.J. and Hersam, M.C., 2017. Chemical vapor deposition of monolayer MoS_2 directly on ultrathin Al_2O_3 for low-power electronics. *Applied Physics Letters*, 110(5), p.053101.

14. Tao, H., Zhang, Y., Gao, Y., Sun, Z., Yan, C. and Texter, J., 2017. Scalable exfoliation and dispersion of two-dimensional materials-an update. *Physical Chemistry Chemical Physics*, 19(2), pp.921–960.

15. Yang, S., Jiang, C. and Wei, S.H., 2017. Gas sensing in 2D materials. *Applied Physics Reviews*, 4(2), p.021304.

16. Cho, B., Hahm, M.G., Choi, M., Yoon, J., Kim, A.R., Lee, Y.J., Park, S.G., Kwon, J.D., Kim, C.S., Song, M. and Jeong, Y., 2015. Charge-transfer-based gas sensing using atomic-layer MoS_2. *Scientific reports*, 5, p.8052.

17. Joshi, N., Hayasaka, T., Liu, Y., Liu, H., Oliveira, O.N. and Lin, L., 2018. A review on chemiresistive room temperature gas sensors based on metal oxide nanostructures, graphene and 2D transition metal dichalcogenides. *Microchimica Acta*, 185(4), p.213.

18. Lu, G., Qu, L. and Shi, G., 2005. Electrochemical fabrication of neuron-type networks based on crystalline oligopyrene nanosheets. *Electrochimica Acta*, 51(2), pp.340–346.

19. McGovern, S.T., Spinks, G.M. and Wallace, G.G., 2005. Micro-humidity sensors based on a processable polyaniline blend. *Sensors and Actuators B: Chemical*, 107(2), pp.657–665.

20. Prasad, G.K., Radhakrishnan, T.P., Kumar, D.S. and Krishna, M.G., 2005. Ammonia sensing characteristics of thin film based on polyelectrolyte templated polyaniline. *Sensors and Actuators B: Chemical*, 106(2), pp.626–631.

21. Ulman, A., 2013. *An introduction to ultrathin organic films: From langmuir-blodgett to self-assembly.* Academic press.

22. Richardson, J.J., Björnmalm, M. and Caruso, F., 2015. Technology-driven layer-by-layer assembly of nanofilms. *Science*, 348(6233), p.aaa2491.

23. Agbor, N.E., Petty, M.C. and Monkman, A.P., 1995. Polyaniline thin films for gas sensing. *Sensors and Actuators B: Chemical*, 28(3), pp.173–179.

24. White, N.M. and Turner, J.D., 1997. Thick-film sensors: Past, present and future. *Measurement Science and Technology*, 8(1), p.1.

25. Ruangchuay, L., Sirivat, A. and Schwank, J., 2004. Selective conductivity response of polypyrrole-based sensor on flammable chemicals. *Reactive and Functional Polymers*, 61(1), pp.11–22.

26. Varghese, O.K., Kichambre, P.D., Gong, D., Ong, K.G., Dickey, E.C. and Grimes, C.A., 2001. Gas sensing characteristics of multiwall carbon nanotubes. *Sensors and Actuators B: Chemical*, 81(1), pp.32–41.

27. Tanvir, N.B., Yurchenko, O., Laubender, E., Pohle, R., Sicard, O.V. and Urban, G., 2018. Zinc peroxide combustion promoter in preparation of CuO layers for conductometric CO_2 sensing. *Sensors and Actuators B: Chemical*, 257, pp.1027–1034.

28. Mourya, S., Kumar, A., Jaiswal, J., Malik, G., Kumar, B. and Chandra, R., 2019. Development of Pd-Pt functionalized high performance H_2 gas sensor based on silicon carbide coated porous silicon for extreme environment applications. *Sensors and Actuators B: Chemical*, 283, pp.373–383.

29. Neri, G., Bonavita, A., Micali, G., Rizzo, G., Callone, E. and Carturan, G., 2008. Resistive CO gas sensors based on In_2O_3 and $InSnO_x$ nanopowders synthesized via starch-aided sol-gel process for automotive applications. *Sensors and Actuators B: Chemical*, 132(1), pp.224–233.

30. Franke, M.E., Koplin, T.J. and Simon, U., 2006. Metal and metal oxide nanoparticles in chemiresistors: Does the nanoscale matter?. *Small*, 2(1), pp.36–50.

31. Gurlo, A., Bârsan, N. and Weimar, U., 2005. Gas sensors based on semiconducting metal oxides. In J. L.G. Fierro, *Metal oxides*, pp.705–760. CRC Press.

32. Hansman, R.J., Figliola, R.S., Sydenham, P.H. and Dieck, R.H., 1999. Measurement characteristics. In J. G. Webster , *Measurement, instrumentation & sensors*, pp.1–62. IEEE Press.

33. Padilla, M., Perera, A., Montoliu, I., Chaudry, A., Persaud, K. and Marco, S., 2010. Drift compensation of gas sensor array data by orthogonal signal correction. *Chemometrics and Intelligent Laboratory Systems*, 100(1), pp.28–35.

34. Srinives, S., Sarkar, T., Hernandez, R. and Mulchandani, A., 2015. A miniature chemiresistor sensor for carbon dioxide. *Analytica Chimica Acta*, 874, pp.54–58.

35. Korotcenkov, G., 2014. Bulk doping of metal oxides. In R. A. Potyrailo, *Handbook of gas sensor materials*, pp.323–340. Springer, New York.

36. Korotcenkov, G., 2014. Bulk and Structure Modification of Polymers. In R.A. Potyrailo, *Handbook of gas sensor materials*, pp.341–357. Springer, New York.

37. Xu, J., Han, J., Zhang, Y., Sun, Y.A. and Xie, B., 2008. Studies on alcohol sensing mechanism of ZnO based gas sensors. *Sensors and Actuators B: Chemical*, 132(1), pp.334–339.

38. Moon, J., Park, J.A., Lee, S.J., Zyung, T. and Kim, I.D., 2010. Pd-doped TiO_2 nanofiber networks for gas sensor applications. *Sensors and Actuators B: Chemical*, 149(1), pp.301–305.

39. Sadek, A.Z., Wlodarski, W., Kalantar-Zadeh, K., Baker, C. and Kaner, R.B., 2007. Doped and dedoped polyaniline nanofiber based conductometric hydrogen gas sensors. *Sensors and Actuators A: Physical*, 139(1–2), pp.53–57.

40. Sin, M.L.Y., Chow, G.C.T., Wong, G.M.K., Li, W.J., Leong, P.H.W. and Wong, K.W., 2007. Ultralow-power alcohol vapor sensors using chemically functionalized multiwalled carbon nanotubes. *IEEE Transactions on Nanotechnology*, 6(5), pp.571–577.

41. Korotcenkov, G., 2014. Surface functionalizing of carbon-based gas-sensing materials. In R. A. Potyrailo , *Handbook of Gas Sensor Materials*, pp. 359–372. Springer, New York.

42. Tran, T.H., Lee, J.W., Lee, K., Lee, Y.D. and Ju, B.K., 2008. The gas sensing properties of single-walled carbon nanotubes deposited on an aminosilane monolayer. *Sensors and Actuators B: Chemical*, 129(1), pp.67–71.

43. Espinosa, E.H., Ionescu, R., Bittencourt, C., Felten, A., Erni, R., Van Tendeloo, G., Pireaux, J.J. and Llobet, E., 2007. Metal-decorated multiwall carbon nanotubes for low temperature gas sensing. *Thin Solid Films*, 515(23), pp.8322–8327.

44. Alshammari, A.S., Alenezi, M.R., Lai, K.T. and Silva, S.R.P., 2017. Inkjet printing of polymer functionalized CNT gas sensor with enhanced sensing properties. *Materials Letters*, 189, pp.299–302.

45. Hossein-Babaei, F. and Orvatinia, M., 2003. Analysis of thickness dependence of the sensitivity in thin film resistive gas sensors. *Sensors and Actuators B: Chemical*, 89(3), pp.256–261.

46. Yamazoe, N., Kurokawa, Y. and Seiyama, T., 1983. Effects of additives on semiconductor gas sensors. *Sensors and Actuators*, 4, pp.283–289.

6 3D Nano Printing

*Current Status and Emerging
Trends of a Novel
Fabrication Technique and
Its Industrial Applications in
Biomedicines*

F. Ahmad, A. Nazeer, and S. Ahmad

CONTENTS

6.1 Introduction...74
6.2 Conventional 3D Printing – Medical Applications.......................................75
    6.2.1 Inkjet 3D Printing..76
    6.2.2 Extrusion 3D Printing 76
    6.2.3 Light-Assisted 3D Printing 77
    6.2.4 DOPsL 3D Printing...77
    6.2.5 TPP 3D Printing..77
6.3 Biomedical Applications – 3D Printing 78
    6.3.1 Surgical Applications 78
    6.3.2 Disease Modelling..79
    6.3.3 Regenerative Biomedicine 79
6.4 Materials for 3D Printing..80
    6.4.1 Prerequisite Parameters 80
    6.4.2 Appropriate Biomaterial Choice 81
        6.4.2.1 Melt-Cure Polymers 81
        6.4.2.2 Hydrogels..81
6.5 Novel 3D Printing and Materials..84
    6.5.1 Novel SLA Materials 85
    6.5.2 Multi-Material 3D Printing...85
    6.5.3 Embedded 3D Printing...86
    6.5.4 4D Printing 87

DOI: 10.1201/9781003220602-6

    6.5.5   Electrically Controlled 3D Printing..90
6.6   Discussion and Conclusions..91
References..92

6.1   INTRODUCTION

A physical part (3D) is produced by 3D printing using CAD programs by adding layers of a material called additive manufacturing (AM) or layered manufacturing. 3D printing employs some techniques such as selective laser sintering (SLS) material jetting, stereolithography (SLA), material extrusion and binder jetting used for different materials and areas for producing complex parts with saving material and time. 3D printers being a part of a developing technology are now more accessible and affordable than before. This technology, which is preferred especially for many applications in the field of health, provides great benefits especially for medical and dental imaging, by handling medical device design and production that define the patient-specific anatomical structure. Applications using biocompatible materials such as the creation of tissue without any damage to living cells, blood vessel production, dental implants and special medical prostheses are just some of the contributions of the 3D printer to the biomedical field. This technology is also being explored to fix or replace defective organs such as kidneys and heart in which these organs will perform the same biological functions as the original organs. 3D-printer technology has become a preferred application in many sectors, especially in recent years its use in biomedical applications. This method is used in surgical applications, medical imaging, pharmaceutical industry, production of patient-specific medical prostheses and implants, vet medicine applications, skin engineering and stem cell studies and organ printing [Bozkurt, and Karayel, 2021].

3D objects are fabricated in the conventional 3D-printing process by adding layers of materials onto a planar surface as a line or a point via material extrusion, vat photopolymerization, material or binder jetting, powder-bed fusion, sheet lamination and directed energy deposition. A commercial 3D printer that entered the market in 1986 was a refined version of the first printer introduced four decades ago. A vat photo-polymerization process first converts a liquid plastic-like acrylate into a solid object, through a laser scan of liquid photocurable material. Later, others started creating alternatives to the original UV light-based system. The selective laser sintering method was introduced in 1989, in which laser power was used for fusing or sintering powdered materials, typically made of plastic, metal, ceramic and glass, to create solid 3D objects in a layer-by-layer manner by selecting appropriate powdered materials. In case of fused deposition modelling (FDM), a thermoplastic filament feed into a heated nozzle was used in depositing filaments on a printing substrate in a layer-by-layer manner by extruding the heated plastic filaments through a nozzle to build up the objects. FDM-based 3D printers pioneered a new way of manufacturing products since their invention and provided a new method of creating prototypes at a lower cost. SLA being the first system of high-speed and high-resolution additive manufacturing (AM), it has emerged as a cost-effective method of 3D-printing method. Despite the rapid advancements in AM, its low printing speed, scalability and quality have hampered the adaptation of 3D printing in large-scale manufacturing applications. Nevertheless, 3D printing capable of printing complex 3D objects with high customizability has attracted the interest of many researchers, because its features are extremely useful for rapid prototyping, creating concept models and manufacturing end-products ready to be sent to the market [Tetsuka, and Shin, 2020].

The recent developments in machine learning-based processes, computer-aided design (CAD) software and in novel materials, ranging from plastic and metals to ceramics and even food products, are further expanding the stage for 3D printing. Subsequently, efforts were made to fabricate even complex parts of the human body using biomaterials-based inks for 3D printing in the same way. Many advantages of 3D printing in biomedical applications are paving the way for possible medical solutions such as transplantation of human tissues or organs for regenerative medicine, and the 3D printing of human tissues and organs is now an emerging research topic. For the first time in 2001 came the successful report of transplanting a 3D-printed bladder, which was fabricated using a dome-shaped scaffold of the size of a human bladder constructed from a biodegradable polymer followed by coating the patient's own bladder cells layer-by-layer on it using a 3D printer. Two different types of bio-inks comprising of urothelial cells on the inside and muscle cells on the outer surface of the scaffold. Subsequently, the fabrication of other complex organs like the heart and liver, a method to mimic the vascular networks was required for keeping the organs alive. In 2004, a 3D printer was used in creating tubular structures for fabricating the blood vessels and vascular networks. Constructing 3D-biological hollow tubes and culturing the appropriate cells on the outer surface of 3D printed hollow tubes. The printer used in these fabrications had three print-heads that deposited bio-inks onto a gelatine sheet serving as the extracellular matrix (ECM). Until 2010, this technology was the basis for 3D bioprinting. Subsequently, 3D bioprinting has been developed for fabricating various artificial biological tissue constructs for various biomedical applications like tissue regeneration. 3D bioprinting has also been widely used in the fabrication of bio-mimetic tissue models for studying the pathogenesis of various diseases, identifying and optimizing potential drugs, and inventing useful novel medical applications because it has emerged as a promising technology to create complex tailor-made biological constructs with desired physical and biological properties [Gu, et al., 2018; Tetsuka, and Shin, 2020].

An overview of the recent advancements of new materials and 3D-printing techniques developed and reported for taking care of unfulfilled needs of the conventional 3D-printing methodologies is included in the followings, especially in biomedical applications, such as printing speed, cell growth feasibility and complex shape achievement.

6.2 CONVENTIONAL 3D PRINTING – MEDICAL APPLICATIONS

Nozzle-based layer-by-layer deposition using bio-ink has been used as a 3D-printing method in biomedical applications to create biological constructs. The primary 3D-printing methods for medical applications include inkjet-based, extrusion-based and light-assisted methods. The most used platform is based on the extrusion method followed by the light-assisted and inkjet-based printing approaches. All these methods can print scaffolds for cell culture or biological constructs using cell-laden bio-inks. However, there are some differences in the printing resolution, materials, speed and mechanism among these methods [Capel, et al., 2018].

6.2.1 INKJET 3D PRINTING

The first inkjet bioprinter was a modified version of a commercially available benchtop inkjet printer with few picoliter droplets of bio-ink composed of bio-materials or cell mixtures dispensed on an electronically controlled stage along the z-axis. Multiple actuation mechanisms have been used in modern inkjet printers involving thermal, piezoelectric, electromagnetic, electrostatic or acoustic, to produce a precise droplet. Inkjet-based 3D-printing methods can print at a speed of ~100 mm/s and a minimum resolution in the range of 20–100 mm. The nozzle diameter and the physical or chemical properties of the bio-ink de-termine the resolution of the printed constructs. Typically, higher printing re-solution can be obtained with a smaller diameter of the nozzle heads. Inkjet-based methods generally require bio-inks with lower viscosity for offering a relatively fast printing speed compared to other techniques. However, they provide low cell densities and decreased cell viability and have problems caused by the inherent inability of the printing head to provide a continuous flow, limiting their cap-ability to 3D-print biological constructs compared to extrusion-based techniques [Tetsuka, and Shin, 2020].

6.2.2 EXTRUSION 3D PRINTING

Extrusion-based 3D printing in contrast to inkjet printing can control the flow of continuous bio-inks and has been more widely employed. A dispensing system involving pressure, mechanical or solenoid valves is used to drive the 3D-printing system. Extrusion 3D printing can print cell-laden biomaterials as bio-inks onto a target substrate or material in a layer-by-layer regime. In extrusion-based methods, bio-inks should have a viscosity in the range of 0.001–1000 mPa. s. A wide variety of bio-inks involving biomaterials such as gelatine, alginate, hyaluronic acid (HA) and polyethylene glycol (PEG)-based hydrogels, de-cellularized extracellular matrix (dECM) and cell spheroids are applicable, which make extrusion-based methods highly advantageous compared to other printing methods. However, they have lim-itations in printing speed and resolution. Their printing speed is in a wide range between 0.1 and 150,000 mm/s, typically 10–50 mm/s, and is the lowest among the three types of printing approaches. In the case of a conventional single nozzle, it requires a long time to create large size tissue constructs with bio-inks with good viability. A resolution of minimum 5–100 mm and generally over 100 mm has been reported. This resolution makes it difficult to mimic the architecture of native com-ponents of the body such as micro-vessels, aligned myofibers and neuronal networks. In comparison to inkjet 3D printing methods, extrusion 3D-printing methods can handle bio-inks with higher cell densities but at lower printing speeds and resolutions [Gopinathan, and Noh, 2018; Tetsuka, and Shin, 2020].

6.2.3 LIGHT-ASSISTED 3D PRINTING

Light-assisted 3D printing offers significant improvements in printing speed and resolution, accompanied by smooth features, different from inkjet and extrusion 3D printing. For light-assisted methods, bio-inks with a wide range of viscosities, even fluids, are suitable enabling to use a larger range of biomaterials but these are restricted to photo-crosslinkable bio-inks, typically composed of synthetic and natural biomaterials with photo-crosslinkable groups including gelatine methacryloyl (GelMA), poly(ethylene glycol) diacrylate (PEGDA) and others. To ensure an efficient light penetration depth affecting the quality of the final constructs and the printing resolution, these biomaterials should be transparent against the light source used. Two types of light-assisted 3D printing involve namely: the digital light processing (DLP) method and the two-photon polymerization (TPP) method used in fabricating biological constructs [Hippler, et al., 2019; Tetsuka, and Shin, 2020].

6.2.4 DOPsL 3D PRINTING

SLA is performed using a digital micromirror-array device (DMD) comprising several millions of micro-sized mirrors controlled independently. In this method, the construct is created in a layer-by-layer regime, where one layer is fabricated and then the printing stage is lowered or raised to create a new layer. The entire layer is cured simultaneously. Based on this method, a dynamic optical projection stereolithography (DOPsL) system, which enables the rapid fabrication of complex 3D constructs, was developed. The DOPsL method provides a higher printing speed than other techniques, using a few million micromirror chips simultaneously, which makes it easy to fabricate large-scale complex constructs with submicron resolutions. The printing speed reaches 500 mm/s and the printing resolution is as low as ~10 mm. This superior performance enabled researchers to build complex constructs: tissue constructs with fractal geometries, microfluidic mixing chambers, high-precision microwells constructed with tuneable Poisson ratios, aligned cardiac scaffolds, vasculature networks and liver microarchitectures. The DOPsL method also uses a wide variety of photopolymerizable hydrogels: GelMA, PEGDA, glycidyl methacrylate hyaluronic acid (GMHA) and others, but capable biomaterials are limited to materials that can be photopolymerized [Tetsuka, and Shin, 2020].

6.2.5 TPP 3D PRINTING

TPP 3D printing was developed from SLA as a kind of laser-based direct-writing technique. A femtosecond laser is used to polymerize the photo-linkable monomers repeatedly and selectively to generate constructs. A femtosecond laser can induce two-photon absorption, which is the basic mechanism of the TP method. In a two-photon absorption process, the simultaneous absorption of the two photons induces the excitation of a molecule to a higher-energy electronic state. The probability that a molecule undergoes the two-photon absorption process relies on the square of the light intensity of the incident light. The photons can be confined inside a voxel of size below 1 mm, which enables the printing resolution of TPP to reach only 100 nm.

Thus, TPP is an ideal platform for printing 3D objects with nanoscale to microscale features. The printing speed of the TPP reaches 20 mm/s, which is much faster than those of the nozzle-based 3D printings. TPP also accepts various polymers such as hydrogels, PEGDA, HA, collagen, bovine serum albumin and laminin as bio-inks. Although light-assisted 3D printing techniques have some limitations in the size of the printable constructs, they are now used in various tissue engineering applications and have great potential for fabricating complex 3D biological constructs within a short time [Sakai, et al., 2018; Wang, et al., 2018; Perevoznik, et al., 2019; Tetsuka, and Shin, 2020].

6.3   BIOMEDICAL APPLICATIONS – 3D PRINTING

This technology offers benefits in biomedical applications and devices owing to its ability to manufacture the product according to patient-specific needs. Many instruments used in surgery are produced by forging or casting using moulds with special surface coating meeting certain surface and mechanical properties. The procedures involved are generally not cost-effective especially in case of patient-specific implants. Machining of titanium alloys, for instance, is more difficult as it has low elastic modulus, high mechanical strength and low thermal conductivity compared to 314L stainless steel. Patient-specific implants (Ti-alloy) are not economical to manufacture as they would generate large material waste, and it is not possible to manufacture functional grade implants. Artificial kidneys, hearts, dental implants, knee prostheses, lenses and pacemakers are also needed in biomedical applications. 3D printing for these medical constructs allows producing customized geometry of implants at lower costs. The unit cost remains constant for all the products since no special tooling is necessary in 3D printing. Because of these advantages of this technology, its biomedical applications are increasing day by day especially in implants and tissue engineering and these applications are projected to increase more in the near future. Despite some remarkable achievements, the development of organ tissue with this method goes on posing several challenges. From cancer treatment to patient-specific prostheses; inventions strengthened with 3D printing are sought to improve the quality of life or save patients' life [Yeo, et al., 2018; Bozkurt, and Karayel, 2021].

6.3.1   SURGICAL APPLICATIONS

3D printing is extremely useful in surgical planning to ensure a better visualization of the patient's anatomical structure. A template made by 3D printing helps to precisely guide the surgical procedure, estimate appropriate angles and have a prior opinion of the direction and size of the bone. Under normal conditions, it is difficult to assess the location of blood flow and predict the structure of the bones. Therefore, this technology is used today as a guide to provide correct planning and supervision during surgery. Surgical applications also include cardiovascular, neuro, orthopaedic, general, plastic and aesthetic surgeries. This method is preferred in vascular surgery, tumour resections, orthopaedic surgery and neurosurgery. Improved surgical results with reduced errors increase patient safety. Additionally, it is beneficial both during and before surgery in training the medical students and surgical assistants.

In one of the studies of 3D-printed models, an obvious benefit was achieved in examining the patient's anatomy. It enables the anatomical structure of the patient to be viewed during the preoperative planning stage, to simulate the surgical intervention, as well as to test surgical instruments with the help of a 3D model [Bozkurt, and Karayel, 2021].

6.3.2 DISEASE MODELLING

3D-printing technology is used to create a copy of the patient's anatomy before complex operations or to understand the disease. In this way, the copy of the organ can be examined more clearly than imaging methods like tomography with reduced errors. In addition, it can be adopted as an easy and fast method in the diagnosis and treatment of the disease by modelling kidney diseases.

Realistic 3D models printed with different materials can be used to replicate anatomical structures and pathologies with high accuracy. 3D-printed models generated from medical imaging data acquired with computed tomography, magnetic resonance imaging or ultrasound augment the understanding of complex anatomy and pathology, assist preoperative planning and simulate surgical or interventional procedures to achieve precision medicine for improvement of treatment outcomes, train young or junior doctors to gain their confidence in patient management and provide medical education to medical students or healthcare professionals as an effective training tool. 3D printed models have been used in studying congenital heart disease, coronary artery disease, pulmonary embolism, aortic aneurysm and aortic dissection and aortic valvular disease.

There are studies that provide help not only in the kidney but also in cardiovascular and liver diseases as mentioned above. Traditional imaging methods may not be sufficient, especially in complex cardiovascular surgery because these methods are limited to a flat screen. 3D modelling for many surgeries can provide a complete representation of the anatomy, prevent unexpected findings and provide personalized treatment. In this way, it can reduce the duration of the operation and the possibility of error and provide patient safety [Ma, et al., 2018; Kelly, et al., 2019; Sun, 2020; Bozkurt, and Karayel, 2021].

6.3.3 REGENERATIVE BIOMEDICINE

3D printing is rapid prototyping using an additive technique to fabricate complex architecture with high precision through a layer-by-layer building process. This automated, additive process facilitates the manufacturing of 3D products having precisely controlled architecture including external shape, internal pore geometry and interconnectivity in a highly reproducible and repeatable manner. Therefore, in the regenerative medicine, it can provide an excellent alternative for biomimetic scaffold fabrication by accurately positioning multiple cell types and biofactors simultaneously into complex multi-scale architectures that represent the structural and biochemical complexity of living tissues or organs. In the past three decades, 3D-bioprinting has been widely developed to directly or indirectly fabricate 3D cell scaffolds or medical implants for the field of regenerative medicine. It offers very precise spatiotemporal control on the placement of cells, proteins, DNA, drugs, growth factors and other bioactive substances to better guide tissue formation for patient-specific therapy. 3D printing of bioactive scaffolds contains two types namely: acellular functional scaffolds which incorporate biological components, and cell-laden constructs aiming to replicate native analogues aiming to produce biocompatible, implantable constructs for tissue/organ regeneration [Cui, et al., 2017; Jang, J., et al., 2018].

6.4   MATERIALS FOR 3D PRINTING

Bio-inks used in 3D printing for biomedical applications are composed of bioma-terials and cells. For 3D printing of biological constructs, biomaterials act as an ECM for cells, providing sufficient structural support and promising cellular attachment, to pattern the cells and the tissues. They also regulate cellular functions and behaviours. The ideal bio-inks should not only be printable but also be nontoxic and biocompa-tible to facilitate the biological behaviour of the seed cells or tissues. In order to sustain the functions of printed biological tissues, bio-inks should fulfil certain characteristics required for each specific 3D-printing technique [Zhang, et al., 2018].

6.4.1   PREREQUISITE PARAMETERS

There are three main classes of biomaterials that are utilized in 3D printing: melt-cure polymers, hydrogels and dECM. For biomaterials adopted in 3D printing, the most important prerequisite parameter is the biocompatibility besides preserving cell at-tachment and cell migration. For hydrogel-based biomaterials, photo-initiators are needed to crosslink the hydrogels by light exposure (UV and visible light), but these photo-initiators should not affect the cell viability. It has been reported that some kinds of photo-initiators and monomers show cytotoxicity if left unreacted during the crosslinking process of hydrogels. The degradation rate of biomaterials should also be matched with the regeneration rate of tissue in order to offer sufficient structural support for cell activities to complete tissue regeneration. The elasticity of hydrogels also affects the attachment and proliferation of cells, which depends on the glass transition state of hydrogels. Water-swelling causes a lower polymer glass transition and results in a decreased transition temperature. Subsequently, the hy-drogels become a rubbery elastic state because the hydrogels are plasticized through the incorporation of excess water molecules. Therefore, the viscoelasticity of the hydrogels might be increased below Tg because the rearrangement of the polymer segments is restricted below Tg. Also, in case hydrogels are sufficiently water-swollen during cell culturing, they can contain a large number of bioactive molecules that exist in cell culture media, resulting in improved cellular behaviours such as proliferation, differentiation and elongation [Tetsuka, and Shin, 2020].

For biomaterials used in extrusion-based methods, another important parameter is non-Newtonian behaviour, determining the viscosity and flow behaviour of the bio-materials during dispensing. When pressure is applied to biomaterials during dis-pensing, they exhibit a variety of responses including shear thinning. The viscosity decreases with the increase in the shear rate. The shear force got the conformation of polymer chains reorganized in the hydrogel and enhances the alignment of the polymer chains from a randomly oriented conformation by reducing the viscosity of hydrogels during dispensing called shear thinning. Shear thinning has an impact on high molecular-weight biomaterials. This effect enables the easy dispensing of fluid materials under pressure and causes the fluidic biomaterials to restore to their gel state by relaxing their stress. Because the fluidic event of biomaterials is initiated by the yield stress, which is an instantaneous stress, minimum stress should be applied before they are dispensed. The yield stress affects the shear stress, and the structural network of biomaterials breaks when the applied shear force is greater than the yield stress. The yield stress also helps maintain the homogeneous distribution of cells within the bio-ink. The yield stress required for specific biomaterials can be estimated by extrapolating the flow curve at a low shear rate against a zero shear rate [Tetsuka, and Shin, 2020].

6.4.2   Appropriate Biomaterial Choice

The most important concern in 3D printing for biological constructs is the proper choice of biomaterials, which enables the design of target tissue scaffolds with desired chemical and physical properties.

6.4.2.1   Melt-Cure Polymers

Melt-cure polymers have high mechanical strength and durability and can act as effective structural supports for tissues and cells. Typical melt-cure polymers are polycaprolactone (PCL), polylactic acid (PLA), and polyurethane (PU). Compared to PU and PLA, PCL is favourable as scaffolds because of its low melting point of ~60°C, which can reduce the temperature-induced cell damage. Several groups have shown using PCL in a liver-on-a-chip, cartilage reconstruction, bone generation, muscle analogues and vascular networks. Similarly, PLA and PU have also been used in a heart-on-a-chip and neural tissues such as nerve grafts. However, melt-cure polymers typically require either high process temperature or the use of toxic solvents, which brings less cytocompatibility with cells compared to that of other biomaterials. In the printing process, the integration of these melt-cure polymers into cell-supportive hydrogels is also difficult [Tetsuka, and Shin, 2020].

6.4.2.2   Hydrogels

Hydrogels are one of the most important biomaterials since they inherently contain a large amount of water molecules and show good swelling features. Hydrogels can be categorized into two main classes: naturally-derived hydrogels such as collagen, gelatine, HA, alginate, etc., and synthetic hydrogels including PEG, poly (lactic-glycolic)acid (PLGA), PEGDA, etc. Hydrogels can form gel-like structures through physical, chemical or enzymatic crosslinking. Either a permanent or a reversible hydrogel is formed depending on the type of crosslinking state. Typically, irreversible permanent hydrogels are formed by introducing covalent bonds. Conversely, physical interactions such as hydrogen bonds and ionic forces produce reversible hydrogels. Although chemical crosslinking requires post-curing, the resultant permanent hydrogels show higher mechanical strength than physically cross-linked reversible hydrogels. However, hydrogels typically lack mechanical strength and shape fidelity compared to melt-cure polymers. In order to improve their mechanical strength and shape fidelity, the integration of hydrogels with melt-cure polymers such as PCL and PLGA has been investigated.

The most commonly used natural hydrogels such as gelatine, collagen, alginate and HA are deployed in 3D printing which is biodegradable and can promise native ECM-like environments required for cellular activities because they have similar mechanical properties and biological activities to the natural ECM. Natural hydrogels show a defined structure and a distinct molecular weight, owing to their biological production methods. Collagen, being the main component of the natural ECM and the most abundant protein in mammalian tissues, has been used in applications like a liver-on-a-chip and tissue constructs such as cartilage constructs. Partial hydrolysis of collagen causes a helix-to-coil transition and thus produces another soluble protein-based polymer, gelatine possessing lower antigenicity than collagen and also undergoes gelation with a change in temperature. Although it remains in the gel state below 37°C, an elevated temperature converts it to a liquid allowing it to be utilized as a sacrificial material for cells to construct organs-on-a-chip such as a liver-on-a-chip. After cell incubation, only liquid gelatine is removed at decreased temperature leaving the cells. Gelatine is also applied to produce a photopolymerizable hydrogel, GelMA, by the introduction of a methyl acrylate group as a synthetic part, which is a potential hydrogel for 3D printing. GelMA has been extensively used in various tissue engineering fields such as organ-on-a-chip, the construction of vascular networks, etc [Yin, et al., 2018].

Alginate and HA are also used as scaffolds for cartilage, chondrocytes, vascular networks with branch structures, skin tissue and muscle constructs. The physical properties of HA once modified through chemical modifications by PEG, thiolate, guest–host supramolecular complexes enhance printability and stability. Alginate can be modified with RGD motifs to offer 3D printability features needed for printing human pluripotent stem cells to generate mini-livers.

Other natural hydrogels, matrigel, fibrinogen, thrombin, chitosan and agarose are used in drug conversion in liver tissue, skin and muscle constructs, high-cell-density bioinks, the reconstruction of cartilage and bones and the construction of vascular networks, respectively. Despite all these natural hydrogels, their main limitations remain as relatively low mechanical strength, immunogenicity and stability compared to synthetic hydrogels.

In contrast, synthetic hydrogels have a well-defined structure, and their properties such as the degradation rate, mechanical strength and structural characteristics can be more easily controlled reproducibly to enhance cell adhesion. Since the late 1960s, the poly(2-hydroxyethyl methacrylate) (PHEMA) hydrogel has been widely used as an implantable material. However, currently, the most commonly used synthetic hydrogels are PEG and Pluronic F-127. PEGDA, a photo-crosslinkable hydrogel, is generated by the addition of photoinitiators.166 PEGDA has been utilized in vascular construction 40,167 and in-ear construction as a sacrificial material. Other PEG-based hydrogels: poly-(ethylene glycol)methacrylate (PEGMA), poly(ethylene glycol)-tetra-acrylate (PEGTA), etc. have also been studied for the reconstruction of bone and cartilage and vascular network construction.

Pluronic F-127 is a temperature-responsive hydrogel that can be converted to a liquid state at low temperatures allowing its application as a sacrificial material for the reconstruction of bone and cartilage, tissue engineering of muscle and construction of vascular channel networks.

PVA can be photo-crosslinked to fabricate hydrogels and used in vascular tissue and cartilage constructs. PVA hydrogels typically show a higher mechanical strength than most other synthetic hydrogels and are copolymerized with PEG to produce biodegradable hydrogels, and their degradation rate is between that of the PVA hydrogel and the PEG hydrogel.

While natural hydrogels possess better compatibility with cells, synthetic hydrogels have better processability such as printability and shape fidelity enable them to produce hybrids. In hybrid hydrogels, synthetic hydrogels enhance the mechanical strength while natural hydrogels retain the cell viability and functionality by offering an ECM microenvironment. A technique for fabricating hydrogel bio-inks with both high printability and cytocompatibility allows the hybridization of gelatine with low-concentration GelMA to produce reversible hydrogels by changing the bio-ink temperature to regulate the processability during 3D printing. The hybrid hydrogels showed higher cell compatibility than GelMA hydrogels. PEG and GelMA copolymerized hydrogels have been developed to tune their degradation rate and stiffness profiles.

The PEG-GelMA hydrogels exhibit improved cell viability and attachment compared to PEG hydrogels. A hybrid hydrogel composed of PVA-gelatine and PEG exhibited modulus strength in the range of 10–100 kPa by changing the concentration of PVA and gelatine and the molecular weight of PVA and used in cartilage regeneration. Hybrid hydrogels of alginate and Pluronic F-127 were printed at high resolution using the extrusion method, and effectively crosslinked to produce constructs with high cytocompatibility and long-term structural fidelity. As alternative approaches, synthetic materials such as PCL and polydimethylsiloxane (PDMS) have been deposited as supportive scaffolds and mixed into natural hydrogels such as alginate, collagen, gelatin and fibrinogen.

An appropriate biomaterial choice is made by considering a combination of the following factors: the printing method to be used, the target biological tissues and constructs, the cell types and the biological processes to apply. Regardless of the selected bio-ink, biomaterials have a quick crosslinking ability either in a chemical or physical manner to form a hydrogel network structure after or during the printing of 3D constructs. For instance, further development of water-soluble photoinitiators combined with high UV-visible absorption ability is necessary for 3D printing of hydrogels. Recently, highly efficient water-soluble nanoparticle-based UV curable inks were reported allowing the 3D printing of hydrogels in an aqueous solution. The water-soluble nanoparticles were made from 2,4,6-trimethylbenzoyl-diphenylphosphine oxide (TPO) with light-absorption in 385–420 nm range with an extinction coefficient over 300 times that of commercially available water-soluble photoinitiators such as PIs. An n - p^* transition in the aroyl-phosphinoyl chromophore with strong conjugation between the phosphonyl group and the carbon atom of the adjacent carbonyl group is the origin of the strong, long wavelength absorption. Thus, the polymerization rate is magnificently enhanced. This enabled the 3D printing of hydrogels without adding any solvents [Xu, et al., 2018; Hippler, et al., 2019; Wang, et al., 2019].

The development of a biomaterial enabling the creation of complex biomimetic tissues is not possible as most of the natural biomaterials cannot recreate the complexity of natural ECMs because they only have a single component from natural ECMs and lack important major components including proteoglycans, elastin, growth factors and cell-binding glycoproteins such as laminin and fibronectin. This is insufficient to mimic complex living tissues where a microenvironment with cell-to-cell connection and 3D cellular organization is typical. Consequently, dECM derived from living tissues and organs was chosen for 3D printing as well as tissue engineering and regenerative medicine. In the decellularization process to fabricate dECM, all cellular components are removed from living tissues of organs through a combination of chemical, mechanical and enzymatic treatments, which yield collagen while retaining important components of the native ECM. A cellular scaffold was explanted from mice to effectively support the formation of myofibers. Bio-inks made of decellularized tissues derived from pepsin solubilized cartilage, adipose and cardiac tissues exhibited the practicability of these tissue-specific dECMs as bio-inks for use in nozzle-based 3D printing. In fact, the constructs printed from bio-inks made of these dECMs exhibited enhanced functionality of encapsulated mesenchymal stem cells derived from human inferior turbinate tissue, human adipose-derived stem cells and rat myoblasts compared to bio-inks made from collagen. The fabrication of functional skeletal muscle constructs was also reported by other research groups using skeletal-derived dECM bio-inks. The significant increase in the osteogenic genes of human adipose-derived stem cells within dECM-PCL constructs manifested the effectiveness of dECM in bone regeneration compared to PCL scaffolds. These studies imply the versatility of dECM in 3D printing for creating complex biological tissue constructs with a living tissue and organ-like microenvironment. However, dECM-based bio-inks still have inferior printing capability of shape fidelity, which should be addressed. dECM has also inherent ethical usage problems because of its origin. Other biomaterials such as cell spheroids and tissue strands also have potential for replicating the functions and developing processes of native living tissues and organs. The direct 3D printing of cell spheroids or tissue strand-laden bio-inks has been reported through a scaffold-free method [Ashammakhi, et al., 2019; Highley, et al., 2019; Zhang, B., et al., 2019; Tetsuka, and Shin, 2020].

6.5 NOVEL 3D PRINTING AND MATERIALS

Interdisciplinary nature of the research involved 3D printing and advanced functional materials for constructing 3D objects by accumulating a layer leading to a step structure along the edges called a stair-step effect. Limitations of speed, geometry and surface quality exist in material layering methods. They also have difficulties creating 3D objects with both complexity and multifunctionality. The core problem is to improve the printability and formability of novel biomaterials without losing the features of the original material during the 3D-printing process and the production of complex and multi-functional constructs.

6.5.1 Novel SLA Materials

Recently, a novel 3D-printing technique was reported using computed tomography (CT), called computed axial lithography (CAL) based on the image reconstruction procedure, which is a technique widely used in medical imaging and non-destructive testing. Recent developments in CT for use in cancer treatment provided an intensity-modulated radiation therapy (IMRT) method, which enables the tar-geted tumour areas of the patient's body to be exposed to a critical radiation dose in 3D. Instead of the patient, a photo-responsive material is subjected to CT scans in CAL to obtain stair-step-free, smooth, flexible and complex 3D objects. The re-searchers used a viscous liquid, made from polymers with photocurable grafts and dissolved oxygen molecules, and designed the materials to react against a certain threshold of patterned light for solidification. The desired 3D shape was formed by projecting light onto a rotating cylinder of the liquid. Using CAL, the formation of a centimetre-scale geometry was completed in less than 1 min. It has the potential to produce a large array of geometries with a lateral size of up to ~55 mm within a time range of 30 to 300 s. It is also possible to add new parts into an already existing object. The printing materials do not have to be transparent. Even opaque 3D objects can be created using a dye molecule that absorbs visible light in a wide wavelength range except for the curing wavelength. A versatile photopolymerizable hydrogel was de-veloped to enable fabricating of complex 3D objects for projection stereolithography. To date, it has been difficult to create complex 3D-transport systems where organs transport blood via bio-physically and bio-chemically entangled complex vascular networks. To solve this problem, an intravascular and multi-vascular design used photopolymerizable hydrogels by incorporating a food dye as a biocompatible photo-absorber. Monolithic transparent hydrogels with intravascular 3D-fluid mixers and bicuspid valves were produced in minutes using polyethylene glycol diacrylate with the food dye. A hydrogel model of a lung-mimicking air sac with airways was re-ported to enable the delivery of oxygen to the surrounding blood vessels. Successful implantation of bio-printed constructs including liver cells into mice was also de-monstrated [Grigoryan, et al., 2019; Tetsuka, and Shin, 2020].

6.5.2 Multi-Material 3D Printing

Alternatively, the challenges in producing 3D complex vascularized cellular net-works were sorted out by using multi-material 3D-printing systems. The develop-ment of an integrated tissue organ printer (ITOP) was reported for fabricating any shapes of human-scale tissue constructs by designing multi-dispensing systems for extruding and patterning multiple cell-laden hydrogels in a single construct: the poly(e-caprolactone) polymer as a supporting construct and the Pluronic F-127 hydrogel as a sacrificial layer. Multiple materials and techniques were developed for optimizing the carrier material capable of positioning cells in the liquid form on distinct locations inside the 3D structure, sophisticated nozzle modules with a re-solution of 2 mm for biomaterials and 50 mm for cells, and photo cross-linkable cell laden hydrogels with photocurable ability even after cell passage. Simultaneous printing of an outer sacrificial acellular hydrogel mould served as a supporting layer. The lattice of microchannels permitted the diffusion of nutrients and oxygen into the printed tissue constructs. The ITOP successfully generated various 3D constructs with multiple cell types and biomaterials showing potential for fabricating various types of vascularized tissues. The fabrication of novel organ-on-a-chip devices has also been demonstrated using a multi-material 3D-bioprinting system using biocompatible soft material-based multiple-function inks. High conductance and piezo-resistive characteristics of the inks induced self-assembly into physio-mimetic laminar cardiac tissues. The cardiac micro-physiological devices were printed in a single step and applied to study the drug responses and the contractile mechanism of laminar cardiac tissues. Furthermore, very recently, an extrusion-based new multi-material printing technique was reported that allowed printing with up to eight different inks within a single nozzle termed as multi-material multi-nozzle 3D printing. They designed a printhead with a Y-shaped junction that enabled the injection of multiple inks into a single nozzle, where each ink with different viscosities can be adjusted by varying the length of the ink channels. Precisely controlled high-speed pneumatic valves were utilized to achieve rapid and seamless switching between different inks, which drastically enhanced the printing speed to fabricate complex 3D objects in a fraction of the time of conventional extrusion-based techniques. Using the MM 3D printing, successful fabrication of 3D objects with a centimetre-scale, such as foldable origami structures and locomotive soft robots, composed of two alternating epoxy or silicon inks with different stiffnesses, was demonstrated within minutes at a speed of 10–40 mm/s [Cui, et al., 2017; Skylar-Scott, Mueller, et al., 2019; Tetsuka, and Shin, 2020; Shapira, and Dvir, 2021].

6.5.3 EMBEDDED 3D PRINTING

Embedded 3D printing is another strategy for obtaining complex tissue-like constructs demonstrated by printing a 3D network of interconnected channels within a matrix composed of an acellular hydrogel and silicone using a viscoelastic, sacrificial ink. After curing the matrices and removing the sacrificial ink, a 3D construct with an interconnected channel network was realized. Embedded 3D printing involves extruding a viscoelastic ink into a reservoir with a high plateau shear elastic modulus, a low yield stress, and a photo-crosslinking ability. For this, a Pluronic F127 triblock copolymer with a hydrophobic poly(propylene oxide) segment and two hydrophilic poly(ethylene oxide) segments as a reservoir was synthesized using a chemical modification of the terminal hydroxyl groups of the hydrophilic poly (ethylene oxide) segments with diacrylate groups. Another embedded printing strategy was realized based on supramolecular assembly of shear-thinning hydrogel inks through guest–host complexes, where a mixture of two different supramolecular hydrogels, adamantane modified HA serving as a guest and b-cyclodextrin modified HA serving as a host, was injected into a supporting hydrogel to create cell-laden 3D-structures like spirals and channels. The formation of intermolecular guest-host non-covalent bonds between the adamantane modified HA and cyclodextrin modified HA allowed for the rapid formation of assemblies. This technique was also extended for biomedical applications such as drug delivery. Recently, a technique for generating complex, freeform and liquid 3D architectures was reported using formulated aqueous two-phase systems (ATPSs) using a poly-ethylene oxide matrix and an aqueous bio-ink made of a long carbohydrate mole-cule, called dextran to provide several orders of magnitude lower tension compared to typical aqueous/organic phases, which suppressed the deformation of printed structures. The chemical interaction between hydrogen bonding within the polymers provided sufficient resistance against deformation and the aqueous-in-aqueous re-configurable 3D architectures printed on the interface of the noncovalent membrane that could stand for weeks. Tailor-made micro-constructs with perfusable vascular networks were created by separately combining different cells with compartmen-talized bio-inks and matrices for embedded 3D-printing, synthetic and biopolymer matrices with a viscoplastic response and self-healing features. Organ building blocks (OBBs) composed of patient-specific-induced pluripotent stem cell (iPSC)-derived organoids and a technique called sacrificial writing into functional tissue (SWIFT) were deployed in the assembly of thousands of OBBs into living matrices at a high cellular density and introduced into perfusable vascular channels. The OBB matrices exhibited the desired viscoelastic and self-healing behaviour to allow the rapid therapeutic-scale assembly of patient and organ-specific tissues [Luo, et al., 2019; Skylar-Scott, Uzel, et al., 2019; Tetsuka, and Shin, 2020].

In another embedded system including electronic functionalities, a novel system using multiple materials printing simultaneously was reported by Glasgow University research team for speedy fabrication of biomedical wearables and prosthetics at affordable costs. Medical prosthetics and wearables, needed for in-dividual's needs, were manufactured by integrating the sensors and electrical components, as opposed to affixing them on the outside. This configuration sealed off the components from the outside contact leading to minimizing wear and tear. Using biocompatible patient-specific medical implants with fully embedded com-ponents are possible to be interfaced by wireless network to IoT for better man-agement using cell phones. In this context, a bendable sheet of plastic, with LEDs and strain sensors, using a modified 3D printer was prepared to meet the patient's needs. After modifying the printer, the team used printed layer of plastic with spaces for placing the LEDs. The 3D printer applied a silver palladium paste to connect the LEDs to strain sensors. Finally, another plastic sheet was printed on top of the sensors to seal the package. These steps were completed in one print, with only a brief pause to place the LEDs. The completed structure was subjected to bending strain to ensure that the LEDs still functioned as they would be required to function into wearable or prosthetic devices. By directly integrating sensors and electrical components into devices, this approach greatly reduced the manufacturing time of complex structures and prevents wear and tear that would otherwise be unavoidable. Moving forward, the researchers are considering to include dexterous sensory robotic limbs and have already developed a prosthetic hand with embedded sensors as a prototype [WP-01; Hamzah, et al., 2018].

6.5.4 4D PRINTING

The restriction of microstructures fabricated using 3D printing being static was overcome by considering 4D printing for fabricating a complex structure that changes with time in response to an external stimulus in intended manner. 4D printing is a branch of additive manufacturing, where the fabricated constructs are no longer static. They transform into complex structures by changing the size, shape, property and functionality under external stimuli, which makes 3D printing alive [Chu, et al., 2020].

An idea of fabricating 3D objects reacting against an external environment stimulus was introduced as the 4D-printing method using stimuli-responsive smart materials. This could prepare self-assembling and self-regulating constructs by changing their shape upon external stimuli. Currently, many studies are going on to fabricate 4D-printed constructs with shape-changing abilities such as bending, twisting, elongating and corrugating against external stimuli such as temperature, humidity, or light. The success of 4D printing relies on the development of new smart materials, novel printing techniques and mathematical modelling of deformation mechanisms. Most widely studied smart materials for 4D printing are temperature-sensitive materials. The deformation mechanism of temperature-sensitive materials must be accompanied by the shape memory effect. Shape memory polymers (SMPs) are used because of their ease of printability and capability of recovering their original shape under an external stimulus after undergoing deformation. The Tg value of SMPs is typically higher than their operating temperatures. Their shapes can be programmed through subsequent heating (4Tg) and cooling treatments. At the operating temperature, they adopt a temporarily deformed shape. After the temperature increases to 4Tg, they return to their original shape. For example, SMP fibres were incorporated into an elastomeric matrix to create a hinge structure that could be d with a maximum deformation angle depending on the Tg value of the SMPs. A 4D-active shape-changing structure was realized using direct-write printing of UV photo cross-linkable poly(lactic acid)-based bio-inks based on SMPs and shape memory nanocomposites (SMNCs). The printed constructs exhibited superior shape memory behaviour, which allowed 3D-1D-3D, 3D-2D-3D and 3D-3D-3D configurational transformations. To improve their motion freedom further, a six-petal leaf with a bilayer structure of paper laminated with polylactic acid was fabricated that uniformly curled into a flower shape upon changing the environment temperature. This strategy could create complex structures with corrugated and helical configurations. The fabrication of temperature-responsive multi-fingered grippers consisting of rigid segments of poly(propylene fumarate) and stimuli-responsive hinges made from poly(N-isopropylacrylamide-co-acrylic acid) using the stereolithography technique was reported using the grippers to grip drugs at >32°C and release them into the targeted tissue at body temperature of 37°C. Fabrication of containers made from photoresist panels and thermo-responsive PCL hinges was also demonstrated using photolithography. Similar approaches used temperature as an external stimulus. Humidity-responsive materials that deform on taking up or releasing moisture were used for 4D printing [Tetsuka, and Shin, 2020].

Initially, 3D objects printed from inks composed of rigid polymers and humidity-responsive materials were demonstrated and upon changing the moisture level, the volume of the printed object was extended and folded by 200% from its original state. However, the obtained object was relatively fragile against repeated motion of folding and unfolding. Structures with anisotropic swelling properties by confining hydrogels in one direction using stiff materials were used in 4D-printed structure with a four times higher transverse swelling strain characteristic than that of longitudinal strain using a hydrogel ink with cellulose fibrils that were aligned by the shear forces generated from the contact between the ink and the print bed. Humidity-responsive natural hydrogel constructs were reported using carboxymethyl cellulose hydro-colloids with quick response properties by using hydrophobic thin films derived from cellulose stearoyl esters (CSEs). Their actuation properties could control the changes in the temperature of the surrounding aqueous environment. The soft ac-tuators, humidity-responsive sensors and drug delivery systems were reported using humidity-responsive hydrogels (e.g., PEGDA) and biodegradable elastomers (e.g., poly(glycerol sebacate)). The use of light-responsive materials offers a basis to de-velop novel stimuli-responsive constructs and printing techniques because light as a stimulus has the ability to focus energy only on the desired area, enabling rapid and local control or switching of light-responsive materials. The photo-responsive ma-terial is locally heated by the absorbed light [Tetsuka, and Shin, 2020].

A sunflower-like 3D object composed of carbon black and a PU-based SMP was reported with sequential bud-to-bloom deformations driven by heat generated from the absorbed light, in which light caused deformations of various self-folding structures. The use of light source stimuli for bent 4D-printed constructs was ex-plored using a poly(lactic-co-glycolic) acid capsule loaded with plasmonic gold nanorods. In this system, the capsules are ruptured by laser irradiation at the re-sonance wavelength of the gold nanorods. Electric and magnetic fields are used in 4D printing as heat sources. A soft artificial muscle (mixture of silicone elastomer and ethanol) was reported using a phase shift characteristic from the liquid state to the gas state in ethanol under an applied current to control the volume of the elastomer matrix. Polypyrrole (PPy) films were used to create an origami micro-robot, which was controlled by influencing the water absorption/desorption through an on/off current. Incorporation of magnetic-NPs into a hydrogel-based micro-gripper successfully controlled the microrobot remotely by applying magnetic fields. The fabrication of silicone rubber-neodymium-iron-boron (Nd-Fe-B) hybrid 3D structures with programmable ferromagnetic domains were used by applying a magnetic field during printing. A shape change by magnetic actuation was reported apart from physical stimuli such as temperature and light, chemical stimuli like pH, ion concentration and biological stimuli like glucose and enzymes have also been used in advancing 4D printing and related materials [Kim, et al., 2018; Mulakkal, et al., 2018; Lui, et al., 2019; Momeni, et al., 2019; Tetsuka, and Shin, 2020].

4D-printed constructs have the capability of changing their shape and functionality with time. This time-dependent shape-change ability can provide tremendous po-tential applications for use in biomedical actuators such as self-bending/tightening valves, staples and stents, biomedical microrobots to deliver and release drugs upon external stimulation for targeted therapy, and biosensors for medical diagnostics. Another application is the fabrication of scaffolds for tissue regeneration, which al-lows the scaffolds to mimic the complexity of human tissues that possess a dynamic change in their tissue conformations during the tissue regeneration process. 4D-printed tissue constructs responding to fluctuations in the external environment and geometry change can offer a favourable dynamic microenvironment for tissue regeneration that could not be precisely mimicked in conventional 3D-printed tissue constructs [Khanmohammadi, et al., 2018; Morimoto, et al., 2018; Wu, et al., 2018; Zhang, Z., et al., 2019; Chu, et al., 2020; Tetsuka, and Shin, 2020].

6.5.5   ELECTRICALLY CONTROLLED 3D PRINTING

Electrohydrodynamic jetting allows the generating submicrometer jets that could attain speeds above 1 m/s, but such jets couldn't be precisely collected by too slow mechanical stages. In one of the studies it was reported controlling the voltage applied to electrodes located around the jet, its trajectory could be continuously adjusted with lateral accelerations up to 10^6 m/s^2. Through electrostatically deflecting the jet, 3D objects with submicrometer features were successfully printed by stacking nanofibers on top of each other at layer-by-layer frequencies as high as 2000 Hz. The fast jet speed and large layer-by-layer frequencies achieved printing speeds up to 0.5 m/s in-plane and 0.4 mm/s in the vertical direction, three to four orders of magnitude faster than techniques providing equivalent feature sizes [Liashenko, et al., 2020].

3D-hierarchical architectures mimicking a natural nacre were reported by developing a novel electrically assisted 3D-printing technique by fabricating the complex 3D-constructs with superior mechanical and electrical properties using 3-aminopropyltriethoxysilane grafted graphene nanoplates (GNs) with a thickness of ~8 nm, and surface area as large as 120 to 150 m^2/g to strengthen the interface with the polymer matrix of epoxy diacrylate and glycol diacrylate. An electric field of 433 V/cm was applied to align GNs in the polymer matrix during the printing process in which the ink was polarized under the electric field and higher dipole moment was realized in the direction parallel to the GNs because of the shape anisotropy, resulting in the alignment of the GNs. Their superior mechanical toughness originated from the synergistic effects of the hydrogen bonding and pi–pi-interactions between the GNs and the polymer and the covalent Si-O-Si bonding between the aminopropyltriethoxysilane grafts on the GNs. This technique is capable of fabricating a lightweight and strong smart object for use in not only biomedical applications but also transportation, aerospace and military applications. Another novel method for patterning liquid hydrogels with a resolution as low as 100 mm by introducing a capacitor edge effect called PLEEC. The PLEEC system consists of five layers: a pair of silver adhesive electrodes isolated by a dielectric polyimide film layer, an insulating bottom acrylate film layer and a top Teflon film layer that acts as an insulator to keep the liquidus hydrogels on the top surface isolated from the upper electrode. The top layer should be hydrophobic to send any liquid hydrogels away when the electric field is off. Upon applying the electric field, the electrostatic force traps the liquidus hydrogels on the top of the surface layer by the capacitor edge effect. Printed hydrogel objects could effectively respond to the environment temperature. The fabrication and the operation of ionic high-integrity hydrogel display devices were also demonstrated using hydrogels as inks relying on the physical and chemical properties of hydrogels, which place some constraints on their formability. However, the PLEEC system combined the 3D-patterning and stacking processes of hydrogels offer great opportunities in rapid fabrication of prototyping of hydrogel constructs with complex geometries and devices with multiple components [Jang, T.-S., et al., 2018; Yang, et al., 2019; Liashenko, et al., 2020; Tetsuka, and Shin, 2020].

6.6 DISCUSSION AND CONCLUSIONS

3D-printing techniques are progressively being used in medical applications because of their robust capabilities to produce biomimetic biological structures with ease. A brief discussion provided in this chapter covers the conventional and recent advancements made in 3D-printing techniques and materials for biomedical applications. Technological challenges faced currently in 3D-printing technologies are related to achieve higher resolution, higher printing speed and larger scale with retaining good biocompatibility. Conventional 3D-printing techniques have already succeeded in generating biological constructs such as cartilage, bone, heart, brain and muscle, but achieving complex, reproducible, large biological constructs with vascularized architectures suitable for biomedical applications is still posing problems. Recent light-assistance-based 3D-printing techniques, such as computed axial lithography (CAL), show great promises for achieving a microscale resolution and a speed up to ~1 mm/s. The combination of such projection stereolithography and dye-added photopolymerizable hydrogels also enabled the fabrication of a lung-mimicking air sac with airways that enable the delivery of oxygen molecules into surrounding blood vessels. Constructs involving liver cells are successfully implanted into mice. The recently developed integrated tissue-organ printer (ITOP) technique also has potential to produce large-scale biological tissue constructs with a complex geometry, human-scale and high structural integrity. ITOP was found feasible to create sizable biological constructs that mimicked the structure of native living tissues: human ear-shaped tissue constructs integrated with cartilage tissues, vascularized functional constructs and skeletal muscle constructs.

Embedded 3D-printing techniques have additional potential to obtain complex tissue-like constructs with some challenges, but these technologies greatly advance the field of tissue engineering. Development of novel biomaterials merged with the desired mechanical properties and high cytocompatibility is needed for the success of 3D printing in biomedical applications. There are still limitations on the variety of biomaterials that can be used in conventional 3D printing of biomedical constructs. Because of the prerequisite parameters of biomaterials to possess the specific features of biocompatibility and formability, hydrogels are commonly used as biomaterials to have biological 3D constructs. Efforts have been made to develop multi-functional biomaterials and bio-inks that can mimic the natural ECM with the application of decellularized ECM in 3D printing using extrusion-based and light-assisted methods. The dECM-based bio-inks have heterogeneous constituents such as cell-binding proteins and growth factors present in the ECM of native tissue compared to natural hydrogels, which are highly purified forms of a single ECM component that enabled creating patient/organ-specific cell-laden constructs that possess native ECM-like microenvironments. This strategy is useful for developing 3D-printed biological tissues and organs because dECM is capable of modulating the biological activities such as cell proliferation, differentiation, migration and maturation. However, there are still problems with the printing shape fidelity and ethics of its widespread use in 3D printing. Recently developed, multi-functional biomaterials such as stimuli-responsive hydrogels and reversible crosslinking polymers for 4D printing are also promising for creating programmable 3D constructs with complex geometries for biomedical applications and new design systems. The future of biomaterials and biomaterial-based 3D/4D printing is very promising. Further improvements in biomaterials and printing technologies will bring many successes in the fabrication and engineering of tailor-made functional 3D-biological constructs with more complex geometries and artificial organs.

The current state of developments taking place in 3/4D bioprinting has just started proliferating to number of human organs. It is therefore too early to conclude very much specific. But it is certain that the use of biomaterials and more detailed knowledge of microbiology derived from using nanoscale materials will provide much wider field for number of biomedical breakthroughs in the time to come.

REFERENCES

Ashammakhi, N., Ahadian, S., Xu, C., Montazerian, H., Ko, H., Nasiri, R., Barros, N., and Khademhosseini, A. (2019), Bioinks and bioprinting technologies to make heterogeneous and biomimetic tissue constructs. Mater Today Bio., 1, 100008.

Bozkurt, Y., and Karayel, E. (2021), 3D printing technology; methods, biomedical applications, future opportunities and trends, J. Mater. Res. and Technology, 14 September–October 2021, 1430–1450.

Capel, A.J., Rimington, R.P., Lewis, M.P., and Christie, S.D.R. (2018), 3D printing for chemical, pharmaceutical and biological applications. Nature Rev., Chemistry, 2, 422–436.

Chu, H., Yang, W., Sun, L., Cai, S., Yang, R., Liang, W., Yu, H., and Liu, L. (2020), 4D Printing: A review on recent progresses. Micromachines, 11, 796.

Cui, H., Nowicki, M., Fisher, J.P., and Zhang, L.G. (2017), 3D Bioprinting for organ regeneration. Advanced Healthcare Materials, 6(1), 10.1002/adhm.201601118. https://doi.org/10.1002/adhm.201601118

Gopinathan J., and Noh, I. (2018), Recent trends in bio-inks for 3D printing. Biomater Res., 22, 11.

Grigoryan, B., Paulsen, S.J., Corbett, D.C., Sazer, D.W., Fortin, C.L., Zaita, A.J., Greenfield, P.T., Calafat, N.J., Gounley, J.P., Ta, A.H., Johansson, F., Randles, A., Rosenkrantz, J.E., Louis-Rosenberg, J.D., Galie, P.A., Stevens, K.R., and Miller, J.S. (2019), Multivascular networks and functional intravascular topologies within biocompatible hydrogels. Science, 364(6439), 458–464.

Gu, G.X., Chen, C.-T., Richmond, D.J., and Buehler, M.J. (2018), Bioinspired hierarchical composite design using machine learning: Simulation, additive manufacturing, and experiment. Mater. Horiz., 5, 939–945.

Hamzah, H.H., Shafiee, S.A., Abdalla, A., and Patel, B.A. (2018), 3D printable conductive materials for the fabrication of electrochemical sensors: A mini review. Electrochem. Commun., 96, 27–31.

Highley, C.B., Song, K.H., Daly, A.C., and Burdick, J.A. (2019), Jammed microgel inks for 3D printing applications. Adv. Sci., 6, 1801076.

Hippler, M., Blasco, E., Qu, J., Tanaka, M.,Kowollik, C.B., Wegener, M., and Bastmeyer, M. (2019), Controlling the shape of 3D microstructures by temperature and light. Nat. Commun., 10, 232.

Jang, J., Park, J.Y., Gao, G., and Cho, D.W. (2018), Biomaterials-based 3D cell printing for next-generation therapeutics and diagnostics. Biomaterials, 156, 88–106.

Jang, T.-S., Jung, H.-D., Pan, H.M., Han, W.T., Chen, S., and Song, J. (2018), 3D printing of hydrogel composite systems: Recent advances in technology for tissue engineering. Int. J. Bioprint., 2018, 4, 126.

Kelly, B.E., Bhattacharya, I., Heidari, H., Shusteff, M., Spadaccini, C.M., and Taylor, H.K. (2019), Volumetric additive manufacturing via tomographic reconstruction. Science, 363(6431), 1075–1079.

Khanmohammadi, M., Dastjerdi, M.B., Ai, A., Ahmadi, A., Godarzi, A., Rahimi, A. and Ai, J. (2018), Horseradish peroxidase-catalysed hydrogelation for biomedical applications. Biomater. Sci., 6, 1286–1298.

Kim, Y., Yuk, H., Zhao, R., Chester, S.A., and Zhao, X. (2018), Printing ferromagnetic domains for untethered fast-transforming soft materials. Nature, 558(7709), 274–279.

Liashenko, I., Rosell-Llompart, J., and Cabot, A. (2020), Ultrafast 3D printing with sub-micrometer features using electrostatic jet deflection. Nature Communications, 11(1), 10.1038/s41467-020-14557-w

Lui, Y.S., Sow, W.T., Tan, L.P., Wu, Y., Lai, Y., and Li, H. (2019), 4D printing and stimuli-responsive materials in biomedical aspects. Acta Biomater, 92, 19–36.

Luo, G., Yu, Y., Yuan, Y., Chen, X., Liu, Z., and Kong, T. (2019), Freeform, reconfigurable embedded printing of all-aqueous 3D architectures. Adv. Mater., 31, e1904631.

Ma, X., Liu, J., Zhu, W., Tang, M., Lawrence, N., Yu, C., Gou, M., and Chen, S. (2018), 3D bioprinting of functional tissue models for personalized drug screening and in vitro disease modelling. Adv Drug Deliv Rev., 132, 235–251.

Momeni, F., Sabzpoushan, S., Valizadeh, R., Morad, M.R., Liu, X., and Ni, J. (2019), Plant leaf-mimetic smart wind turbine blades by 4D printing. Renewable Energy, 130, 329–351.

Morimoto, Y., Onoe, H., and Takeuchi, S. (2018), Biohybrid robot powered by an antagonistic pair of skeletal muscle tissues. Sci. Robot., 3, eaat4440.

Mulakkal, M.C., Trask, R.S., Ting, V.P., and Seddon, A.M. (2018), Responsive cellulose-hydrogel composite ink for 4D printing. Mater. Des., 160, 108–118.

Perevoznik, D., Nazir, R., Kiyan, R., Kurselis, K., Koszarna, B., Gryko, D.T., and Chichkov, B.N. (2019), High-speed two-photon polymerization 3D printing with a microchip laser at its fundamental wavelength. Opt. Express, 27, 25119–25125.

Sakai, S., Kamei, H., Mori, T., Hotta, T., Ohi, H., Nakahata, M., and Taya, M. (2018), Visible light-induced hydrogelation of an alginate derivative and application to stereolithographic bioprinting using a visible light projector and acid red. Biomacromolecules, 19, 672–679.

Shapira, A., and Dvir, T. (2021), 3D-Tissue and organ printing-hope and reality. Advanced Science, 8(10), 2003751.

Skylar-Scott, M.A., Mueller, J., Visser, C.W., and Lewis, J.A. (2019), Voxelated soft matter via multi material multi nozzle 3D printing. Nature, 575(7782), 330–335.

Skylar-Scott, M.A., Uzel, S.G.M., Nam, L.L., Ahrens, J.H., Truby, R.L., Damaraju, S., and Lewis, J.A. (2019), Biomanufacturing of organ-specific tissues with high cellular density and embedded vascular channels. Sci Adv., 5(9), eaaw2459.

Sun, Z. (2020), Clinical applications of patient-specific 3D printed models in cardiovascular disease: Current status and future directions. Biomolecules, 10(11), 1577.

Tetsuka, H., and Shin, S.R. (2020), Materials and technical innovations in 3D printing in biomedical applications. J. Mater. Chem. B, 8, 2930.

Wang, J., Lu, T., Yang, M., Sun, D., Xia, Y., and Wang, T. (2019), Hydrogel 3D printing with the capacitor edge effect. Sci. Adv., 5, eaau8769.

Wang, Z., Jin, X., Tian, Z., Menard, F., Holzman, J.F., and Kim, K. (2018), A novel, well-resolved direct laser bioprinting system for rapid cell encapsulation and microwell fabrication. Adv. Healthcare Mater., 7, e1701249.

WP-01; Text @ https://innovate.ieee.org/innovation-spotlight/wearables-biomedical-sensor-embedded-3d-printer/

Wu, J., Zhao, Z., Kuang, X., Hamel, C.M., Fang, D., and Qi, H.J. (2018), Reversible shape change structures by grayscale pattern 4D printing. Multifunct. Mater., 1, 015002.

Xu, C., Lee, W., Dai, G., and Hong, Y. (2018), Highly elastic biodegradable single-network hydrogel for cell printing. ACS Appl. Mater. Interfaces, 10, 9969–9979.

Yang, Y., Li, X., Chu, M., Sun, H., Jin, J., Yu, K., Wang, Q., Zhou, Q., and Chen, Y. (2019), Electrically assisted 3D printing of nacre-inspired structures with self-sensing capability. Sci. Adv., 5, eaau9490.

Yeo, J., Jung, G.S., Martín-Martínez, F.J., Ling, S., Gu, G.X., Qin, Z., and Buehler, M.J. (2018), Materials-by-design: Computation, synthesis, and characterization from atoms to structures. Physica scripta, 93(5), 053003. 10.1088/1402-4896/aab4e2

Yin, J., Yan, M., Wang, Y., Fu, J., and Suo, H. (2018), 3D bioprinting of low-concentration cell-laden gelatin methacrylate (GelMA) Bio-inks with a two-step cross-linking strategy. ACS Appl. Mater. Interfaces, 10, 6849–6857.

Zhang, Z., Jin, Y., Yin, J., Xu, C., Xiong, R., Christensen, R.K., Ringeisen, B.R., Chrisey, B.D., and Huang, Y. (2018), Evaluation of bioink printability for bioprinting applications. Appl. Phys. Rev., 5, 041304.

Zhang, B., Gao, L., Ma, L., Luo, Y., Yang, H., and Cui, Z. (2019), 3D Bioprinting: A novel avenue for manufacturing tissues and organs. Engineering, 5, 777–794.

Zhang, Z., Demir, K.G., and Gu, G.X. (2019), Developments in 4D-printing: A review on current smart materials, technologies, and applications. Int. J. Smart Nano Mater., 10, 205–224.

7 Nanomanufacturing of Biomedicines
Current Status and Future Challenges

F. Ahmad, A.K. Ogra, and S. Ahmad

CONTENTS

7.1 Introduction..95
7.2 Nanomanufacturing Areas..102
    7.2.1 Nanomedicine Formulations 102
        7.2.1.1 Drug Syntheses...103
        7.2.1.2 Micro/Nanonization Processes 105
    7.2.2 Translation – Laboratory Experiments to Nanomanufacturing 112
    7.2.3 Quality Control...114
7.3 Nanoproduction of Biomedicines – Emerging Areas 115
    7.3.1 Anti-Inflammatory Nanomedicines...115
    7.3.2 Anti-Diabetic Nanoformulations...120
    7.3.3 Alzheimer's Disease and LNP-Carriers...125
7.4 Discussions and Conclusions 127
References..129

7.1 INTRODUCTION

Significantly notable differences, established between the physicochemical properties of the nanoscale materials species compared to their bulk counterparts, provide the necessary guidelines to take into account while considering the setting up of production plants for nanoscale materials for their applications in manufacturing the nanoproducts/products. Accordingly, the nanoproduction system envisages more elaborate considerations compared to those involved in classical manufacturing technology using bulk materials for producing goods and services for the users. Accordingly, the processes involved in nanoproduction or nanomanufacturing are anticipated to take care of full-fledged value-chain system encompassing the complete "product life-cycle," starting from the initial R&D, product design, manufacturing and further extending to additional considerations such as software development, applications, besides disposal and/or recycling in terms of environmental interactions. The related know-how is developed using knowledge/experience from different fields

DOI: 10.1201/9781003220602-7

of science and engineering to design and implement the processes that deliver the products with specified tolerances at affordable costs, and ultimately meeting the consumer demands. Due to more involved scopes of nanomanufacturing, the continuing evolutions of the newer technologies are found helpful to incorporate compared to those in traditional manufacturing where changes are only introduced if found necessary to affect the product quality and cost effectiveness. The additional burden in terms of evaluating the influence of nanomaterials unintentionally escaping into the environment and causing pollution and affecting human health in terms of their impacts on the ecosystems must be carefully incorporated while the policymakers and manufacturers must revise various associated standards, and regulations to keep in view especially environment pollutions, health hazards and overall safety issues in an integrated manner. Having recognized the importance of nanomanufacturing and its impact on developing novel products and processes, several programs were set up in the US under the "National Nanotechnology Initiatives" (NNI) briefly described below.

Nanomanufacturing, an inevitable link connecting different phases of R&D and product feasibility studies, and subsequent translation of the laboratory scale results into high-volume production, requires solving numerous issues of the manufacturing system. These include product design, reliability and quality, process design and control, shop floor operations and supply-chain management according to the consumer market demands in a networked manner. Nanomanufacturing, thus, is expected to handle various problems related to manipulating the nanoscale material structures, components, devices/machines and systems for large-scale reproduction of value-added components and devices. Nanomanufacturing, adapting either bottom-up or top-down approach, is comprised of molecular systems engineering, and their hierarchical integration leading to larger scale systems. In the nanoscale materials and molecular systems, the classical rules for describing their properties are modified significantly. However, the behaviour of the final product would still depend on the collectively integrated behaviours of all those nanoscale building blocks put together. Consequently, the challenges faced in nanomanufacturing represent an inherently multi-disciplinary set of problems related to nanoscale structures (1–100 nm range) with proper inclusion of top-down and bottom-up processes in providing multi-scale systems integration. To attain the desirable economy of scale for large-scale production, new concepts and principles must be discovered to complement the existing manufacturing capabilities and the infrastructures. A number of scientific disciplines are therefore necessary to invoke in understanding and control of nanoscale phenomena – in terms of the concepts of physics, chemistry, biology, material and information sciences, engineering and polymer sciences. Newer techniques so developed in this context must finally bridge the gaps between the innovations in research laboratory on one end and the economically viable nanoproducts development on the other. The challenges faced by nanomanufacturing involving controlling assembly of three-dimensional heterogeneous systems would be to process nanoscale structures in volumes to support the applications without altering their basic physicochemical features, and to ensure the long-term reliability of nanostructures through testing and characterizations. These challenges do require further R&D in the field of materials and device characterizations involving nanoscale materials as the building blocks of fabrication and synthesis integration. Advanced instrumentations, thus, become inevitable to characterize the nanostructures for providing not only the predictive simulation of their anticipated behaviour but also to contribute to the design and integration of nanodevices and systems. Finally, the knowledge sharing and outreach established via several rounds of user interactions would be more helpful in overcoming the problems of technology transfer by creating public awareness as reported in the cited references [Feng, et al., 2019].

Nanomanufacturing, thus, needs free access to nanoscale materials synthesized in adequate volume, processing techniques and handling tools that enable the assembly of nanoscale components and devices to realize the functional system meeting the target goals. Nanomanufacturing, in particular, has been driven by the current success achieved in the semiconductor Industry, where the ever-growing push to create smaller, faster and more efficient microprocessors that are commercialized by aggressively reducing the feature sizes in the range of 10 nm and below. On the side of federal support, the National Nanotechnology Initiative (NNI) set up in 2001, undertook the major responsibility of supporting the development of nanotechnologies in the US by identifying nanomanufacturing as one of seven Program Component Areas, for which ~US$50 million was allocated for R&D activities. The National Science Foundation (NSF) took initiatives to support four major "Nanoscale Science and Engineering Centres" particularly to address the nanomanufacturing requirements. With nearly 60 sponsored Research Centres and over 1200 Nanotech Companies in the US under NNI were identified already in 2006 to initiate the program of translating the outcome of laboratory experimentations to product manufacturing for meeting the emerging market needs and the activities in this context. The major application areas, being benefited by nanomanufacturing would include: electronics and semiconductor, IT, Telecommunications, Aerospace, and Automotive Industries, Energy Sectors, Pharmaceuticals, Biomedical and Biotechnology, Environmental Remediation and Green Technologies along with National Security as included in the various related documents released from time to time [RR-01].

Developing manufacturing capabilities of nanoscale materials and devices, and incorporating their characteristic features into the final products, has been the main objective of the domain of Nanotechnology. Recognizing the excellent prospects of nanomanufacturing from the outset, the US-sponsored program of NNI was set up after conducting several workshops. In one of the NNI reports, published way back in 2007, nanomanufacturing was listed as one of the major areas for accelerating the overall progress. This report emphasized considering different aspects of the promotional activities related to the R&D for hierarchical nanomanufacturing; infrastructure development for creating distributed nanomanufacturing research centres, and user facilities involving modelling, simulation and design; tool development; environmental and occupational health and safety issues; education, as well as Societal Impacts. The subsequent development of R&D needs for producing nanomedicines and health-care products was found attracting immediate attentions globally. Successes achieved in the areas like "gene therapy" and "newer antiviral vaccine developments," thus, became most sought-after in alleviating the human sufferings, although, most of these novel disease-fighting formulations are still undergoing extensive clinical trials due to stringent regulations to be met before their entry into the consumer market [WP-01].

Three new initiatives were, accordingly, announced in 2016 to focus on treating the cancer as global threat to human population. The development of advanced tools and techniques for nanoscale syntheses, and characterization, thus, became essential for supporting the R&D activities under such initiatives for providing nanoenabled therapeutic and diagnostic products. In addition, the NNI assessments recommended more emphasis on refining or replacing the laboratory procedures with reliable, consistent and more viable manufacturing processes to take care of the clinical supplies. Sufficient start-ups were, thus, subsequently encouraged to develop and adopt these methods on the product development timeline as extensive trials on disease models were found necessary to know whether these nanotherapies would affect the body as these details were ultimately required in proceeding for the clinical development as reported in the associated references [WP-02]. Another critical step in nanomanufacturing of biomedicines was, thus, identified as preclinical toxicity studies and subsequent clinical trials using the drugs produced under good manufacturing practice (GMP) conditions. These studies were found essential before knowing whether the results, so arrived at, would provide an effectively usable nanomedicine. Even the options of roping in the help from outsourcing were explored and encouraged as and when available. However, the start-ups did step in after acquiring the capabilities of conducting preclinical studies, completing method validations and deploying GMP scale-up protocols.

Although many hurdles, in general, were envisaged during the scale-up of any new drug development, testing the viability for nanomanufacturing even posed bigger and more challenging problems such as the essential need for nanocharacterization at every stage of discovery, development and commercialization. Nanomaterials, meeting medical-grade purity and reproducibility, have rather been difficult to produce earlier. It was quite often the case that expensive nanomedicines were found lacking in complying with FDA's cGMPs regulations showing poorly reproducible efficacies. Safety considerations did raise serious concerns not only at the stage of preparing dosages and clinical results but also in avoiding the occurrences of the in-process contaminations compromising the product toxicity.

For retaining the lead position in nanomanufacturing of biomedicines, US found it imperative to create sustainable nanomanufacturing infrastructures for undertaking not only innovative research studies (i.e., cell and gene therapies) but also extending to the other commercial uses. For instance, there are only very few manufacturers with cGMP compliance for nanomedicines. Inadequate quality assurance testing protocols permitted only very few services incorporating small molecule drugs in FDA-approved, bio-compatible nanoparticle-based formulations.

The National Cancer Institute (NCI) was created and supported by the Alliance for Nanotechnology in Cancer, dedicated to accelerating the cure and treatment from the laboratory to the patient bed. Consequently, Nanotechnology Characterization Laboratory (NCL) was established to fulfil the demands of translating the R&D experiences into the user applications. By developing a standard set of appropriate test protocols for NMs used in cancer diagnosis or treatment, NCL could thus aid in developing the animal trials, and regulatory review processes. NCI was assigned with the tasks of advancing nanotechnology into cancer-relevant applications in clinical practices; help in NM-characterization and standardization protocols before transferring the technologies from laboratory to manufacturing that could make these results available to the patients and work for the next generation of cancer therapies. The NCL, in collaboration with National Institute of Standards and Technology (NIST), and the Food and Drug Administration (FDA), facilitated to accelerate the transition of basic nanobiotechnology research into clinical applications. NCL, while striving hard to provide an analytical cascade for NM characterization, facilitated the clinical development and regulatory review of NMs for cancer trials along with identifying and characterizing critical parameters related to NM absorption, distribution, metabolism, excretion and toxicity profiles; and examining the multicomponent and combinatorial aspects. NCL has particularly been sharing database of "NM-structure and property relationships" derived from the pre-clinical trials in the academic and industrial research institutions [WP-01].

Despite exhaustive mandate of NCL, a major limitation was found in using only sub-gram quantities of nanomaterials against having access to kilogram quantities required for preclinical safety assessments and Phase I clinical trials. These situations prevailing in the US were challenged by the European Union's Framework for Research and Innovation Programme, known as "Horizon 2020," which clearly articulated a set of goals addressing to the scale-ups for the production of nanomedicines as discussed in the program [Horizon-2020].

The updated NIH/NCI "Nanotechnology Cancer Plan" released in 2015 included a section on commercialization of nanoproducts for Cancer and the manufacturing challenges faced therein. It was observed that the most frequent shortcoming encountered by the manufacturers lied in the advancement of therapeutic NPs in absence of thorough characterization techniques of the product and the identification of the critical quality attributes. This required early emphasis on the availability of appropriate analytical methods, which was something very frequently neglected. While the plan did a good job of outlining the challenges related to manufacturing NPs for biomedical uses, NIH was, somehow, unable to support the manufacturing of these nanoenabled products.

The Translation of Nanotechnology in Cancer (TONIC) Consortium was established in 2011 to bring together Alliance-funded research centres, pharmaceutical and biotechnology companies, and patient advocacy groups to promote collaborations between academia and industry by sharing knowledge about the best practices in translating nanotechnology from the laboratory benches to the marketplace. Consequently, a working group was setup to develop the clinical protocols for testing nanodrugs in patients by addressing the process limitations and gaps specific to the nanotherapeutics. Another program of Nano-Bio Manufacturing Consortium (NBMC) was subsequently launched to integrate a suite of nano-bio manufacturing technologies and transition to the industries by involving the convergence of nanotechnology, biotechnology, advanced manufacturing and flexible electronics for real-time, remote physiological and medical monitoring. These efforts led to provide a technology platform for human performance monitoring in military and civilians involved in strenuous professions of pilots, special operations personnel, firefighters and trauma care providers to name a few.

Nanomanufacturing of medicine is an essential step in realizing the benefits of the considerable investment made under NNI. Nanomedicine manufacturing poses specific challenges that are not being met by other NNI efforts. In nanomanufacturing, it is necessary to understand the process details as arrived at by the Association of Nonmanufacturers in consultation with R&D and market research teams. Having witnessed considerably successful syntheses of a variety of NMs possessing novel properties compared to their bulk counterparts, their commercialization on mass scale was not only envisaged but also guidelines and plans were evolved to fulfil the basic requirements of setting up of nanomanufacturing of biomedicines. Endeavours were thus made in implementing several nanoinitiative programs consequently. Some of the recommendations made from time to time in this context are briefly mentioned below [WP-03].

Commercially viable production of nanostructured materials and products was defined as the outcome of nanomanufacturing. Ultra-small devices, structures, features and miniaturized systems were subsequently created using the nanoenabled processes that were very useful in the fields such as chemistry, physics, molecular biology, medicine, aerospace engineering and many more. These newer NMs and products were manufactured using various processes such as material removal/ deposition, electronic and medical device fabrications, electrostatic coating, fibres and lithography to name a few. Nanomanufacturing, being a quite recent development, represents a mix of upcoming fields of science and technology, and has already started creating altogether newer markets.

Nanoscale science, engineering and technology put together helps in understanding, characterization and control of the nanoscale materials i.e., at atomic and molecular dimensions. Advances in nanotechnology promise new materials and structures that offer solutions for improving human health, optimizing available energy and water resources, supporting a vibrant economy, raising the standard of living and increasing national security. The NNI has thus been found playing the role of coordinating the multiagency efforts of expediting discovery, development and deployment of nanoscale science and technology for the benefit of human society. Subsequently, 21st Century Nanotechnology Research and Development Act of 2003 examined and commented on the NNI mechanisms to advance areas of nanotechnology development and commercialization via improving the physical and human infrastructure needs for their successful realization as reported in cited references [WP-03].

Next came the "InterNano" initiative, an important service provided through "National Nanomanufacturing Network" (NNN) by collecting information and connecting the nonmanufacturer community from the researchers and practitioners by creating, collating, contextualizing and disseminating the relevant resources such as news highlights, reviews, process details and assessments of the ongoing nanomanufacturing practices. One could, thus, either use these resources or contribute information to the "InterNano" knowledge base. Cooperative working of "InterNano" deployed complementary informatics to facilitate the data-sharing among the user groups. "InterNano" also provided an editorial overview with the help of "NNN." "InterNano," thus, created a virtual community of nanomanufacturing researchers and practitioners through information-sharing related to nanomanufacturing. The role of "NNN" involved a combination of electronic resource, community of practice and network of experts working in the areas of nanomanufacturing. This established an effective mechanism for information sharing, facilitating and organizing events with special focus on critical and emerging issues involving the areas of informatics, standards, education and workforce training. This was anticipated to provide an open platform for archiving information, in which, stakeholders either contribute or have access to the relevant information for accomplishing their nanomanufacturing tasks. The areas of activity in this context included developing the requisite processes of nanomanufacturing using appropriate tools for realizing nanoscale objects and NMs. These objectives were accomplished by employing adequate characterization techniques for assessing the impact on environmental, health and safety considerations besides social and economic implications, informatics and standards for nanomanufacturing and commercialization, regulation and intellectual property associated therein [WP-02].

Nanoproduction of biomedicines, for example, involves a multi-step process that requires the constant but stringent control of the nanomaterial properties such as size, shape, charge, structure, composition, physicochemical, pharmacokinetics and biopharmaceutical properties. The challenges in nanomedicine product development come from the transition from small laboratory batch size to large industrial volumes and the selection of excipients required in producing high-quality pharmaceuticals. R&D efforts involved in translating low-volume production for scale-up posed challenges to produce some specific NMs. The development of nanomedicines relies on the progress made in manufacturing technology to accommodate scalable processes complying with GMP-based quality guidelines to ensure the quality of the processes and manufactured products, involving detailed written procedures for each process that affect the finished product's quality. Moreover, the application of GMP guidelines minimized the risks involved in production starting from the selection of raw materials, equipment and extending to train the workforce along with providing documented proof of the implementation of the correct procedures at each step of the manufacturing [Đorđević, et al., 2021].

In this endeavour, the pharmaceutical industry worked with the EC to set up the European Technology Platform on Nanomedicine (ETPN) in 2005, as an initiative to address the application of nanotechnology in healthcare sector [WP-03]. The objectives of ETPN were to create conditions for the successful translation of nanomedicines by shaping and supporting public funding in the promising areas of research and designing a unique technical infrastructure in the form of Nanomedicine Translation Hub. The ETPN Nanomedicine Translation HUB offered custom mentoring through Translation Advisory Boards, product characterization and GMP manufacturing through pilot lines. These free-of-charge services were open to all, including entrepreneurs, SMEs, industry and academic laboratories with the sole purpose of removing the specific roadblocks to nanomedicine product development and support/accelerate clinical development of promising nanomedicine research as discussed in detail in the enclosed references [Martins, et al., 2020; Souto, et al., 2020].

A brief description of a number of above-mentioned initiatives and monitoring programs clearly highlights the relevance of intensive efforts to be made in advancing nanomanufacturing biomedicines and health care products. More relevant information will be included in the upcoming text of this chapter here.

7.2   NANOMANUFACTURING AREAS

Starting from the basic concepts of nanomanuctacturing, a clear paradigm shift appeared to be inevitable before entering into the processing technologies providing *in situ* device, process and materials analyses. The first step in this context begins with the design for reliability by considering the critical issues for optimizing the product nanomanufacturing. Considering trade-off between cost and process complexity must be included while striving for fault-tolerant, redundant designs where defects are not catastrophic. Combining critical inspection procedures introduced between each process step provides the compensation to improve process yields in subsequent steps. Inherent to the manufacturing of future nanosystems, it would require a specific design phase associated with the variability of the nanosystem. In other words, a new set of materials, devices and nanocomponents would be designed into the system, which might inevitably involve creating problems downstream. Instead of designing with compromised reliability, in terms of introducing more tolerances in individual parameters for improving process yield, the design for variability would seek to improve processes and yields by focusing on manufacturing designs, the introduction of new nanomanufacturing practices would be preferred to set up revolutionary process methodologies. This approach including redundant designs, critical process controls and incorporation of new nanomaterials in the system design to improve reliability, and functionally tailored materials properties, would be necessary to meet the requirements. Ultimately, the ability to incorporate unlike materials and uncommon building blocks to enable unique functionality will be necessary to evolve for nanomanufacturing [RR-01].

With the ongoing developments in nanoscience and technology for several decades, some areas of nanomanufacturing have already achieved maturity and many others are at different stages of development as discussed in this chapter. It is worth noting that all the processes chosen for nanomanufacturing are not only related to altogether new NMs for industrial applications, but some are also dedicated to replace the well-established conventional processes and materials using conventional materials.

The development of nanomanufacturing of biomedicines has implicit objective of acquiring commensurate advantages over the existing systems. In each case, it is essential to mature the associated laboratory scale synthesis protocols with viable scale-up capability. Overall, the requirement of huge capital investment to meet the costs of raw materials, process developments and characterization facility, indeed, makes it necessary to have a very detailed study and careful planning for arriving at the setting up of the nanomanufacturing plants to meet the demands.

Brief descriptions of some successful nanomanufacturing efforts already made are briefly described to highlight their inherent potentials not only in creating new markets but also in contributing effectively in upgrading the living standards of the human populations globally.

7.2.1   NANOMEDICINE FORMULATIONS

It has been possible to make use of a variety of ligand molecules to conjugate with nanoscale carriers, identified and evolved recently in nanotechnology development, to imbibe hydrophilic and hydrophobic features in addition to acquiring improved drug loading capacity for the purpose of site-specific deliveries appropriate for treating human diseases. These nanocarriers deployed in targeted deliveries are found more effective compared to conventional treatments. Although, still some issues like side effects and cytotoxicity are to be studied in detail for minimizing them appropriately by designing the suitable ligand molecules. Better understanding about the site-specific therapeutic deliveries is certainly going to improve the efficacies of the therapies being developed for further cure of different ailments [Yetisgin, et al., 2020].

Having better insight into the syntheses of a large variety of nanoparticulate species of inorganic, organic and biomolecular origins, a number of their conjugated forms have started maturing for large-scale production and subsequent commercial applications. Further research and developments in the area of identifying the biochemical pathways followed by these species in the living organisms, plants and environment (i.e., air, water and soil) have enriched the understanding of the basic phenomena involved in a number of cases and it is, therefore, expected to create still larger-scale demands, which can be met only by nanoproduction. The current situation regarding realization of multifunctional nanoparticulate species, by combining different site-specific targeting molecules, creating sufficient drug loading sites on the surface, imparting capability to reach the target site and enabling imaging molecules to provide evidence of their arrivals, site-specific delivery of the drug molecules stimulated externally followed by their ultimate removal from the system, has reached to maturity in few cases and more are expected in near future. Having mastered the processes involved in synthesizing the engineered nanoparticulate species with desired structural and physico-chemico-biological features imparted by surface coatings and conjugations are very much sought-after ensembles today.

7.2.1.1 Drug Syntheses

Most of the drugs are intentionally produced in the crystalline form to take care of their stability, self-segregation of impurities and ease of handling. These crystalline drugs are actually in polymorphic states undergoing various phase transitions that alter their dissolution rates and transport characteristics. Choosing the most appropriate and stable form of the drug becomes imperative in the initial stages of development. The current studies are now specifically directed towards knowing the origin of molecular level polymorphism for designing the most stable polymorphic drug using computational tools to identify the most appropriate molecular structure [Wang, H., Zhou, et al., 2021].

The assembly and stability of drug NPs are, in principle, controlled through supersaturation and precipitation leading to nucleation and growth competing with simultaneous dissolution that decide under dynamic equilibrium the final particle size and its distribution. The crystalline and amorphous phases present in the drug-NPs have major influences on the stability and bioavailability *in-vivo*. Biopharmaceuticals need to take care of two main factors like drug solubility in the biofluid and the drug permeability while passing through the gastrointestinal membrane. It is important to note that almost 40% of the marketed drugs and up to 90% of the new drugs being introduced in the market are poorly soluble in aqueous media. It has also been noted that significantly large percentage of all the approved drugs suffer from low permeability. Having extremely large variety of drug molecules under consideration, the objective of tuning the required efficacy in each case would be rather a difficult target to achieve.

Alternately, it was found much easier to fill the poorly water-soluble drugs into polymeric nanocarriers with the desired surface modifications resulting in the required physicochemical properties. Several formulations of this type are available commercially including aprepitant, fenofibrate and sirolimus. Subject to deploying the production technology and processing of the drug materials, the resultant nanoparticulate assemblies are mostly polymorphic in structure. Drug nanocrystals (NCs) of desired shape and size are possible to prepare separately by controlling the synthesis parameters used. Nanocrystalline drugs have been prepared by taking into account three basic approaches of precipitation, homogenization and milling as discussed sometime back [Paliwal, et al., 2014].

Filling of crystalline/amorphous drug-NPs inside the polymeric nanocarriers, being better to control during production, has currently been developed subsequently. The polymeric nanocapsules are produced by dissolving the polymer in water-miscible organic solvent and forming a colloidal suspension after the mixture is allowed to diffuse in aqueous medium in the presence of a surfactant. This process, known as solvent displacement technique, demonstrated higher drug loadings. The utility of nanocapsules is, however, limited to water-miscible solvents. It is an inefficient process to encapsulate larger number of hydrophilic drugs [Qian, et al., 2021].

To take care of the associated practical challenges, various formulation strategies have been explored using amorphous solids, nanocrystals, liposomes, micro/nanoemulsions and co-crystals for targeted deliveries. Amenability of amorphous solids and their dispersions that are more acceptable due to their commercial success makes it easy to enhance drug solubility and dissolution rates. For example, amorphous glibenclamide exhibited 14x higher solubility than its crystalline form in an aqueous buffer. Similarly, solubility of amorphous indomethacin was found much higher than the γ-crystal API over a temperature range of 5°C to 45°C. Amorphous pranlukast exhibited ~6x improved stability in water and ~20x in phosphate-buffered saline, compared with the crystalline substance. Despite all these reported successful observations, the intrinsic instability of the amorphous solid drugs has still been a major concern in terms of its wider adoption in nanomanufacturing of the modern drugs as reported in the cited references [Qian, et al., 2021].

In a novel solution of taking care of the instability of amorphous drugs was reported deploying anti-plasticizer polymers and/or the stabilizer to maintain the amorphous structure of the drug molecules during their storage. Polymers generally help in enhancing their stability via mechanisms such as creating a physical barrier, chemical potentials and glass transition temperature and drug-polymer interactions. Nontoxic polymers such as polyvinylpyrrolidone, hypromellose, polyvinylpyrrolidone/vinyl acetate and hypromellose acetate succinate have been FDA (US) approved as excipients for the oral dosages of drugs as reported recently [Qian, et al., 2021].

Of late, the entry of natural polymer-modified drug delivery nanocarriers has been showing better options by using chitosan (CS), cellulose, starch, collagen, gelatine, alginate and elastin. CS and its derivatives are currently being recognized as promising nanocarriers due to their biodegradability, better biocompatibility, non-toxicity, low immunogenicity, great versatility and beneficial biological effects. CS, either alone or its composites, are suitable for fabricating products such as hydrogels, membranes, beads, porous foams, NPs, in-situ gel, microparticles, sponges and nanofibers/scaffolds to name a few. CS-based nanocarriers are employed in producing controlled and targeted drug delivery systems for different kinds of drugs comprising anti-inflammatory, antibiotic and anticancer properties as well as for proteins/peptides, growth factors, vaccines, small DNA and short interfering RNA (siRNA). A recent comprehensive review examined CS-based nanocarriers in various forms such as NPs, nanofibers, nanogels and chitosan coated liposomes and their applications in medical and pharmaceutical fields using different routes of administration involving oral, transmucosal, pulmonary and transdermal with reference to classical formulations [Iacob, et al., 2021].

7.2.1.2 Micro/Nanonization Processes

Numerous methods of preparing micro-/nano-sized material ensembles have been developed over a period of last few decades as reported in the publications. However, out of all these protocols, only some of them are worth scale-up and therefore knowing the nature of drug NPs and their favourable/adverse interactions will enable selecting the right combination for achieving the targeted specifications.

Some of the generic processes are briefly described in the following to highlight their applicability under different circumstances.

7.2.1.2.1 Ball-Milling

A simple top-down process of ball-milling has been in use since long for preparing the ultra-fine drug suspensions, in which, size reduction is achieved via shear force created out of the impact force exerted during the milling process. Process parameters like the milling type and dispersion media influence the size and stability of the resultant nanoparticulate ensemble. Ball-milling generally uses a hollow cylindrical container rotating around its axis, which is partially filled with the balls of different materials such as steel, stainless steel, ceramic or rubber. Relying on the energy released from the impact and attrition between the balls and the powder, this method turned out very cost-effective, reliable and easy to operate. Reproducible results have been achieved by controlling energy and speed in wet and dry conditions over a wide range of materials including metal oxides, pigments, catalysts, cellulose, fibres, chemicals, polymers and hydroxyapatite besides many others. Only disadvantages associated are noted in terms of unavoidable contaminations, irregular shaped NMs, noise, long milling and cleaning times. Out of two options of direct and indirect milling methods, the direct one employs the rollers directly acting on the particles to transfer the kinetic energy. In the indirect case, the kinetic energy is first transferred to the mill body and then to the grinding medium. Ball-mills are also divided into three groups namely: tumbler, vibratory and planetary ball-mills. A tumbler mill consists of a cylinder partially filled with steel balls rotating about its longitudinal axis, in which, the efficiency of the process mainly depends on the diameter of the milling container. Larger diameter containers provide greater heights of the fall consequently transferring a higher energy to the balls. In the vibratory mills, the vessel containing the sample and the grinding medium are shaken back and forth at higher vibrational frequencies. The factors like frequency and amplitude of vibration and the mass of the milling medium are important factors to take into account. In planetary mill, the vessels are placed on a rotating supporting disk, which rotates around its own axis as well. Again, the size of the vessels is an essential parameter for higher efficiency as a larger distance allows a higher kinetic energy and therefore stronger impacts as discussed in a very recent publication [Piras, et al., 2019].

7.2.1.2.2 Extrusion

The extrusion is another low-cost and scalable process of converting hydrophobic drugs into nanoparticulate forms. For example, a manual extrusion uses a gas-tight syringe and polycarbonate membrane. However, it might develop heterogeneity, especially when the pore sizes are <100 nm due to variations in manual pressure applied each time. A recent method was reported using a nanoporous membrane extrusion for NP-production of hydrophobic drugs by inducing precipitation of drug-loaded NPs at the exit of nanopores. Another novel one-step process of converting a liquid stabilized nanosuspension into a solid formulation *via* hot-melt extrusion was reported combining an internal devolatilization process, in which, the polymer was fed into the extruder and allowed to melt. Subsequently, a stable nanosuspension was added to this *via* side inlet. After removing the water by devolatilization, the solidified polymer came out through the outlet. Various aspects of extrusion-based nanomanufacturing have been discussed extensively in the cited references [Prabha, et al., 2021].

7.2.1.2.3 Supercritical Fluid Processing

Supercritical fluids (SCF) based technology offers another extremely promising route of NP-production using mild temperature conditions and avoiding organic solvents by selecting pressure and temperature generally above the critical point. These techniques have been operated in different modes such as – solvent-rapid expansion from supercritical solution; swelling and plasticizing agent saturated solutions; antisolvent-gas antisolvent or supercritical antisolvent process or aerosol solvent extraction system and solution-enhanced dispersion; and solvent-based polymerization in the dispersed media. These formulations are found offering several significant advantages of inducing particle de-aggregation, improving solubility, and dissolution rates, controlled release and drug absorption enhancement. To design NPs with controlled particle size and size distribution, morphology, inner core structure and minimal residual solvent concentration, this method turns out to be one of the best. However, the poor solvent power of supercritical CO_2, higher cost and use of large amount of the CO_2 are few limitations as noted by several research groups. The specific advantages of using SCF technology include reproducible productions of novel polymorphic/metastable solid forms of APIs that are not possible under normal crystallization. Free-flowing powders with reduced adhesion and cohesion are found appropriate for aerosol preparations for pulmonary drug delivery. Thermally labile materials like recombinant proteins and antibodies are produced using SCF-based micronization for commercial-scale clinical applications [Chakravarty, et al., 2019; Franco, and De Marco, 2021].

The APIs lacking in appropriate physicochemical, pharmacokinetic and pharmacodynamic properties are better encapsulated in such a way that the drug delivery carriers not only acquire improved stability after getting protected from harsh conditions such as light, oxygen, temperature, pH, enzymes and others but also exhibit enhanced dissolution rates and bioavailability. Conventional production of these drug carrier formulations has several drawbacks such as thermal and chemical stability of the APIs, excessive use of organic solvents, higher levels of solvent residues, difficult control of particle size and size distributions, drug loading-related challenges, besides high energy consumptions over a long duration.

7.2.1.2.4   Micro/Nano-Emulsion Processes

The first method of synthesizing the polymeric-NPs called solvent evaporation used an oil-in-water (O/W emulsion for production of solid nanospheres. The drug and the organic phase of polymer, dissolved into a polar organic solvent, were dispersed together in ethyl acetate in place of more toxic dichloromethane and chloroform used earlier. Adding aqueous surfactant of polyvinyl acetate to the organic phase prior to emulsification using a high-speed homogenization or ultrasonication, resulted in a dispersion of the nanodroplets. Nanoparticulate suspensions were formed after evaporation of the polymer solvent diffused through the continuous phase of the emulsion. The solvent was evaporated either by continuous magnetic stirring at room temperature or in a slow process of reduced pressure in case of using dichloromethane and chloroform. The resultant solid NPs were washed out and collected by centrifugation, followed by freeze-drying for long-term storage. Biodegradable polymers employed in drug nanocarrier formulations are saturated poly(α-hydroxy esters): poly (lactic acid) (PLA), poly(glycolic acid) (PGA) and poly(lactic-co-glycolide) (PLGA), as approved by the US FDA and the EMA because of their safety profile, confirmed biocompatibility, low levels of immunogenicity and toxicity, as well as their biodegradation during *in vivo* studies [Zieli'nska, et al., 2020].

It was alternatively noticed that polymeric nanogels prepared by emulsification showed minimum level of toxicity besides stability in the presence of serum and stimulus responses since they possess a high drug encapsulation capacity, tuneable size and are relatively easy to prepare. Therefore, they have widely been used in biosensors, drug delivery, tissue engineering and biomimetic material designs. Their special features such as size, surface charge, hydrophilicity and hydrophobicity, or even type of polymer, are found governing their applications by providing a successful polymer functionalization using alcohols, amines and thiols. This was possible due to the fast reaction kinetics, stability of isocyanates toward radicals and good yields, although their further uses were limited due to toxicity of isocyanate or the instability of the final mixtures containing isocyanate and polymer mixtures as reported recently [Chenthamara, et al., 2019; Jain, and Thareja, 2019; Lombardo, et al., 2019; Pinelli, et al., 2020].

Emulsification and solvent diffusion-based formulations employ generally oil in water (O/W) emulsions of a partially water-miscible solvent containing polymer and drug, and aqueous solution with surfactant. These emulsions, comprising of a partially hydro-miscible organic solvent (benzyl alcohol or ethyl acetate) saturated with water, provide an initial thermodynamic equilibrium condition for both the phases. Subsequent dilution with a large quantity of water induces the solvent diffusion from the dispersed droplets into the external phase resulting in the formation of colloidal particles. Generally, this is the method used to produce nanospheres, but nanocapsules were also obtained once a small amount of triglyceride was added to the organic phase. Finally, the organic solvent is eliminated by evaporation or by filtration producing 80–900 nm size NPs. These emulsification/solvent diffusion-based methods have been modified with the emulsification/reverse salting-out method, which deployed the separation of a hydro-miscible solvent from an aqueous solution, through a salting-out effect that produced the nanospheres. Examples of suitable salting-out agents include electrolytes, such as $MgCl_2$, $CaCl_2$ or $Mg(CH_3COO)_2$, as well as sucrose. The miscibility of acetone and water is reduced by saturating the aqueous phase allowing the formation of an O/W emulsion from the other miscible phases. The O/W emulsions were obtained under intense stirring at room temperature followed by dilution in an appropriate volume of DI-water or an aqueous solution in order to allow the diffusion of the organic solvent to the external phase resulting in polymer precipitation, and consequently producing the nanospheres. The remaining solvents and salting-out agents were subsequently eliminated by cross-flow filtration. The dimensions of the nanospheres obtained by this method vary between 170 and 900 nm. The average size could be adjusted between 200 and 500 nm, by varying polymer concentration of the internal phase/volume of the external phase. A similar method was also adopted in the preparation of lipid nanoparticles (LNPs) in which careful selection of the salting-out agent is extremely important since it can affect drug encapsulation efficiency. The advantages of this technique include high efficiency, easy to scale-up, useful for thermolabile substances and protein encapsulations with minimal stress. However, this method is mostly applicable in lipophilic drugs, where extensive washing is required. Most of these observations were reported in the enclosed references [Wang, et al., 2016; Crucho, and Barros, 2017; Lim, and Hamid, 2018; Vasile, 2018; Souto, et al. 2019; Zieli'nska, et al., 2020].

7.2.1.2.5   *Microfluidization*
Microfluidization, of late, has been reported as an ideal route for producing drug delivery carriers appropriate for nanomedicines. A frontal collision between the two fluids under high-pressure conditions (i.e., up to 1700 bar) generates shear and cavitation appropriate for forming the NPs in the presence of surfactants. In order to control the size of NPs, several cycles (i.e., from 50 to 100) are required for reducing their diameters – a serious limitation. Microfluidic liposomes, prepared recently, are far superior than traditionally prepared ones in extrusion and sonication. Microfluidic hydrodynamic focusing was also used to synthesize NPs and vesicles of various lipids as discussed in a recent publication [Đorđević, et al., 2021].

Understanding the physicochemical behaviour of polymeric-NPs was realized when attempts were made to characterize them. One of the critical challenges was faced in identifying the drug association mechanisms to the polymeric-NPs. The study of NPs and their ecotoxicology is another necessary consideration for continuing the development of efficient nanocarriers, showing no risk to the environment or human health in their potential applications. It might be possible to obtain quantitative and qualitative measurements that might help to establish methods, principles and procedures that would guarantee the results with analytical quality as discussed in the cited references [Gagliardi, et al., 2021].

7.2.1.2.6 Nanogel (NG) Formulations

Biodegradable hydrophilic polymers such as chitosan, gelatine and sodium alginate, when cross-linked, generate NPs out of a mixture of two aqueous phases having counter ionic materials in an ionic gelation technique. For example, chitosan, a cationic polymer, when cross-linked with sodium tripolyphosphate, the positively charged amino groups of chitosan interacting with negatively charged tripolyphosphate form nanosize coacervates as a result of electrostatic interactions followed by ionic gelation involving the transition of the material from liquid to gel form due to ionic interactions at room temperature. NGs are three-dimensional networks of polymers capable of adsorbing a large amount of water and are used in preparing most of the NPs for drug delivery applications. With a rational design, it is possible to have highly hydrophilic NGs with tunable size and porosity, deformability and degradability. The NGs of an appropriate size (10–200 nm) might be used in selective tumour accumulation for anticancer drug delivery by taking advantage of the enhanced permeation and retention effects while attached to the targeting ligands. The hydrophilic polymer forming NGs with the degree of cross-linking controlling the softness is important for the interaction of NPs with the biological system. For example, NGs with varying degrees of softness demonstrate a variable cellular uptake. The deformed NGs easily pass through the pores that are even smaller than the hydrodynamic size of the NGs and this is important in controlling *in-vivo* circulation and accumulation of NGs in the tumour cells. The porous NGs provide large spaces for drug loading comprising small molecules, nucleic acids and proteins. Tuned drug deliveries are realized by controlling the cross-linking degree. Hydrophilic NGs degrade easily using degradable or natural polymers as matrix. Polysaccharides are the most popular biomacromolecules to construct the NGs for drug delivery as discussed recently [Wang, H., Deng, et al., 2021].

Optimization of nanoemulsification process was attempted using sonication of polymeric nanodispersions by simply dissolving the polymer in the organic phase followed by slowly removing the solvent from the polymer containing nanodroplets to have the NPs. Sonication causes growth, and collapse of very small size bubbles in the liquid phase in the presence of high-intensity ultrasound. All sonicators de-agglomerate particles and make them dispersed in the liquid medium. Improved homogeneity is achieved by applying sonication resulting in a much narrower particle size distribution. An important aspect to consider in the dispersion is the resultant final stability of the dispersion exhibiting a situation where the particles do not settle down from their dispersed phase and the average hydrodynamic diameter do not vary by more than 10% between the five repeated measurements taken during 10 min. Kaur, et al., proposed ways to measure dispersion stability by estimating zeta potential from the measurement of electrophoretic mobility and the characteristic absorption of NPs in the UV spectral range as reported sometime back. It was further recommended that for every combination of NM-type and liquid medium, it requires a separate exercise of optimizing the basic protocol, which can be done by careful adjustment of various factors such as sonication time, strength and sonicator type. A detailed method comprising step-wise sequences of the dispersion procedure is there for improving the comparability. The current methodology and control parameters reported are helpful to use in other dispersing media apart from the water and comparisons can be drawn accordingly [Kaur, et al., 2017].

7.2.1.2.7   Liposomal Nanocarriers

Liposomes are extremely versatile for preparing the nanocarrier platforms by transporting hydrophobic or hydrophilic molecules, including small molecules, proteins and nucleic acids. Liposomes were used as the earliest nanomedicine delivery platforms used in the clinical applications. The next generation of LNPs, including solid lipid nanoparticles, nanostructured lipid carriers and cationic lipid-nucleic acid complexes, exhibit more complex internal architectures with enhanced physical stabilities. With their ability to control the location and timing of drug delivery in the body, LNPs are very useful in treating human diseases.

Having noted the importance of LNPs in early explorations of preparing vaccines or treating a number of diseases via RNA delivery. A large number of products are in pipeline for clinical trials. The method of microfluidics was extensively deployed for their manufacturing due to better scalability, reproducibility and faster preparation. In a current study of various operational formulation parameters, few of them like flow rate ratio (FRR) and the total flow rate (TFR) were found affecting physicochemical properties of the final product. For instance, increased TFR or FRR could produce reduced particle size. The amino lipid, buffer choice and type of nucleic acid payloads affect the characteristics of these LNPs with a resultant high loading (>90%) in case of sub-100 nm particles. These results are quite encouraging to develop manufacturing of the LNPs using microfluidics. Having observed the potential of LNPs in drug delivery, it will be helpful to understand various protocols involved in preparing LNPs in the recent past as discussed in current reviews [Roces, et al., 2020; Tenchov, et al., 2021].

Out of hot and cold high-pressure homogenizations, the hot process is operated at temperatures above the melting point of the lipid and is applied to lipophilic drugs as reported by [Montoto, et al., 2020]. High shear mixers produce drug-loaded lipid and aqueous phases. It is noteworthy that the final form of NPs depends upon quality of the nanoemulsion. The dispersed phase in nanoemulsion congeals upon cooling down to room temperature forming the LNPs. However, this method is not suitable for the thermolabile drugs besides its limitation due to drug distribution and losses into the aqueous phase during the processing. An alternative method of LNP-production is to use cold homogenization known as high-pressure milling of suspension, in which, the lipid with the drug is melted and solidified quickly to get LNPs. The prepared microparticles are stirred in cold aqueous surfactant followed by homogenization below room temperature to convert them into NPs. Comparing hot and cold high-pressure homogenization processes, it was noted to have high lipid concentrations with a narrow particle size distributions, and larger particles with broad size distribution, respectively. This method also limits thermal exposure to the drug and offers negligible degradation of temperature-sensitive biomolecules including the drugs. Factors affecting size distribution of LNPs in both hot and cold processes included temperature, type of homogenizer, pressure and homogenization cycles as reported elsewhere [Bagul, et al., 2018].

Hot microemulsion of low melting lipids, emulsifier, co-surfactant and water, when added to stir excess cold water causes precipitation of the lipid-NPs. These protocols of lipid-NPs without high energy inputs were found suitable for loading thermolabile drugs. The parameters such as composition, temperature gradient and pH of the medium did influence the final product quality. Only limitation was the removal of excess water, high concentration of surfactant and co-surfactant for meeting the regulatory compliances. LNP dispersions are obtained using solvent emulsification and evaporation technique. In this, lipid and lipophilic drugs are dissolved in organic solvent and the organic phase is emulsified in an aqueous phase resulting in O/W emulsion. Organic solvent is then removed under pressure to obtain LNPs due to precipitation. The size of these LNPs is controlled by the type and concentration of the lipid, surfactant and co-surfactant in the organic phase. For instance, ~5 wt.% lipid content could produce particle size in the range of 30–100 nm, while higher lipid contents increased the particle size due to higher viscosity of the dispersed phase with a decrease in the homogenization efficiency. This method was also suitable for thermolabile drug loadings. This is an improved version of the "solvent emulsification evaporation" method as described earlier. The LNP-dispersions were prepared from the emulsion containing solvents (partially water-miscible) with low toxicity. Similar to the above-described method, drug dissolved in the organic solvent precipitated out instantly as a result of diffusion of the organic solvent resulting in sub-100 nm NPs with low polydispersity. Use of a surfactant is important in optimization of desired size. For example, non-ionic surfactants produce larger NPs than ionic surfactants. It was thus suggested that a combination of two or more surfactants could control the particle size better for such combinations affect synergistically on the emulsion stability in terms of coalescence rate. Combination of surfactants produces mixed films at the interface with high surfactant coverage *vis a vis* sufficient viscosity for the desired stability as discussed by several research groups [Fernandes, et al., 2021].

LNPs are mass-scale produced using molten lipids pressing against a porous membrane that produces the nanosized droplets. Inside the membrane, the aqueous phase is circulated to sweep the droplet away from the pore outlets. These swiped droplets, once cooled, get solidified. Factors such as bath temperature of aqueous and lipid phases, cross-flow velocity of aqueous phase, lipid phase pressure and membrane pore size, ultimately decide the particle size of LNPs. The best feature of this method includes simple apparatus, controllable particle size by careful selection of process parameters, and scaling-up possibilities as discussed in the cited references [Bochicchio, et al., 2021].

Varying the frequency and intensity of ultrasonic energy is successfully used in producing LNPs from the preformed microparticles controlling the size of resultant NPs. A combination of high-speed stirring and ultrasonication at higher temperature has also been explored. However, the broader particle size distributions along with potential metal contaminations arising from the sonication probe are there as basic limitations.

Multiple emulsion method based on solvent emulsification-evaporation was also used for incorporating the hydrophilic drugs in the lipid matrix of the NPs. However, the particle sizes are in the micrometer range and hence are designated as "lipospheres" in place of NPs. A stabilizer is also added to encapsulate a hydrophilic drug into the internal aqueous phase. This minimizes the partitioning of the drug from internal to external phase when solvent evaporates from the multiple emulsion as reported [Pineda-Reyes, et al., 2021].

7.2.1.2.8 *Engineered Metal and Metal Oxide Nanoplatforms*

Engineered metal and metal oxide NPs, appropriate for drug delivery, have been used with IR irradiation or magnetic field excitation for thermal drug release. Site-specific intracellular targeting of mitochondria has been successful with the fullerene molecules. Au-NPs (1–150 nm) are promising drug delivery platforms due to their inert, non-toxic and biocompatible features. Smaller diameter Au-NPs are easily surface functionalized with higher loading of targeting ligands. Using proper stabilizing agents, it has been possible to synthesize varying particle sizes of Au-NPs. Other metallic NPs used in biomedical applications are Co, Ni, Fe, Au and their respective oxides such as magnetite, maghemite, cobalt ferrite and chromium dioxide that are synthesized and modified with functional chemical groups for enabling them to be decorated with various molecules including therapeutic agents, biological molecules such as peptides, proteins and DNA. They offer unique characteristics including magnetic properties besides stability and biocompatibility. Thus, magnetic-NPs can be targeted to a specific location in the body by using an external magnetic field. For example, super-paramagnetic iron oxide-NPs with large magnetic susceptibility, are widely used in clinics as contrast agents in MRI-based bioimaging. Likewise, super-paramagnetic properties of NPs facilitate the stable delivery of therapeutic agents to the body/cell and proper accumulation at the target tissue providing a reproducible and safe treatment. When metallic-NPs are subjected to an alternating magnetic field, they produce heat (magnetic hyperthermia) enabling them to ablate the cancer tumours. Au-NPs are widely used in cancer diagnosis and therapy due to their unique optical and localized surface plasmon resonance with relatively lower cytotoxicity. Radiation of appropriate wavelength falling on Au-NPs exhibits photothermal conversion that heats up the targeted tumour tissues to kill cancer cells. Besides, Au-NPs are used for drug delivery, where light irradiation can trigger the drug release. It has been possible to tailor the optical and LSPR properties for imaging, optical and electrochemical detection, diagnosis and photothermal therapy as reported recently [Ahmad, 2015; Yetisgin, et al., 2020].

7.2.2 Translation – Laboratory Experiments to Nanomanufacturing

The translation of laboratory experiments to industrial-scale manufacturing needs to address several additional challenges that include developing cost-effective production technology with viable yield, high precision control of the assembly of nanostructures based on evolving reliable test procedures for defect control. Currently, non-selective defect control in the semiconductor industry, e.g., takes 20%–25% of the total manufacturing time. Removal of defects from nano-scale system is projected to take up more time as it requires selective and careful removal of the impurities. It is further necessary to ensure maintaining the physicochemical properties and quality of nano-system during high-rate and high-volume production as well as during the lifetime of the product after manufacture. Ultimately, it is essential to assess the environmental, ethical and social impacts of meeting the large-scale requirement of materials, devices and systems during nanomanufacturing.

Drug syntheses are, in general, possible either in crystalline or amorphous solid forms by modifying their syntheses protocols. However, the routes of using bare drug forms in crystalline and amorphous forms are rather difficult to control. Instead, an alternate route of encapsulating these drugs inside a polymeric nano-capsule is found easier from nanoproduction angle. Organic nanocapsules prepared using polymeric and lipid compounds are preferred by the pharmaceutical industries as they have better affinity for encapsulating active drug molecules as conjugates of proteins, DNA delivery vehicles, liposomes and co-polymer micelles. Several reports are there on organic nanocapsules such as dendrimers, micelles, liposomes and ferritin that possess the desired features like non-toxicity and biodegradability. Hollow core micelles and liposomes nanocapsules are more sensitive to thermal and electromagnetic radiation and offer themselves for controlled drug delivery applications. The drug-carrying capacity, stability and delivery of organic nanocapsules, either in the form of entrapped or adsorbed drugs, determine their applications according to their efficacies. Organic nanocapsules are most widely used in preparing the biomedicines.

Almost spherical solid lipid nanoparticles (SLNPs) with solid core are stabilized by suitable surfactant. These core lipids are chosen from the fatty acids, acylgly-cerols, waxes and varying mixtures of the surfactants. Biological membrane lipids include phospholipids, sphingomyelins, bile salts and sterols (i.e., cholesterols) as stabilizers. Biological lipids possess minimum cytotoxicity and solid cores permit better-controlled drug release due to enhanced mass transfer resistance. LNPs used in mRNA vaccines for SARS-CoV-2 (i.e., virus causing COVID-19) were made using four types of lipids: an ionizable cationic lipid with positive charges binding to negatively charged mRNA, a PEGylated lipid, a phospholipid and cholesterol as described by several research groups mentioned in the cited references [Ulbrich, et al., 2016; Ahuja, et al., 2020; Yao, et al., 2020].

Several laboratory-scale concepts and ideas pursued over last several years in the recent past for developing the pharmaceutical nanocarriers are now entering into the stage of maturity. The nanoformulations evolved in this context are proven to be useful for their applications in site-specific targeted drug delivery. However, the current challenge is how to optimize them to ensure their safety, efficacies and scalability, so that they are mass-produced for industrial applications leading to their clinical uses. In this context, LNPs being primarily non-toxic, biocompatible and easy-to-produce are used extensively in numerous formulations. The LNPs are increasingly being used for transport and delivery of different therapeutic agents, from biotechnological products to small drug molecules. In a recent review, the biopharmaceutical aspects of LNPs were addressed to using descriptive statistics of the state-of-the-art solid-lipid NPs research covering a span of 7 years by focussing on the therapeutics, absorption and distribution processes and current efforts for translating the related formulations into their clinical applications [Montoto, et al., 2020].

7.2.3 QUALITY CONTROL

Embedded metrology for *in situ* quality control of nanoscale systems was noted as one of the critical issues involved in verifying the features and properties of materials, components and devices during nanomanufacturing. From processing and manufacturing angles, metrology is generally found disruptive and complicated resulting into either relatively low cycle time and throughput, or expensive in terms of costly equipment, as well as cost to the overall manufacturing process. Available metrology and analytical tools with sufficient nanoscale resolution for processes, materials and devices, face restrictions in imaging or analyzing large areas and, thus, render these tools ineffective for nanomanufacturing. Alternately, the concept of embedded or *in situ* metrology is suggested to be more appropriate as it does not interrupt the processing during measurements and analysis. For deploying embedded metrology, the approaches used must provide sufficient accuracy, calibration and repeatability, however, it needs to identify the areas where sensor technology requires further development. The need for interpreting the nanoscale properties of materials, devices, components and nanosystems by analyzing some macroscale property, accesses to advanced models are necessary to translate such measurements to nanoscale features and properties. Combining statistical and physical models would further improve the interpretation of both materials and device properties and metrology indicative of the nanosystem performance and manufacturing yield as discussed in detail sometime back in the cited references [RR-01].

The size, shape, morphology, size distribution, targetability and functionality of developed NP-species are the key parameters for engineering their physicochemical properties appropriate for their biomedical applications. Such desired characteristics should be reproducible and scalable. A desired reproducible drug release profile from NPs is required to further establish batch-to-batch uniformity and quality performance assured by *in-vitro and in-vivo* performance correlations. At present, very little data is available to establish such critically important parameters of quality control of nanomedicines. Even, till date, no officially established drug release method is available for its evaluation. Mostly, conventional official dissolution methods (established for solid dosage forms) such as paddle or basket method are generally used to perform the same. Dialysis is also used in various variants (reverse dialysis, rotating dialysis, double dialysis, etc.) to estimate drug release from the NPs. However, these methods need validation. Since nanomedicines are specialized formulations delivering the bio-actives precisely to the target sites; an effective method is necessarily needed to generate the correlations to ensure their quality production. Despite significant advances in health care science and pharmaceutical industry, selection of the method and material is affected by market demands too. One of the important aspects of scale-up production of NPs is the cost of the finished product and also the consumable market. For more details, readers are suggested to look into the cited references given in a current review paper [Jian, et al., 2020].

7.3   NANOPRODUCTION OF BIOMEDICINES – EMERGING AREAS

Very large and complex area of biomedicines requires a variety of molecular species with their characteristic therapeutic features to reach the disease-inflicted sites. Contrary to the old method of administering classical drug formulations, the building blocks of nanomedicine are engineered NP-species with desired surface modifications (hydrophilic and hydrophobic nature) having provisions of vacant sites for conjugating drug molecules, targeting ligands and imaging radicals. Out of huge variety of nanostructured material species loaded with numerous possible characteristic features, it is rather a tough task to choose from the available list of material species studied so far. However, with ongoing developments in nanoscience and technology, the search is slowly becoming more manageable using the compiled data of structure-property relationships and using machine learning, artificial intelligence and advanced simulation tools. With time, the compiled data is growing in size and therefore it is inevitable to disregard the help rendered by the mathematical tools.

In highlighting the challenges faced during development of disease-specific targeted drug delivery nanoplatforms, three typical examples of diseases out of a huge list are taken here. Going through a brief overview of the salient features of developing target-specific drug delivery nanoplatforms is discussed in case of inflammatory, diabetes and Alzheimer's diseases. For keeping the discussion limited to the basic theme of targeted drug delivery platform, the mention of other diseases is not included purposely.

7.3.1   ANTI-INFLAMMATORY NANOMEDICINES

Inflammation is an immune response opposing the harmful stimuli and is an essential part of numerous diseases encountered in humans. Consequently, anti-inflammatory treatments have also been found helpful in treating a variety of diseases in general. For instance, in life-threatening vascular diseases, atherosclerosis causes myocardial infarction due to macrophage-mediated inflammation as the basic mechanism. The use of statins with anti-inflammatory properties confirmed that ~44% of the major cardiovascular problems could be eliminated with intensive cholesterol-lowering therapies as discussed by several groups [Katsuki, et al., 2017; Ma, et al., 2017].

The infiltration of leucocytes and inflammatory chemokines assists the tumour progressions. Interleukin-6 and tumour necrosis factor-α promote the tumour metastasis. In commonly used chemotherapy, immunotherapy was found quite effective in treating tumours. However, these therapies have limitations such as low specificity, bone marrow suppression and drug resistance because dying cells stimulate not only the inflammatory effects but also reduce the therapeutic effects and enhance the drug resistance as reported [Hori, et al., 2018].

Current drug delivery nanoplatforms are reported to increase the drug concentration and reduce drug resistance due to tumour cell targeting. Consequently, combined uses of anti-inflammatory and anti-tumour drug delivery systems have been very effective in using red blood cell membrane vesicles coated onto black phosphorus QDs as drug carriers with doxorubicin (OX) and kirenol (KIR) as anti-tumour and anti-inflammatory drugs, respectively. The RBC@BPQDs-DOX/KIR platforms possessing both anti-inflammatory and anti-tumour properties still need further studies for better stability and biotoxicity as reported by several teams [Huang, et al., 2019; Yao, et al., 2020].

Similarly, in case of ocular disorders like xerophthalmia and allergic diseases attempts were made using intraocular implants in glaucoma and uveitis. Most ocular diseases and procedures are related to an inflammatory response category. However, most cases of local inflammation reaching inner eye structures are treated via intraocular injections, but with several limitations such as drug bioavailability, local side effects and unstable drug concentrations. Nanocarriers allow anti-inflammatory drugs to reach the target sites. Chronic inflammatory diseases like age-related macular denegation and uveitis require the drug to be maintained at certain concentration for successful treatment. However, biodegradable polymer carriers can cause intra-ocular inflammation. Synthesis of self-assembling block co-polypeptide-NPs, once loaded with dexamethasone showed higher loading efficiency and lasting drug release in the affected eyes. Liposomal nanocarriers were found improving the intimate contact between anti-inflammatory drug and corneal surfaces allowing improved drug absorption. Dexamethasone, ibuprofen and other anti-inflammatory drugs loaded onto these nanocarriers were found improving the drug efficiency and maintaining a certain concentration into ocular tissues as reported in the cited references [Chang, et al., 2020].

Sepsis is another life-threatening disease caused by the dysregulated host response to infection, and is associated with high morbidity and mortality globally. The pathogenesis of sepsis being still unclear, it is suggested that the body's immune system receives stimulation from the bacteria to attack the invasive bacteria. Part of the inflammatory process is assumed resulting from leucocytic adhesion and inflammatory cytokine-induced endothelial cell activation. Uncontrolled and excessive inflammatory caused sepsis is treated using antibiotics and anti-inflammatory drugs to eliminate bacteria. However, limitations are still there in terms of severe adverse effects, dysfunction of liver and kidney, and poor bioavailability. Recent studies have been successful to overcome some of these issues by deploying nanoformulations to deliver antibiotics inside the cells in mouse models. A co-delivery system of antibiotic and anti-inflammatory agents was reported targeting infectious microenvironments for effective management of sepsis in a mouse model. However, more studies are required to reveal its full potential as discussed in the cited references [Singer, et al., 2016; Zhang, et al., 2018; Wang, C., et al., 2021].

Additional efforts are becoming necessary to take care of the emerging identifications of more and more number of inflammatory diseases. For instance, chronic diseases such as osteoarthritis, rheumatoid arthritis and skin diseases are treated with Non-Steroidal Anti-Inflammatory Drugs (NSAIDs) or Glucocorticoids in frequent/continuous treatments. Diseases like trauma, acute airway and ischaemic inflammation require high dosages of anti-inflammatory drugs. The development of nanoenabled pharmaceuticals not only expands the scope of site-specific drug deliveries where carriers facilitate the drug delivery to the selected target tissues but also offers controlled release of sufficient drug for improved quantities of anti-inflammatory agents as described in the cited publications [Shao, et al., 2016; McMasters, et al. 2017; Wu, et al., 2019; Shi, Y., et al., 2020; Xu, et al., 2020; Wang, H., Zhou, et al., 2021].

A variety of natural and synthetic polymer-NPs have been found effective in treating inflammatory diseases. Incorporating drugs into hydrophobic PLGA, for instance, was studied for site-specific targeted drug delivery applications in this context. An implantation model employing stromal cell-derived factor-1α (SDF-1α) into PLGA exhibited reduced inflammatory responses when placed subcutaneously. Peptides with amino acids encapsulated into collagen IV in Col IV-Ac2–26 PLGA-NPs showed significant efficacy in an atherosclerosis model. PLGA was also found effective in treating osteosarcomas, osteoarthritis, bone cancer metastasis and other inflammatory bone diseases. Encapsulating DOXY into PLGA nanospheres in-hibited the growth of *E. coli* and *S. aureus*, after controlled release of DOXY. A biodegradable thermo-sensitive implant of poly (ethylene glycol) monomethyl ether (mPEG) and PLGA as a solution-to-hydrogel (sol-gel) drug delivery system offered advantages in treating osteomyelitis, as demonstrated by mPEG-PLGA and teico-planin in the osteomyelitis rabbit model [Mir, et al., 2017].

In inflammatory bowel diseases (IBDs), comprising Crohn's disease and ul-cerative colitis, the chronic inflammations damage the gastrointestinal (GI) tract. Oral decoy oligonucleotide (ODN), when given to mouse models in the form of chitosan (CS)-modified PLGA nanospheres with nuclear factor kB (NF-kB) decoy ODN as oral drug, it improved the ODN stability, reduced the bloody faeces and diarrhoea. Experimental results showed that Decoy ODN-loaded CS-PLGA-NS turned out as an effective strategy for UC treatment. Patients suffering from chronic inflammatory disease known as rheumatoid arthritis showed relief when treated with low doses of methotrexate (MTX). Lutetium-177 was found decreasing sy-novial tissue inflammations in which 1,4,7,10-Tetraazacyclododecane-1,4,7,10-tetraacetic acid (DOTA) was used as an agent of Lutetium-177 by complexing hyaluronic acid (HA) with ^{177}Lu-DOTA and encapsulating into MTX and PLGA ^{177}Lu-DOTA-HA-PLGA (MTX) formulation that demonstrated drug delivery performance in anti-rheumatic therapy. However, still more *in vivo* tests are required to corroborate the results as reported by the research teams involved in this study [Trujillo-Nolasco, et al., 2019, Yazeji, et al., 2017].

Exosomes (30–100 nm), secreted by tumour cells, mesenchymal stem cells and immune cells, turned out to be good carriers with low cytotoxicity, non-immunogenicity and endogenous properties. Recently, exosome-based drug delivery platforms were prepared to show natural targeting because of the native biological functions of the original cells in facilitating the disease treatment [Yan, et al., 2020].

Curcumin possessing anti-inflammatory, antineoplastic and antioxidant proper-ties shows poor systemic bioavailability. Curcumin-loaded exosomes injected into a mice model, protected the animals against lipopolysaccharide (LPS)-induced septic shock. The exosome-curcumin complex enhanced the anti-inflammatory effect. The exosomes derived from the TGF-β1 gene-modified BMDC (TGF-β1-EXO) were found to inhibit the development of dextran sulphate sodium (DSS)-induced murine

IBD. The exosomes extracted from the culture supernatant of granulocytic myeloid-derived suppressor cells (G-MDSC) were found effective against the DSS-induced colitis to restore intestinal immune balance. Molecularly engineered M2 macrophage-derived exosome combined with hexyl 5-aminolevulinate hydrochloride (HAL) showed excellent inflammation-tropism with anti-inflammatory effects in atherosclerosis. Exosome encapsulating dexamethasone sodium phosphate, and surface modified with folic acid (FA)-polyethylene glycol (PEG)-cholesterol, showed anti-inflammatory effect against RAW264.7 cells as discussed by several research groups [Trujillo-Nolasco, et al., 2019; Yang, and Merlin, 2019; Robb, et al., 2020; Wu, et al., 2020].

Lipid-NPs (LNPs) are the earliest nanodrug delivery platforms for treating tumours. Currently, anti-inflammatory effects of LNPs have been studied in interleukin-10 (IL-10) formulation for checking the pro-inflammatory cytokines and exhibiting efficacy in several inflammatory diseases such as IBD, rheumatoid arthritis and organ transplantation. PSL conjugated IL-10 (PSL-IL-10) showed not only anti-inflammatory and anti-obesity effects in mice models but also mimicked the apoptotic cells changing the inflammatory M1 to anti-inflammatory M2 macrophages. PS modified liposomes and DSPE-PEG2000-cRGDfK, forming an apoptotic body liposome (AP-Lipo), when loaded with pioglitazone (PIO) showed better recognition of the activated vascular endothelial monolayer and upregulated anti-inflammatory cytokines *in vitro*. PIO-loaded AP-Lipo targeted atherosclerotic plaque caused anti-inflammatory inhibitions of M1 polarization and promoting M2 macrophage polarization. Polyetheretherketone, modified with dexamethasone plus minocycline-loaded liposomes (Dex/Mino liposomes), showed an enhanced anti-inflammatory, antibacterial and Osseo integrative behaviour suitable for an orthopaedic/dental implant nanomaterials in clinical applications. Liposome containing PS was found directly affecting anti-inflammatory cytokine production with the capability of an anti-inflammatory and immune-modulatory agent.

Target specific drug delivery platforms employing collagen hydrogel, gelatine hydrogel, lactide-co-glycoside (PLG) scaffold, hyaluronic acid hydrogel, amphiphilic poly-N-vinylpyrrolidone (Amph-PVP) and Emulgel were found effective as anti-inflammatory agent exhibiting enhanced drug delivery accuracy. Different morphologies of these drug delivery platforms were found targeting various macrophages that regulate the pro-inflammatory and anti-inflammatory effects in treating inflammation as reported in the cited papers [Kuskov, et al., 2017; Gul, et al., 2018; Wu, et al., 2019; Xu, et al., 2019; Chang, et al., 2020].

Being more stable and endowed with many applications in nanomedicine, inorganic metal oxide NPs (MO-NPs) have been investigated as drug delivery carriers. For instance, ZnO-NPs influenced the transmission of genetic messages. ZnO is a very strong antibacterial agent with bacteriostatic properties arising out of electrostatic interaction between ZnO and the cell wall causing destruction of the bacterial cell, and allowing penetration of the Zn^{2+} ions to produce ROS. The induced membrane damage and interaction with DNAs finally resulted into cell death. In one of the several studies conducted in this context, the targeted release of the drugs by folic acid-modified PEG-ZnO-NPs showed anti-inflammatory and anti-tumour properties for drug delivery. In another experiment, the controlled release of

DOX hydrochloride by meso-porous ZnO-NPs showed anti-inflammatory and anti-tumour co-effect.

Aspirin-laden monocytes inside gold nanocages (As@GNCs) were found responding to the infection area by differentiating the bactericidal macrophages as noted in MRSA-induced osteomyelitis mice model. A peptide-Au-NP hybrid, P12 (G20), with a GNP core (20 nm) could reduce Toll-like receptor signalling influencing the inflammatory response of acute lung injury. The bare ZnO and Au-NPs suffering from aggregation made them otherwise unstable. The metal NPs, modified with polymers or peptides, or changing their structures could possibly be used in improving their biostability for further investigations.

The fluorescent semiconductor QDs are used in theranostic applications along with their drug delivery properties as noted in NSAIDs anti-inflammatory molecules in treating inflammatory disorders. QD-Celecoxib conjugates are particularly noted to target the inflamed paw of the mice, not only by exerting an accurate anti-inflammatory effect but also offering bio-imaging capabilities. Combining N-doped graphene QDs (N-GQDs) sodium 10-amino-2-methoxyundecanoate (SAM) exhibited improved downregulation of COX-2, iNOS, TNF-α, NF-κß, IL-1α, IL-1ß, IL-4 and IL-6 compared with cells treated with SAM alone.

CNTs conjugated with anti-inflammatory drugs have been studied as the main drug vehicle to help or modify drug release from another parent delivery system. NSAIDs, used in alleviating arthritic pain, were administered via transdermal and oral routes offering lower concentration that lacked in precise control, and exhibited systemic side effects, respectively. Electrospun fibres are reported for transdermal drug delivery platforms using poly-ethylene oxide (PEO) and water-insoluble pentaerythritol triacrylate (PETA) along with ketoprofen. Several aspects of the observations mentioned above were discussed in detail in the cited references [Jia, and Wei, 2017; Vimala, and Soundarapandian, 2017; Buszewski, et al., 2018; Kalangi, et al., 2018; Sameer Kumar, et al., 2018; Gao, et al., 2019; Agarwal, and Shanmugam, 2020; Shi, M., et al., 2020].

The engineered NP-formulations of biomedicines have, in turn, been successful in promoting the anti-inflammatory therapy as well. Newer approaches of minimizing the shortcomings of the therapeutic protocols have consequently been employing both organic and inorganic NPs. Some protocols of these upcoming routes are still in the exploratory phase, whereas some have already reached the stage of clinical treatments.

Wider choices of nanomedicines are emerging for future development. A number of Chinese herbs are currently under study in this context. For instance, anti-inflammatory effect of the *Angelica pubescens* root extract with Au-NPs (DH-AuNPs) exhibited favourable response at the inflammation sites by releasing the drug accurately. Exploratory results obtained indicate the necessity of further investigations to confirm the potential use of this Chinese herb as an ideal anti-inflammatory drug [Markus, et al., 2017]. QDs have also been used as strong targeting and bio-imaging entities by modifying them with peptides, inorganic materials and antibodies for targeted inflammation therapy, particularly immune-related inflammation. Using these QDs offers real-time positioning and imaging information for better treatment.

Although NPs showed multiple advantages in drug delivery, there are still shortcomings that need to be taken care of before their clinical applications. While comparing different types of NPs, the performance of PLGA has already been finding wider applications in vehicle synthesis. Because of its variable structure and stability, however, PLGA can cause new inflammation *in vivo*, showing little toxicity in the liver and intestine of the rats. The improved biocompatibility due to modified PLGA with HA, chitosan and other materials could be considered as the future directions to improve PLGA-based drug delivery platforms. Exosomes secreted from the tissues offering good biocompatibility and targeting ability have also been employed for improving the uniformity of exosome for further advantages.

Liposomal drug deliveries generally exhibit reduced toxicities. The metal-based carriers like ZnO NPs and GNP though useful but are not stable and hence limit their applications. CNTs are another upcoming class of nanovehicle materials with good biocompatibility, however, their targeting ability needs to be still improved. QDs are widely studied for bio-imaging applications. Since most of the QDs contain heavy metal, it is essential to reduce their biological toxicity and maintain their targeting and bio-imaging ability – a subject of future investigations in drug delivery. A better understanding of anti-inflammatory diseases and targeted features of NPs for different diseases will fundamentally improve the therapeutic effect and reduce the side effects of the drugs being studied currently [Wang, C., et al., 2021].

7.3.2 ANTI-DIABETIC NANOFORMULATIONS

Commonly encountered diabetes mellitus (DM) changes the blood sugar levels and requires the injection of insulin as the common line of treatment. DM is an epidemic disease compromising the quality of life of the patients, with continuously increasing numbers resulting in ~1.5 million deaths yearly worldwide and DM-related loss of human lives is estimated to reach ~366 million in 2030 (WHO Report). Considering these alarming conditions, and the current glucose-lowering medications offering no cure, the treatment has to offer control of glycaemic effect by balancing the benefits against secondary effects of the drug. Conventional drug delivery systems, though advanced in the meantime, have several limitations, e.g., improper and/or ineffective dosage, low potency, limited target specificity and adverse side effects in other organs/tissues. The loading of insulin, and other sugar-lowering drugs and nutraceuticals onto nanocarriers as a more convenient, non-invasive and safer approach through alternative administration routes as reported in the cited references [Fonte, et al., 2015].

In type I DM, the pancreatic β-cells start producing lesser insulin, and in type 2, prevalent in > 85% of total diabetics, causes impairment of pancreatic β-cells leading to compromised capacity. Though, insulin is used in both the types, except its introduction in Type II enters at a stage wherein the glycaemic control fails to respond to diet control, regular exercise, weight loss and oral antidiabetic agents. Currently, insulin delivery in Type I is replaced by surgical implantation of β-Langerhans, whereas in Type II the oral antidiabetic drugs and insulin replacement are common. Insulin, a polypeptide hormone with chains of 21 and 31 amino acids joined by two disulphide bonds controls the glucose uptake and storage in the liver and muscles.

Oral insulin possessing lower bioavailability and insufficient therapeutic effect due to the physiological instability of the polypeptide was attributed to its degradation in the GIT, and the fast systemic clearance. Consequently, insulin has to be administered daily through several injections, which causes discomfort, pain, stress and trauma, resulting in limited compliances. Insulin injections are also associated with the risk of local infections, hypoglycemia, skin necrosis and nerve damage. While attempting a diversity of approaches including permeation enhancers, chemical modifications, enteric coatings, enzymatic inhibitors and NPs to optimize the oral delivery of insulin, a site-specific targeted delivery was also explored.

Emulsified nanosuspensions of glibenclamide, glimepiride and gymnemic acids, when tested in animal models, did show stimulation of pancreatic β-cells to secrete insulin. For instance, diabetic rat models treated with glimepiride nanosuspensions exhibited improved pharmacokinetic and anti-hyperglycaemic activity. Adding phyto-ingredient like berberine showed an enhanced antidiabetic effect in T2DM animal models with lower dosages in comparison to the classical formulations. These nanosuspensions when tested in humans, showed an enhanced antidiabetic and higher glucose-lowering effects. More details of these formulations are discussed in the cited references [Ganesan, et al., 2017; Souto, et al., 2019].

Alternately, several other strategies of encapsulating the insulin into polymer-NPs were proposed to take care of its instability problems during passage through GIT and its lower intestinal permeation. Polymers used in preparing dosages for oral, nasal, transdermal and intraperitoneal administrations include chitosan (oral/nasal), dextran (oral), alginate (oral), poly(γ-glutamic acid), hyaluronic acid (oral), dextran (oral), gelatine (oral/pulmonary), poly(lactic acid), poly(lactide-co-glycolic acid) (oral/transdermal), polycaprolactone (oral), poly amino acid (oral), PVA (oral/transdermal), pluronic (oral), acrylic polymers and polyallylamine. Despite all these high promising attempts made and reported recently, there is still no polymeric-NP-based platform to deliver insulin orally that is available in the market [Mansoor, et al., 2019].

While exploring the metabolic pathway of non-insulin GLP1 hormone, it was noted that after its production in the L-cells of intestine (proglucagon expression, and release after meal intake, stimuli and other molecular mechanisms), it got metabolized fast due to dipeptidyl peptidase IV inactivation even before leaving the gut. It possibly transmitted the information via sensory neurons in the intestine and the liver to express and produce the receptors. GLP1 influences the sugar-dependent insulin release and lowers the glucagon despite having poor oral bioavailability. It's absorption/desorption into PLGA, SLNPs and porous Si-NPs were subsequently deployed for improving its release. Combined with chitosan-NPs (CS-NPs), GLP1 demonstrated sustained release with improved interactions with the intestinal cells. While examining hypoglycaemia influence of GLP1 in animal models by integrating them with anionic, non-ionic and cationic liposomes, the anionic liposomes showed highest encapsulation efficiency (80%) compared to the other two, leading to 50% reduction in the glucose. Recently, self-assembled two derivates namely cadyglp1e and cadyglp1m peptides were found preventing degradation of GLP1 by DPP4 enzyme resulting in better stabilization and extending the presence of high insulin level and reducing glucose in blood. Other three analogs of PEGylated GLP1 along with Au-NPs, namely – [GLP1(7–37)/-Lys(Ac)-NH$_2$, –Lys

(Cys)-NH$_2$ and –Lys(PEGCys)-NH$_2$], exhibit better biocompatibility and absorptions. The PEGylated conjugate GLP1(7–37)-Lys(PEGCys)-NH$_2$] could cross Caco-2 cell barriers and deliver the peptide at higher concentration than in the others, and therefore it was studied for *in vivo* hypoglycaemic activity in rat models. However, GLP and conjugated GLP1/Au-NPs exhibited more reduction in glucose level after 2 h of administration and maintained a longer insulinotropic activity of incretin. Another approach for treating diabetic patients was the route deploying the non-viral delivery of GLP1 plasmid DNA (pDNA) complex through intravenous or oral administration, in which, the pDNA codes a specific protein. This route provides the protein manufacturing by the cells of the small intestine and delivers into the bloodstream. A nano-sized gene complex (HTCA+GLP1) prevented the gastrointestinal degradation. The incorporation of GLP1 and DPP4-inhibitors together into a single delivery system was also anticipated to offer a higher efficacy than just GLP1 alone, particularly nanoparticulate GLP1 and PLGA functionalized with CS and a cell-penetrating peptide with chitosan-modified porous Si-NPs coated with hydroxylpropyl methylcellulose acetate succinate polymer. In experimental animal models, blood glucose level was reduced by 45% by enhancing the pancreatic insulin [Zijlstra, et al., 2014; Simos, et al., 2021].

Another formulation of pituitary adenylate cyclase-activating peptide (PACAP) was found active in carbohydrate and lipid metabolism by activating VPAC1 and two receptors by mediating the glucagon production and glucose-dependent insulin secretion, respectively. A novel peptide-conjugated formulation of selenium-NPs was developed by Lin, et al., by attaching chitosan encapsulated VPAC2, which was responsible to prolong (~170x) the release with circulating half-life of longer than that of PACAP. A sustained hypoglycaemia effect with enhanced insulin secretion and low blood glucose level was reported outmatching the Exendin-4 [Lin, et al., 2017].

Sulfonylureas-based antidiabetic drugs were found enhancing the insulin release from the pancreatic beta cells. For instance, gliclazide niosomes maximize the related hypoglycaemia activity. Comparison of gliclazide/niosomes with gliclazide alone exhibited slow and maximum reduction after 6 h (~48%) and fast reduction of glucose level (45%, 2 h), respectively. Glipizide, a second member, was found increasing pancreatic insulin with a half-life of 3.5 h, which was extended by encapsulating them into poly-E-caprolactone (PCL). Oral form of metformin hydrochloride (MH) with half-life of 6.2 h, when entrapped into niosomes showed drug release extended over a longer period of time. Positively charged niosomes and MH orally administered to diabetic rats exhibited mild release of MH in the initial period whereas maximum reduction in blood glucose was reached after 4 h and sustained for a longer period while blood glucose returning to the initial level after 6–8 h compared to 2–4 h with the drug solution alone. Metformin/liposomes coated with chitosan cross-linked with the biocompatible β-glycerol phosphate increased their stability in the GIT. MH/liposomes presented enhanced bioavailability in rats almost twice as high as that of MH solution. Linagliptin, alternately, taken orally, was found regulating the insulin levels produced by body after eating. It is generally less preferred in comparison to metformin and sulfonylureas as an initial treatment and used along with exercise and diet.

Another family of thiazolidinediones was found reducing the insulin resistance and increasing glucose uptake in the peripheral tissues. Pioglitazone was found enhancing the transcription of insulin-responding genes to increase sensitivity. For optimal response of pioglitazone, nano-formulations were studied with varying particle size and entrapment efficiency. Cumulative drug release varied in the range of 70%–95% using 145–500 nm particle size with 67%–84% entrapment efficiency. The hypoglycaemia capability of 145 nm particle size having entrapment efficiency of 84% exhibited cumulative release of 95% compared to that from the pure pioglitazone in rats. Conjugated pioglitazone-niosomes showed a greater reduction in glucose level than pure pioglitazone in oral administration.

Repaglinide drug stimulated the insulin release from pancreatic cells with a half-life of 1 h in systemic blood circulation and once loaded on PMMA-NPs, it prolonged *in vitro* drug release with no *in vivo* toxicity. The toxicological study for repaglinide-loaded SLNs showed that the surfactant used during the preparation affected the drug release profile and hence changing the surfactants onto NPs (83–91 nm particles), varied the release rate from 1.12% to 1.63%/h with a maximum drug depletion from 47% to 86%. Out of nine different batches of NPs (size from 144 to 497 nm, and entrapment efficiency from 54% to 88%), the batch with 144 nm particles and 82% entrapment efficiency exhibited anti-hyperglycaemia in rats. Free repaglinide achieved maximum reduction of 70% in mean blood glucose level 4 h post-administration in contrast to niosome formulations having 68% maximum 8 h post-administration. Both drugs caused 30% reduction in blood glucose than mean level in rats with normoglycemia. Exenatide, an incretin analog found in lizard saliva, was effective when given by subcutaneous injection. Exenatide could be given orally also by encapsulating into the modified CS-NPs conjugated with a goblet cell target peptide in order to enhance its effectiveness. Higher loading (6x) of exenatide into CS-NPs compared to that of injected subcutaneously led to a similar change in blood glucose level over a period of 12 h. CS-NPs generate a gradual release of exenatide into the blood circulation whereas subcutaneous injection produces an immediate sharp rise at 0.5 h post-injection. Recent studies used DPP4 inhibitor encapsulated vildagliptin into polymer microspheres or into DNA nanospheres. Pure vildagliptin and microsphere were given orally to diabetic rats and induced similar blood glucose reduction up to 70%; however, achieving at different times (4 h and 8 h for pure vildagliptin and for vildagliptin containing microsphere, respectively). Vildagliptin microspheres maintained lower blood glucose concentration for a longer period compared to pure vildagliptin. A genetically modified mice model behaving closer to human T2DM was orally administered with eudragit (cationic polymer)-DNA nanospheres with or without vildagliptin, exhibiting postprandial concentration of GLP1. Vildagliptin nanospheres preserved stable blood glucose level for up to 4 h. Specifically at 3 h GLP1 concertation increased more than 6x particularly attributed to the eudragit-DNA nanospheres properties of small size, high vildagliptin loading efficiency and ability to bypass the acidic pH of the stomach and inhibit the degradation of GLP1. The scope of a positive control (pure vildagliptin) however limits the range of the conclusions that can be drawn as reported by several research groups included in the cited references [Andreani, Kiill, et al., 2014; Andreani, de Souza, et al., 2014; Andreani, et al., 2015; Wakaskar, 2017; Andreani, et al., 2019; Cao, et al., 2019].

Various kinds of trace elements such as Zn, V, Cr, Se and Li, act as cofactors in enzymatic reactions exhibiting their responses to glucometabolic disorders. Comparative study of the hypoglycaemia effects of ZnO and Ag-NPs in comparison to that of insulin treatment on rats confirmed that although Zn and Ag-NPs reduce blood glucose, they were unable to completely restore normoglycemia. Orally administered three dosages of ZnO-NPs (1, 3 and 10 mg/kg BW) were found similar in reducing fasting blood glucose level but insulin level increased significantly only at the higher doses. ZnO-NPs restored β-cells including silencing of microRNA-103 and 143 resulting in improved glucose homeostasis. ZnO-NPs in combination with vildagliptin provides a synergistic effect that enhanced the positive outcome [Ahmad, et al., 2020].

Some metallic-NPs were found possessing enzyme-like antioxidant properties of scavenging the free radicals and decreasing the ROS concentrations. Such NPs, when functionalized with antioxidants or antioxidant enzymes were found behaving more like an antioxidant delivery system in conjunction with inorganic-NPs. These properties of SOD, CAT, oxidase and peroxidase activities were demonstrated by metal oxide-NPs of cerium, iron, cobalt, copper, manganese and vanadium, as well as noble metals such as gold, silver and platinum as reported in the cited publications [Gao, et al., 2017; Nadaroglu, et al., 2017; Pedone, et al., 2017; Lushchak, et al., 2018].

The molecular mechanisms determining the antioxidant capabilities of metallic-NPs are, though, not yet understood fully, however, this behaviour could possibly be attributed to their ability to transform into different multi-oxidation states. For example, antioxidant properties of ceria-NPs show promising catalytic behaviour due to oxygen vacancies on their surfaces and participate in auto-regenerative cycle of their two oxidation states namely – Ce^{3+} and Ce^{4+}. The increase in concentration of Ce^{3+} relative to Ce^{4+} is associated with the reductions in the particle size, and such loss of oxygen due to reduction of Ce^{4+} to Ce^{3+} is accompanied by the formation of an oxygen vacancy on the surfaces of the NPs is evidenced modulating the antioxidant pathways. The interactions of cellular macromolecules including proteins, nucleic acids and lipids with NPs become significantly important in this context. A variety of sites of the NP-protein interaction combined with the varying structures of the NPs combined the availability proteins and duration of interaction – combined together affect the cellular redox environment by either stimulating or inhibiting ROS generation under certain conditions. Under oxidative stress, NPs might prevent some biomolecules from oxidative damage, thereby causing health benefits and disease preventions. This mechanism of antioxidative NPs could possibly be used in preventing diabetes-induced oxidative stress as reported in the enclosed references [Akhtar, et al., 2017; Lushchak, et al., 2018].

Antidiabetic nanoformulations, evaluated in terms of potency, and efficacy, are found better than conventional drugs. However, the nanodrugs have limitations related to the carrier stability and their reproducible characteristics. Having observed in animal trials, a small number of them applied long-term therapeutics protocols. Thus, the need for comparative studies especially in slow T2DM progression animal models is imperative to fully exploit the pros and cons of the nanoformulations over the pure drugs.

Implementing strategies for the management of diabetes mellitus is becoming a necessity as the cost of diabetic treatments burdens the patients, their family, the society and eventually the public health system. Most of the current antidiabetic drugs enhance either insulin release or glucose uptake in certain peripheral tissues. Often, these drugs fail to control or even delay hyperglycaemia and eventually insulin therapy becomes imperative. Antidiabetic treatment for T2DM usually comes in the form of tablets or capsules but most of the times they are accompanied with unwanted side effects depicting the limited effectiveness of current drugs and specifically their inability to access the site of action at desired concentrations in combination with a narrow therapeutic window as they fail to sustain prolonged release. Complex dosing schedules required for maintaining the constant drug concentration become necessary to implement. The therapy goals of diabetes are to improve the quality of the patients' life, to prevent or delay the onset of disease complications and to ultimately decrease mortality. The aim of the antidiabetic treatment is to achieve optimal glycemic control that can be summarized as target values for FPG 70–130 mg/dL, 2-h postprandial glucose <180 mg/dL and bedtime glucose 90–150 mg/dL. Control of glycemia can be imperative for the prevention of microvascular complications such as neuropathy, nephropathy and retinopathy; however, glycemic control alone cannot be effective for the prevention of macro-vascular complications including ischaemic heart disease, and peripheral arterio-pathy. An optimal therapeutic profile of a nanodrug should, in principle, aim to maintain glucose levels as close to normal as possible for an extended period. Steep decrease with an analogous increase in short time being undesirable, the drug should maintain a progressive increase of blood glucose over extended time duration, although this task is extremely difficult in diabetes due to drug clearance from the circulation. The ideal diabetes nanocarrier, therefore, should respond to blood glucose concentration and accelerate or decelerate the drug release accord-ingly; staying in blood circulation for prolonged periods for maintaining glucose levels as low as possible and gradual drug release with minimum drug burst. For improving the health quality of diabetic patients, the contribution of biomedical devices like closed-loop delivery and novel glucose-responsive microneedle patches were also explored with tight control of blood glucose. Some of these considera-tions are discussed in cited references [Souto, et al., 2016; Cefali, et al., 2018].

7.3.3 Alzheimer's Disease and LNP-Carriers

In a recent review, the importance of LNP-nanocarriers-based nasal route of drug delivery was discussed to cure CNS diseases [Akel, et al., 2020; Khan, et al., 2021]. SLNPs are excellent carriers, in this context, showing inhibitory effects against amyloid aggregation for treatment of Alzheimer's Disease (AD). Recent method of inducing expression of p-glycoprotein and breast cancer resistance protein transpor-ters on brain endothelial cells via targeting the MC11 ligands, employed transferrin-functionalized LNPs for inducing the expression of these proteins, as a potential method toward AD therapy. SLNPs loaded with donepezil showed possibility of enhancing the drug delivery to the brain through the intranasal route. SLNPs and donepezil formulations prepared by the solvent emulsification diffusion technique were found having promising drug efficacy compared to other formulations. Curcumin-loaded lipid-core nanocapsules were designed successfully showing neuroprotective effects against Aβ 1–42-induced behavioural and neurochemical changes in AD mice. LNPs with nanocurcumin were found helping in reducing the oxidative stress in AD brain to recover the memory by suppressing Aβ in AD. In another case, LNPs and SLNPs loaded with nanocurcumin showed improved bioavailability of curcumin in the brain as reported by several groups [Sadegh Malvajerd, et al., 2018; Yasir, Chauhan, et al., 2018; Yasir, Sara, et al., 2018; Sadegh Malvajerd, et al., 2019; Yavarpour-Bali, et al., 2019; Arduino, et al., 2020; Sathya, et al., 2020].

Self-assembled amphiphilic nanovesicular liposomes are found better drug delivery carriers to brain tissues as they are easily functionalized and surface modified using several polyether, functional proteins and cell-penetrating peptides that aid in target-specific drug transport across the BBB. For instance, PEGylated liposomes succeeded in evading the RES opsonization. Glutathione-PEGylated liposomes were found enhancing the cellular drug uptake across endothelial BBB. Curcumin-loaded liposomes enhanced the CNS drug delivery via receptors on the BBB cells. Functionalized liposomes were found safe and compatible for gene deliveries to brain tissues. Osthole as an anti-AD compound showed protective effects on hippocampus neurons with anti-Aβ properties reported recently [Arora, et al., 2020].

The influence of artemisia-absinthium-loaded niosome was studied recently against amyloid aggregation to preclude the development of amyloid in AD. Pentamide-loaded chitosan and glutamate-coated niosomes could enhance the drug exposure to AD-brain via the intranasal route exhibiting a significant increase in pentamide efficacy across the BBB resulting in low blood level of folates. The folic acid hindrance met while crossing the BBB revealed that niosomes with span 60 and cholesterol in the ratio of 1:1 showed higher entrapment efficacy with more exposure to the affected brain parts. Rivastigmine is an acetylcholine esterase inhibitor that improves brain functions in CNS disorders. A niosome formulation was reported using sorbitan esters and cholesterol showing improved drug efficacy in the targeted brain tissues [Estabragh, et al., 2018; Rinaldi, et al., 2018; Ansari, and Eslami, 2020].

A nanoemulsion (ultrasonic homogenization) was used for intranasal delivery to bypass the BBB for AD treatment. To improve the clinical efficacy, naringerin nanoemulsion was explored revealing that it could overcome Aβ neurotoxicity and amyloidogenesis as reported in the cited literature [Md, et al., 2018; Kaur, et al., 2020; Nirale, et al., 2020]. Cubosomes, showing potential applications in drug delivery to the brain, took the delivery of donepezil-HCL via cubosomal mucoadhesive. *In-situ* assessment of nasal gel showed promising targeted delivery to the affected brain parts [Patil, et al., 2019; Gaballa, et al., 2020].

Self-assembled amylolipid nanovesicle carriers prepared using lipid-modified starch hybrid showed maximum drug concentration across the BBB for rapid and enhanced influence on the brain cells. Intranasal administration of curcumin-loaded amylolipid nanovesicles was noted to cross the BBB exerting significant effects against AD with promising drug delivery to AD brain tissues [Sintov, 2020].

7.4   DISCUSSIONS AND CONCLUSIONS

Nanomanufacturing of biomedicines generally deploys either of the two approaches namely – bottom-up or top-down as per their commercial viability. In practice, however, more and more top-down methods are in vogue compared to those of the bottom-up ones by the pharmaceutical industry. The major obstacle faced in bottom-up approach at industrial level needs rigorous methods of removing the traces of the remaining solvent that is rather an extremely difficult task.

There are numerous protocols of nanomanufacturing that need scale-up from bench to the market level. For examples, nature of material and its generally regarded as safe (GRAS) status, morphology-specific toxicological properties, *in-vivo* biodegradability, and balancing of multicomponent system at large scale, are some of the features that needs more detailed studies. One has to be careful in selecting the materials, solvents, procedure of NP-syntheses, cost, and acceptability of the finished product both by clinicians and patients. During scale-up of laboratory methods, sometimes the desired features of NPs are lost. For example, in a study of scale-up of emulsification-based NP-preparation, it was observed that with increase in the impeller speed and agitation time, particles size was reduced without altering the entrapment efficiency.

Selection of nanomanufacturing is also important from scale-up point of view to save time during pilot scale batch production. In a comparative study of ibuprofen-loaded NPs, the nanoprecipitation occurred faster than emulsion-based method. After optimization at different levels involving therapeutic needs, market demands, R&D, production steps, scale-up feasibilities, clinical trials and regulatory issues, a nanomedicine finally reaches the market. Some of the commercially available nanomedicines in the market are, e.g., Doxil®, Daunoxome Abraxane®, Ambisome®, Estrasorb Emend, MegaceES, Tricor and Triglide as few examples. In addition, nanoformulations currently available for *in-vivo* imaging include Resovist, Feridex and Gastromark, for example. The nanomedicine products, no doubt, are superior to the conventional drug delivery and imaging systems and hence are in high demand. The overall market of nanomedicines was projected to increase from 12 b US$ in 2012, to 220 b US$ in 2020 and expected to reach 461 b US$ by 2026 (12%CAGR during the forecast period 2021–2026). The nanotechnology of biomedicines offers newer opportunities of developing strategies for prevention, diagnosis and treatment of COVID-19 and other viral infections, which include developing nanoformulations of disinfectants, personal protective equipment, diagnostic systems and nanocarrier systems, for treatments and vaccine development as discussed in detail elsewhere [MR-01]. The nanomedicine market growth is presently driven by various factors including the growing occurrences of cancer; genetic and cardiovascular diseases, advances in nanoenabled diagnostic procedures and growing preference for personalized medicines as discussed and reported [WP-04].

Nanomedicines help in improving the human health by offering treatments for various life-threatening diseases, such as cancer, Parkinson's disease, Alzheimer's disease, diabetes, orthopaedic diseases and diseases related to blood, lungs and the cardiovascular system. Despite several launches of nanomedicine manufacturing made as a whole, the stringent regulatory issues and the high cost of nanoparticulate biomedicines, relative to its traditional counterparts, are hindering the market to grow faster. The nanomedicine market is inclusive of nano formulations of the existing drugs or new drugs or newer nanoformulations. The market research of nanomedicine manufacturing, thus, includes meeting the rising demands of drug delivery platforms, biomaterials, active implants, diagnostic imaging and tissue regeneration for taking care of diseases like cardiovascular, oncological, neurological, orthopaedic, infectious and other diseases.

Antidiabetic nanoformulations were evaluated in terms of efficacy and effectiveness and were found better than conventional drugs. However, the nanodrugs have limitations related to the carrier stability and their reproducible properties. Having gone through the extensive animal trials, a selected number of them were used for arriving at proper therapeutics protocols. Thus, the need for comparative studies, especially in slow T2DM progression animal models, became imperative to know the pros and cons of the nanoformulations over the pure drugs.

After NNI recommendations, extensive efforts were put in refining/replacing the existing laboratory procedures with reliable, consistent and viable manufacturing practices capable of handling the clinical supplies. Sufficient start-ups were, thus, encouraged to develop these methods matching with the product development timeline as trials on disease models were found necessary to know how these nanotherapies would affect the body as these details are required for deciding to proceed further towards the clinical development.

Another critical step in nanomanufacturing of biomedicines is related to the preclinical toxicity evaluations and subsequent clinical trials using drug supply generated under GMP conditions. These studies are conducted before knowing whether the results so arrived at would finally offer an effective nanomedicine. Various possibilities of outsourcing some activities were also explored by sharing the tasks of preclinical trials, validation responsibilities and development of GMP-based manufacturing scale-up protocols.

Although many hurdles came in the way during scale-up and manufacturing like those generally encountered in any new therapeutic development, nanomanufacturing of medicines threw even more severe challenges. Like the necessity of nanocharacterization covering every stage of discovery, development and commercialization. The nanomaterials of medical-grade purity and reproducibility were rather difficult to produce. It was quite often noted that the expensive nanomedicines, produced under FDA's cGMPs, did not comply with the specifications, or showed poor efficacy. Safety concerns not only at the dosage and clinical levels but also in manufacturing needed avoiding contaminations and minimizing toxicological factors.

Nanomanufacturing of medicines would ultimately enable to reaping the benefits of the significantly huge investment made under NNI, as specific challenges met during the development of this important area would not be easy to address by other routes.

Nanomanufacturing and product applications are currently putting pressure on providing strict oversight and mandatory regulations to avoid any adverse effects that may not be known due to the late entry of NPs in the commercially viable production sector. EU's recognition as the next frontier technology with unforeseen applications has put more emphasis thereafter. Consequently, due to inevitable uses of nanoparticulate material species, the related regulations and oversight become unavoidable to let such industry flourish with social and environmental benefits. An international body (OECD) was set up for nano industry providing international standards supported by ISO/TC 229 (Industry Standardisation).

Serious concerns have, nevertheless, been raised against the entry of engineered or incidental NP-species in the environment via several routes. It is necessary to know their possible biochemical pathways followed after the release from the production plant to the environment. Adequate steps must be implemented for ensuring the overall safety of the manufacturing practices. Lack of regulations in this context of NP-release into the environment with no industrial framework makes it still more pertinent to undertake an industry-based review of current practices to assess the industry-specific applications for recommending the resolutions via proper regulations to address the intimately related issues.

Extensive public awareness programs are gaining significance with the rise of several nano industries globally. A communication between the consumers and the products marketed is becoming inevitable to reduce the propagation of negative public opinions. Further studies are needed to investigate about the kinds of the health risks of the workers exposed to these NPs using quantitative estimates of the NPs released to the environment annually by considering the pre-emptive technologies to reduce the negative influence of the incidental NPs from production plants. The biological effect of NPs is emerging as new frontiers in the production industry, enabling technological, health and environmental advancements.

In conclusion, it is recommended that the methods of synthesizing the nano-particulate material species, surface modifications with ample sites for conjugations of therapeutic molecules along with targeting, and imaging radicals, must be kept as simple as possible with minimum use of hazardous chemicals, and also deploying natural polymers more in comparison to the synthetic ones. This will certainly ensure better yield and reproducibility from lot to lot. Structural stabilities with precisely controlled transport properties of the targeted delivery platforms are generally ensured by incorporating the necessary modifications required for accomplishing the final goal. These requirements are need specific and in this context the database compiled in connection with structure-property relationship would be very useful using modern techniques of machine learning-based simulations. Engineered nanomaterials meeting the requirement of target-specific drug delivery and diagnostics involving lower dimension materials are bound to influence in a much bigger way with further development in nanomanufacturing technology. Some of the basic problems have been studied since long [Ahmad, 2015; Kumar, and Ahmad, 2017; Ahmad, 2021].

REFERENCES

Agarwal, H., and Shanmugam, V. (2020), A review on anti-inflammatory activity of green synthesized zinc oxide nanoparticle: Mechanism-based approach. *Bioorg. Chem.*, 94, 103423.

Ahmad, S. (2015), Engineered nanomaterials for drug and gene deliveries – A review. *J. Nano pharmaceutics and Drug Delivery*, 3(1), 1–50.

Ahmad, S. (2021), 2D-metal-NSs based 'Smart drug delivery and imaging platforms' treatment of Neglected tropical diseases (NTDs). *Mod Appro Drug Des.*, 3(2). MADD. 000562.

Ahmad, F., Al-Douri, Y., Kumar, D., and Ahmad, S. (2020), Metal-oxide powder technology in biomedicine. In *Metal Oxide Powder Technologies Fundamentals, Processing Methods and Applications*, Al-Douri, Y. (Ed.), Elsevier: Amsterdam, NL, pp. 121–160.

Ahuja, R., Panwar, N., Meena, J., Singh, M., Sarkar, D. P., and Panda, A. K. (2020), Natural products and polymeric nanocarriers for cancer treatment: A review. *Environmental Chemistry Letters*, 18, 2021–2030.

Akel, H., Ismail, R., and Csóka, I. (2020), Progress and perspectives of brain-targeting lipid-based nanosystems via the nasal route in Alzheimer's disease. *Eur. J. Pharm. Biopharm*, 148, 38–53.

Akhtar, M. J., Ahamed, M., Alhadlaq, H. A., and Alshamsan, A. (2017), Mechanism of ROS scavenging and antioxidant signalling by redox metallic and fullerene nanomaterials: Potential implications in ROS associated degenerative disorders. *Biochimica et Biophysica Acta (BBA) – General Subjects*, 1861(4), 802–813.

Andreani, T., Miziara, L., Lorenzon, E. N., de Souza, A. L., Kiill, C. P., Fangueiro, J. F., Garcia, M. L., Gremiao, P. D., Silva, A. M., and Souto E. B. (2015), Effect of mu-coadhesive polymers on the in vitro performance of insulin-loaded silica nanoparticles: Interactions with mucin and biomembrane models. *Eur. J. Pharm. Biopharm.: Off. J. Arb. Fur Pharm. Verfahr. E.V.*, 93, 118–126.

Andreani, T., Fangueiro, J. F., Severino, P., Souza, A. L. R., Martins-Gomes, C., Fernandes, P. M. V., Calpena, A. C., Gremiao, M. P., Souto, E. B., and Silva A. M. (2019), The influence of Polysaccharide coating on the physicochemical parameters and cytotoxicity of silica nanoparticles for hydrophilic biomolecules delivery. *Nanomaterials*, 9, 1081.

Andreani, T., Kiill, C. P., de Souza, A. L., Fangueiro, J. F., Fernandes, L., Doktorovova, S., Santos, D. L., Garcia, M. L., Gremiao, M. P., Souto E. B., and Silva, A. M. (2014), Surface engineering of silica nanoparticles for oral insulin delivery: Characterization and cell toxicity studies. *Colloids Surf. B: Biointerfaces.*, 123, 916–923.

Andreani, T., de Souza, A. L., Kiill, C. P., Lorenzon, E. N., Fangueiro, J. F., Calpena, A. C., Chaud, M. V., Garcia, M. L., Gremiao, M. P., Silva, A. M., and Souto, E. B. (2014), Preparation and characterization of PEG-coated silica nanoparticles for oral insulin delivery. *Int. J. Pharm.*, 473, 627–635.

Ansari, M., and Eslami, H. (2020), Preparation and study of the inhibitory effect of nano-niosomes containing essential oil from artemisia absinthium on amyloid fibril formation. *Nanomed. J.*, 7, 243–250.

Arduino, I., Iacobazzi, R. M., Riganti, C., Lopedota, A. A., Perrone, M. G., Lopalco, A., et al. (2020), Induced expression of P-gp and BCRP transporters on brain endothelial cells using transferrin functionalized nanostructured lipid carriers: A first step of a potential strategy for the treatment of Alzheimer's disease. *Int. J. Pharm.*, 591, 120011.

Arora, S., Layek, B., and Singh, J. (2020), Design and validation of Liposomal ApoE2 gene delivery system to evade blood–brain barrier for effective treatment of Alzheimer's disease. *Mol. Pharm.*, 18, 714–725.

Bagul, U. S., Pisal, V. V., Solanki, N. V., and Karnavat, A. (2018), Current status of solid lipid nanoparticles: A review. *Mod. Appl. Bioequiv. Availab.*, 3(4), MABB.MS.ID.555617 (2018).

Bochicchio, S., Lamberti, G., and Barba, A. A. (2021), Polymer–Lipid pharmaceutical nano-carriers: Innovations by new formulations and production technologies. *Pharmaceutics*, 13, 198.

Buszewski, B., Railean-Plugaru, V., Pomastowski, P., Rafińska, K., Szultka-Mlynska, M., Golinska, P., et al. (2018), Antimicrobial activity of biosilver nanoparticles produced by a novel *Streptacidiphilus durhamensis* strain. *J. Microbiol. Immunol. Infect.*, 51, 45–54.

Cao, S.-J., Xu, S., Wang, H.-M., Ling, Y., Dong, J., Xia, R.-D., and Sun, X.-H. (2019), Nanoparticles: Oral delivery for protein and peptide drugs. *AAPS Pharmscitech.*, 20, 190.

Cefali, L. C., Ataide, J. A., Eberlin, S., da Silva Goncalves, F. C., Fernandes, A. R., Marto, J., Ribeiro, H. M., Foglio, M. A., Mazzola, P. G., and Souto, E. B. (2018), In vitro SPF and Photostability assays of emulsion containing nanoparticles with Vegetable extracts rich in Flavonoids. *AAPS Pharmscitech.*, 20, 9.

Chakravarty, P., Famili, A., Nagapudi, K., and Al-Sayah, M. A. (2019), Using supercritical fluid technology as a green alternative during the preparation of drug delivery systems. *Pharmaceutics*, 11(12), 629.

Chang, M. C., Kuo, Y. J., Hung, K. H., Peng, C.-L., Chen, K.-Y., and Yeh, L.-K. (2020), Liposomal dexamethasone-moxifloxacin nanoparticle combinations with collagen/gelatine/alginate hydrogel for corneal infection treatment and wound healing. *Biomed. Mater.*, 15, 055022.

Chenthamara, D., Subramaniam, S., Ramakrishnan, S. G., Krishnaswamy, S., Essa, M. M., Lin, F.-H., and Qoronfleh, M. W. (2019), Therapeutic efficacy of nanoparticles and routes of administration. *Biomater. Res.*, 23, 1–29.

Crucho, C. I. C., and Barros, M. T. (2017), Polymeric nanoparticles: A study on the preparation variables and characterization methods. *Mater. Sci. Eng. C Mater. Biol. Appl.*, 80, 771–784.

Đorđević, S., Gonzalez, M. M., Conejos-Sánchez, I., Carreira, B., Pozzi, S., Acúrcio, R. C., Satchi-Fainaro, R., Florindo, H. F., and Vicent, M. V. (2022), Current hurdles to the translation of nanomedicines from bench to the clinic. *Drug Delivery and Translational Research*, 12(3), 500–525. 10.1007/s13346-021-01024-2

Estabragh, M. A. R., Hamidifar, Z., and Pardakhty, A. (2018). Formulation of Rivastigmine niosomes for Alzheimer'disease. *Int. Pharm. Acta.*, 1, 104.

Feng, J., Markwalter, C. E., Tian, C., Armstrong, M., and Prud'homme, R. K. (2019), Translational formulation of nanoparticle therapeutics from laboratory discovery to clinical scale. *J. Translational Medicine*, 17, Article number: 200.

Fernandes, F., Dias-Teixeira, M., Delerue-Matos, C., and Grosso, C. (2021), Critical review of lipid-based nanoparticles as carriers of neuroprotective drugs and extracts. *Nanomaterials*, 2021(11), 563. 10.3390/nano11030563

Fonte, P., Araújo, F., Silva, C., Pereira, C., Reis, S., Santos, H. A., and Sarmento, B. (2015), Polymer-based nanoparticles for oral insulin delivery: Revisited approaches. *Biotechnol Adv.*, 33(6 Pt 3), 1342–1354.

Franco, P., and De Marco, I. (2021), Nanoparticles and nanocrystals by supercritical CO_2-Assisted techniques for pharmaceutical applications: A review. *Appl. Sci.*, 11(4), 1476.

Gaballa, S. A., and Garhy, O.H. El, and Abdelkader, H. (2020). Cubosomes: Composition, preparation, and drug delivery applications. *J. Adv. Biomed. Pharm. Sci.*, 3, 1–9.

Gagliardi, A., Giuliano, E., Venkateswararao, E., Fresta, M., Bulotta, S., Awasthi, V., and Cosco, D. (2021), Biodegradable polymeric nanoparticles for drug delivery to solid tumours. *Front. Pharmacol.*, 12, 601626. 10.3389/fphar.2021.601626

Ganesan, P., Arulselvan, P., and Choi, D.-K. (2017), Phytobioactive compound-based nanodelivery systems for the treatment of type 2 diabetes mellitus-current status. *Int. J. Nanomed*, 12, 1097–1111.

Gao, L., Fan, K., and Yan, X. (2017), Iron oxide nanozyme: A multifunctional enzyme mimetic for biomedical applications. *Theranostics*, 7(13), 3207–3227.

Gao, W., Wang, Y., Xiong, Y., Sun, L., Wang, L., Wang, K., et al. (2019). Size-dependent anti-inflammatory activity of a peptide-gold nanoparticle hybrid in vitro and in a mouse model of acute lung injury. *Acta Biomater*, 85, 203–217.

Gul, R., Ahmed, N., Ullah, N., Khan, M. I., Elaissari, A., and Rehman, A. U. (2018), Biodegradable ingredient-based Emulgel loaded with ketoprofen nanoparticles. *AAPS PharmSciTech*, 19, 1869–1881.

Hori, S. I., Herrera, A., Rossi, J. J., and Zhou, J. (2018). Current advances in aptamers for cancer diagnosis and therapy. *Cancers (Basel)*, 10, 9.

Horizon-2020; www.ukri.org/councils/innovate-uk/guidance-for-applicants/guidance-for-specific-funds/horizon-2020/

Huang, X., Wu, B., Li, J., Shang, Y., Chen, W., Nie, X., and Gui, R. (2019). Anti-tumour effects of red blood cell membrane-camouflaged black phosphorous quantum dots combined with chemotherapy and anti-inflammatory therapy. *Artif. Cells Nanomed. Biotechnol.*, 47, 968–979.

Iacob, A. T., Lupascu, F. G., Apotrosoaei, M., Vasincu, I. M., Tauser, R. G., Lupascu, D., Giusca, S. E., Irina-Draga Caruntu, I.-D., and Profire, L. (2021), Recent biomedical approaches for chitosan based materials as drug delivery nanocarriers, *Pharmaceutics*, 2021(13), 587.

Jain, A. K., and Thareja, S. (2019), In vitro and in vivo characterization of pharmaceutical nanocarriers used for drug delivery. *Artif. CellsNanomed. Biotechnol.*, 47, 524–539.

Jia, X., and Wei, F. (2017). Advances in production and applications of carbon nanotubes. *Top. Curr. Chem.*, 375, 18.

Jian, W., Hui, D., and Lau, D. (2020), Nanoengineering in biomedicine: Current development and future perspectives. *Nanotechnology Reviews*, 9, 700–715.

Kalangi, S. K., Swarnakar, N. K., Sathyavathi, R., Rao, D. N., Jain, S., and Reddanna, P. (2018), Synthesis, characterization, and biodistribution of quantum Dot-Celecoxib conjugate in mouse paw edema model. *Oxid. Med. Cell Longev*, 2018, 3090517.

Katsuki, S., Matoba, T., Koga, J. I., Nakano, K., and Egashira, K. (2017), Anti-inflammatory nanomedicine for cardiovascular disease. *Front. Cardiovasc. Med.*, 4, 87.

Kaur, A., Nigam, K., Srivastava, S., Tyagi, A., and Dang, S. (2020), Memantine nanoemulsion: A new approach to treat Alzheimer's disease. *J. Microencapsul.*, 37, 355–365.

Kaur, I., Ellis, L. J., Romer, I., Tantra, R., Carriere, M., Allard, S., Mayne-L'Hermite, M., Minelli, C., Unger, W., Potthoff, A., Rades, S., and Valsami-Jones, E. (2017), Dispersion of nanomaterials in aqueous media: Towards protocol optimization. *J. Vis Exp: JoVE* (130), 56074.

Khan, N. H., Mir, M., Ngowi, E. E., Zafar, U., Khakwani, M. M. A. K., Khattak, S., Zhai, Y.-K., Jiang, E.-S., Zheng, M., Duan, S.-F., Wei, J.-S., Wu, D.-D., and Ji, X.-Y. (2021), Nanomedicine: A promising way to manage Alzheimer's disease. *Front. Bioeng. Biotechnol.*, 2021.

Kumar, D., and Ahmad, S. (2017), Green intelligent nanomaterials by design (Using nanoparticulate/2D-materials Developments and Future Trends.

Kuskov, A. N., Kulikov, P. P., Goryachaya, A. V., Tzatzarakis, M. N., Docea, A. O., Velonia, K., Shtilman, M. I., and Tsatsakis, A. M. (2017), Amphiphilic poly-N-vinylpyrrolidone nanoparticles as carriers for non-steroidal, anti-inflammatory drugs: *In vitro* cytotoxicity and *in vivo* acute toxicity study. *Nanomedicine: Nanotechnology, Biology and Medicine*, 13(3), 1021–1030.

Lim, K., and Hamid, Z. A. A. (2018), Polymer nanoparticle carriers in drug delivery systems: Research trend. In *Applications of Nanocomposite Materials in Drug Delivery*, Inamuddin, Asiri, A. M., and Mohammad, A., (Eds.), Woodhead Publishing: Cambridge, UK, pp. 217–237.

Lin, C. H., Chen, C. H., Lin, Z. C., and Fang, J. Y. (2017), Recent advances in oral delivery of drugs and bioactive natural products using solid lipid nanoparticles as the carriers. *J. Food Drug Anal.*, 25, 219–234.

Lombardo, D., Kiselev, M. A., and Caccamo, M. T. (2019), Smart nanoparticles for drug delivery application: Development of versatile nanocarrier platforms in biotechnology and nanomedicine. *J. Nanomater.*, 2019. 10.1155/2019/3702518

Lushchak, O., Zayachkivska, A., and Vaiserman, A. (2018), Metallic nanoantioxidants as potential therapeutics for type 2 diabetes: A hypothetical background and translational perspectives. *Hindawi Oxidative Medicine and Cellular Longevity*, 2018, Article ID 3407375.

Ma, L., Manaenko, A., Ou, Y. B., Shao, A.-W., Yang, S.-X., and Zhang, J. H. (2017), Bosutinib attenuates inflammation via inhibiting salt-inducible kinases in experimental model of intracerebral haemorrhage on mice. *Stroke*, 48, 3108–3116.

Mansoor, S., Kondiah, P. P. D., Choonara, Y. E., and Pillay, V. (2019), Polymer-based nanoparticle strategies for insulin delivery. *Polymers*, 11, 1380.

Markus, J., Wang, D., Kim, Y. J., Ahn, S., Mathiyalagan, R., Wang, C., and Yang, D. C. (2017), Biosynthesis, characterization, and bioactivities evaluation of silver and gold nanoparticles mediated by the roots of chinese herbal angelica pubescens maxim. *Nanoscale Res. Lett.*, 12, 46.

Martins, J. P., das Neves, J., de la Fuente, M., Celia, C., Florindo, H., Günday-Türeli, N., Popat, A., Santos, J. L., Sousa, F., Schmid, R., Wolfram, J., Sarmento, B., and Santos, H. A. (2020), The solid progress of nanomedicine. *Drug Delivery and Translational Research*, 10, 726–729.

McMasters, J., Poh, S., Lin, J. B., and Panitch, A. (2017). Delivery of anti-inflammatory peptides from hollow PEGylated poly(NIPAM) nanoparticles reduces inflammation in an ex vivo osteoarthritis model. *J. Control Release*, 258, 161–170.

Md, S., Gan, S. Y., Haw, Y. H., Ho, C. L., Wong, S., and Choudhury, H. (2018), In vitro neuroprotective effects of naringenin nanoemulsion against β-amyloid toxicity through the regulation of amyloidogenesis and tau phosphorylation. *Int. J. Biol. Macromol.*, 118 (Pt A), 1211–1219.

Mir, M., Ahmed, N., and Rehman, A. U. (2017). Recent applications of PLGA based nanostructures in drug delivery. *Colloids Surf. B Biointerfaces*, 159, 217–231.

Montoto, S. S., Muraca, G., and Ruiz, M. S. (2020), Solid lipid nanoparticles for drug delivery: Pharmacological and biopharmaceutical aspects, *Front. Mol. Biosci.*, 7, 587997. 10.3389/fmolb.2020.587997

MR-01; Text available @ https://www.marketdataforecast.com/market-reports/nanomedicine-market

Nadaroglu, H., Onem, H., and Gungor, A. (2017), Green synthesis of Ce_2O_3 NPs and determination of its antioxidant activity, *IET Nanobiotechnology*, 11(4), 411–419.

Nirale, P., Paul, A., and Yadav, K. S. (2020), Nano emulsions for targeting the neurodegenerative diseases: Alzheimer's, Parkinson's and Prion's. *Life Sci.* 245, 117394.

Paliwal, R., Babu, R. J., and Palakurthi, S. (2014), Nanomedicine scale-up technologies: Feasibilities and challenges, *AAPS PharmSciTech*, 15(6), 1527–1534. 10.1208/s1224 9-014-0177-9

Patil, R. P., Pawara, D. D., Gudewar, C. S., and Tekade, A. R. (2019), Nanostructured Cubosomes in an in situ nasal gel system: An alternative approach for the controlled delivery of donepezil HCl to brain. *J. Liposome Res.*, 29, 264–273.

Pedone, D., Moglianetti, M., De Luca, E., Bardi, G., and Pompa, P. P. (2017). Platinum nanoparticles in nanobiomedicine, *Chem. Soc. Rev.*, 46(16), 4951–4975.

Pineda-Reyes, A. M., Delgado, M. H., de la Luz Zambrano-Zaragoza, M., Leyva-Gómez, G., Mendoza-Muñoz, N., and Quintanar-Guerrero, D. (2021), Implementation of the emulsification-diffusion method by solvent displacement for polystyrene nanoparticles prepared from recycled material, *RSC Adv.*, 11, 2226–2234.

Pinelli, F., Perale, G., and Rossi, F. (2020), Coating and functionalization strategies for nanogels and nanoparticles for selective drug delivery. *Gels*, 6, 6.

Piras, C. C., Fernández-Prieto, S., and De Borggraeve, W. M. (2019), Ball milling: A green technology for the preparation and functionalisation of nanocellulose derivatives. *Nanoscale Adv.*, 1, 937–947.

Prabha, K., Ghosh, P., Abdullah, S., Joseph, R. M., Krishnan, R., Rana, S. S., and Pradhan, R. C. (2021), Recent development, challenges, and prospects of extrusion technology. *Future Foods*, 3, 100019.

Qian, K., Stella, L., Jones, D. S., Andrews, G. P., Du, H., and Tian, Y. (2021), Drug-rich phases induced by amorphous solid dispersion: Arbitrary or intentional goal in oral drug delivery? *Pharmaceutics*, 13, 889.

Rinaldi, F., Hanieh, P. N., Chan, L., Angeloni, L., Passeri, D., Rossi, M., Wang, J. T., Imbriano, A., Carafa, M., and Marianecci, C. (2018). Chitosan glutamate-coated niosomes: A Proposal for nose-to-brain delivery. *Pharmaceutics*, 10(2), 38.

Robb, C. T., Goepp, M., Rossi, A. G., and Yao, C. (2020), Non-Steroidal anti-inflammatory drugs, prostaglandins, and COVID-19. *British J. Pharmacology*, 177(21), Themed Issue: The Pharmacology of COVID-19, November 2020; 4899-920.

Roces, C. B., Lou, G., Jain, N., Abraham, S., Thomas, A., Halbert, G. W., and Perrie, Y. (2020), Manufacturing considerations for the development of lipid nanoparticles using microfluidics. *Pharmaceutics*, 12, 1095.

RR-01; Research Challenges for Integrated Systems Nanomanufacturing Report from the National Science Foundation Workshop, February 10–11, 2008, Jeffrey D. Morse (editor); text available @ https://coe.northeastern.edu/Research/nanophm/materialForDistribution/NMSWorkshopReport.pdf

Sadegh Malvajerd, S., Izadi, Z., Azadi, A., Kurd, M., Derakhshankhah, H., Sharifzadeh, M., Akbari Javar, H., and Hamidi, M. (2019), Neuroprotective potential of curcumin-loaded nanostructured lipid carrier in an animal model of Alzheimer's disease: Behavioural and biochemical evidence. *J. Alzheimers Dis.* 69, 671–686.

Sadegh Malvajerd, S., Azadi, A., Izadi, Z., Kurd, M., Dara, T., Dibaei, M., Sharif Zadeh, M., Akbari Javar, H., and Hamidi, M. (2018), Brain delivery of curcumin using solid lipid nanoparticles and nanostructured lipid carriers: Preparation, optimization, and pharmacokinetic evaluation. *ACS Chem. Neurosci.*, 10, 728–739.

Sameer Kumar, R., Shakambari, G., Ashokkumar, B., Nelson, D. J., John, S. A., and Varalakshmi, P. (2018), Nitrogen-doped graphene quantum dot-combined sodium 10-amino-2-methoxyundecanoate: Studies of proinflammatory gene expression and live cell imaging. *ACS Omega*, 3, 11982–11992.

Sathya, S., Shanmuganathan, B., and Devi, K. P. (2020), Deciphering the anti-apoptotic potential of α-bisabolol loaded solid lipid nanoparticles against Aβ induced neurotoxicity in Neuro-2a cells. *Colloids and Surf. B Biointerfaces*, 190, 110948.

Shao, A., Wu, H., Hong, Y., Tu, S., Sun, X., Wu, Q., Zhao, Q., Zhang, J., and Sheng, J. (2016). Hydrogen-rich saline attenuated subarachnoid hemorrhage-induced early brain injury in rats by suppressing inflammatory response: Possible involvement of NF-κB pathway and NLRP3 inflammasome. *Mol. Neurobiol.*, 53, 3462–3476.

Shi, M., Zhang, P., Zhao, Q., Shen, K., Qiu, Y., Xiao, Y., Yuan, Q., and Zhang, Y. (2020), Dual functional monocytes modulate bactericidal and anti-inflammation process for severe osteomyelitis treatment. *Small*, 16, e1905185.

Shi, Y., Xie, F., Rao, P., Qian, H., Chen, R., Chen, H., Li, D., Mu, D., Zhang, L., Lv, P., Shi, G., Zheng, L., and Liu, G. (2020), TRAIL-expressing cell membrane nanovesicles as an anti-inflammatory platform for rheumatoid arthritis therapy. *J. Control Release*, 320, 304–313.

Simos, Y. V., Spyrou, K., Patila, M., Karouta, N., Stamatis, H., Gournis, D., Dounousi, E., and Peschos, D. (2021), Trends of nanotechnology in type 2 diabetes mellitus treatment, *Asian J. Pharmaceutical Sciences*, 16(1), 62–76.

Singer, M., Deutschman, C. S., Seymour, C. W., Shankar-Hari, M., Annane, D., Bauer, M., Bellomo, R., Bernard, G. R., Chiche, J.-D., Coopersmith, C. M., Hotchkiss, R. S., Levy, M. M., Marshall, J. C., Martin, G. S., Opal, S. M., Rubenfeld, G. D., van der Poll, T., Vincent, J.-L., and Angus, D. C. (2016). The third international consensus definitions for sepsis and septic shock (sepsis-3). *Jama*, 315, 801–810.

Sintov, A. C. (2020), Amylolipid nanovesicles: A self-assembled lipid-modified starch hybrid system constructed for direct nose-to-brain delivery of curcumin. *Int. J. Pharm.*, 588, 119725.

Souto, G. D., Farhane, Z., Casey, A., Efeoglu, E., McIntyre, J., and Byrne, H. J. (2016), Evaluation of cytotoxicity profile and intracellular localisation of doxorubicin-loaded chitosan nanoparticles. *Anal. Bioanal. Chem.*, 408, 5443–5455.

Souto, E. B., Silva, G. F., Dias-Ferreira, J., Zielinska, A., Ventura, F., Durazzo, A., Lucarini, M., Novellino, E., and Santini, A. (2020), Nanopharmaceutics: Part I – Clinical trials legislation and good manufacturing practices (GMP) of nanotherapeutics in the EU. *Pharmaceutics.* 2020, 12.

Souto, E. B., Souto, S. B., Campos, J. R., Severino, P., Pashirova, T. N., Zakharova, L. Y., Silva, A. M., Durazzo, A., Lucarini, M., Izzo, A. A., Santini, A. (2019), Nanoparticle delivery systems in the treatment of diabetes complications. *Molecules*, 24, 4209.

Tenchov, R., Bird, R., Curtze, A. E., and Zhou, Q. (2021). Lipid nanoparticles – From liposomes to mRNA vaccine delivery, a landscape of research diversity and advancement. *ACS Nano*, 15, 16982–17015. 10.1021/acsnano.1c04996

Trujillo-Nolasco, R. M., Morales-Avila, E., Ocampo-Garcia, B. E., Ferro-Flores, G., Gibbens-Bandala, B. V., Escudero-Castellanos, A., and Isaac-Olive, K. (2019). Preparation and in vitro evaluation of radiolabeled HA-PLGA nanoparticles as novel MTX delivery system for local treatment of rheumatoid arthritis. *Mater. Sci. Eng. C Mater. Biol. Appl.*, 103, 109766.

Ulbrich, K., Holá, K., Šubr, V., Bakandritsos, A., Tuček, J., and Zbořil, R. (2016), Targeted drug delivery with polymers and magnetic nanoparticles: Covalent and Noncovalent approaches, release control, and clinical studies. *Chem. Rev.*, 116(9), 5338–5431.

Vasile, C. (2018), *Polymeric Nanomaterials in Nanotherapeutics*. Elsevier: London.

Vimala, K., and Soundarapandian, K. (2017), Erbitux conjugated zinc oxide nanoparticles to enhance antitumor efficiency via targeted drug delivery system for breast cancer therapy. *Ann. Oncol.*, 28, 658.

Wakaskar, R. (2017), Types of nanocarriers–Formulation method and applications. *J. Bioequiv. Availab.*, 9, 10000e10077.

Wang, H., Deng, H., Gao, M., and Zhang, W. (2021), Self-assembled nanogels based on ionic gelation of natural Polysaccharides for drug delivery, *Front. Bioeng. Biotechnol.*, 10.3389/fbioe.2021.703559

Wang, Y., Li, P., Truong-Dinh Tran, T., Zhang, J., and Kong, L. (2016), Manufacturing techniques and surface engineering of polymer based nanoparticles for targeted drug delivery to cancer. *Nanomaterials*, 6, 26.

Wang, C., Rosbottom, I., Turner, T. D., Laing, S., Maloney, A. G. P., Sheikh, A. Y., Docherty, R., Yin, Q., and Roberts, K. J. (2021), Molecular, solid-state and surface structures of the Conformational polymorphic forms of Ritonavir in relation to their physicochemical properties, *Pharmaceutical Research*, 38, 971–990.

Wang, H., Zhou, Y., Sun, Q., Zhou, C., Hu, S., Lenahan, C., Xu, W., Deng, Y., Li, G., and Tao, S. (2021), Update on nanoparticle-based drug delivery system for anti-inflammatory treatment. *Front. Bioeng. Biotechnol.*, 9, 630352. 10.3389/fbioe.2021.630352.

WP-01; National Nanotechnology Initiatives; https://www.nano.gov/; www.nap.edu/read/23603/chapter/3; www.nap.edu/read/23603/chapter/5

WP-02; National Nanomanufacturing Network (NNN); www.internano.org/nnn

WP-03; The Nanomedicine Translation Hub, ETPN; https://etp-nanomedicine.eu/about-etpn/nanomedicine-translation-hub/

WP-04; Healthcare Nanotechnology (Nanomedicine) Market – Growth, Trends, COVID-19 Impact, and forecasts (2021–2026), text @ www.mordorintelligence.com/industry-reports/healthcarenanotechnology-nanomedicinemarket#:~:text=Market%20Overview, forecast%20period% 2C% 202021%2D2026

Wu, Y., Sun, M., Wang, D., Li, G., Huang, J., Tan, S., Bao, L., Li, Q., Li, G., and Si, L. (2019), A PepT1 mediated medicinal nano-system for targeted delivery of cyclosporine A to alleviate acute severe ulcerative colitis. *Biomater. Sci.*, 7, 4299–4309.

Wu, G., Zhang, J., Zhao, Q., Zhuang, W., Ding, J., Zhang, C., Gao, H., Pang, D. W., Pu, K., and Xie, H. Y. (2020), Molecularly engineered macrophage-derived exosomes with inflammation tropism and intrinsic heme biosynthesis for atherosclerosis treatment. *Angew. Chem. Int. Ed. Engl.*, 59, 4068–4074.

Xu, Y. B., Chen, G. L., and Guo, M. Q. (2019), Antioxidant and anti-inflammatory activities of the crude extracts of *Moringa oleifera* from Kenya and their correlations with flavonoids. *Antioxidants (Basel)*, 8(8), 296.

Xu, S., Lu, J., Shao, A., Zhang, J. H., and Zhang, J. (2020), Glial cells: Role of the immune response in ischemic stroke. *Front. Immunol.* 11, 294.

Yan, F., Zhong, Z., Wang, Y., Feng, Y., Mei, Z., Li, H., Chen, X., Cai, L., and Li, C. (2020), Exosome-based biomimetic nanoparticles targeted to inflamed joints for enhanced treatment of rheumatoid arthritis. *J. Nanobiotechnol.*, 18, 115.

Yang, C., and Merlin, D. (2019), Nanoparticle-mediated drug delivery systems for the treatment of IBD: Current perspectives. *Int. J. Nanomed.*, 14, 8875–8889.

Yao, Y., Zhou, Y., Liu, L., Xu, Y., Chen, Q., Wang, Y., Wu, S., Deng, Y., Zhang, J., and Shao, A. (2020), Nanoparticle-based drug delivery in cancer therapy and its role in overcoming drug resistance. *Front. Mol. Biosci.*, 7, 193.

Yasir, M., Chauhan, I., Haji, M. J., Abdurazak, J., and Saxena, K. (2018), Formulation and evaluation of glyceryl behenate based solid lipid nanoparticles for the delivery of donepezil to brain through nasal route. *Res. J. Pharm. Technol.*, 11, 2836–2844.

Yasir, M., Sara, U. V. S., Chauhan, I., Gaur, P., Singh, A. P., and Puri, D. (2018). Solid lipid nanoparticles for nose to brain delivery of donepezil: Formulation, optimization by box–Behnken design, in vitro and in vivo evaluation. *Artif. Cells Nanomed. Biotechnol.*, 46, 1838–1851.

Yavarpour-Bali, H., Ghasemi-Kasman, M., and Pirzadeh, M. (2019), Curcumin-loaded nanoparticles: A novel therapeutic strategy in treatment of central nervous system disorders. *Int. J. Nanomed.*, 14, 4449.

Yazeji, T., Moulari, B., Beduneau, A., Stein, V., Dietrich, D., Pellequer, Y., and Lamprecht, A. (2017), Nanoparticle-based delivery enhances anti-inflammatory effect of low molecular weight heparin in experimental ulcerative colitis. *Drug Deliv.*, 24, 811–817.

Yetisgin, A. A., Cetinel, S., Zuvin, M., Kosar, A., and Kutlu, O. (2020), Therapeutic nanoparticles and their targeted delivery applications, *Molecules*, 25, 2193.

Zhang, C. Y., Gao, J., and Wang, Z. (2018), Bioresponsive nanoparticles targeted to infectious microenvironments for sepsis management. *Adv. Mater.* 30, e1803618.

Zieli'nska, A., Carreiró, F., Oliveira, A. M., Neves, A., Pires, B., Venkatesh, D. N., Durazzo, Al., Lucarini, M., Piotr Eder, P., Silva, A. M., Santini, A., and Souto, E. B. (2020), Polymeric nanoparticles: Production, characterization, toxicology and ecotoxicology. *Molecules*, 25, 3731; 10.3390/molecules25163731

Zijlstra, E., Heinemann, L., and Plum-Mörschel, L. (2014), Oral insulin reloaded: A structured approach. *J. Diabetes Sci. Technol.*, 8, 458–465.

8 Experimental Investigation on Spark Behaviour of ECDM for Potential Application in Nanofabrication

Girija Nandan Arka, Mamata Kumari, and Subhash Singh

CONTENTS

8.1 Introduction to ECDM 137
8.2 ECDM Mechanism... 139
8.3 Formation of Gas Film Layer... 141
8.4 Technical Parameters for Experimentation 142
8.5 Result and Discussion 143
8.6 Conclusion 145
References..146

8.1 INTRODUCTION TO ECDM

Electro Chemical Spark generation (discharge) Machining has been developed by combining the two non-conventional (non-traditional) machining one is electro discharge (spark) machining (EDM) process and second is electrochemical machining (ECM). In (ECDM) electrochemical discharge (spark generation) machining process material is removed by two activity first is electric spark energy is responsible for material removal and in second machining is done due to dissolution of material electrochemically between cathode electrode and anode (auxiliary electrode) [1]. This ECDM process is commonly known as electrochemical spark generation machining and sparks generated chemically erosion process. The electrochemical discharge (spark generation) machining (ECDM) is widely applied for conducting as well as non-conducting material for creating the micro holes also micro passage (channel) [2]. The non-traditional machining processes operate by using various forms of energies such as mechanical, thermal, chemical and their combination for the removing the material. The laser beam machining as well as electric discharge (spark generation) machining both comes under thermal (heat)

DOI: 10.1201/9781003220602-8

137

energy-based process, in laser beam machining thermal energy is used performing machining operation on ceramics, metals and polymers. But this machining takes high maintenance cost and expensive equipment, and their industrial application is also limited because of occurrence of heat-affected zone [3]. Electric discharge (spark generation) machining process also takes thermal energy for performing machining operations for micro features on conductive materials only. This electric discharge machining process is heat or thermal assisted erosion operation in this these two-tool electrode's temperature increases due to heat or thermal form of energy till the maximum 20,000 in centigrade [4]. Spark occurs in a gap between tool and work material kept in the dielectric solution, melting and vaporization takes place simultaneously in a very less space between two electrodes. Researchers explained that material which is removing and wearing rate of tool depends on the quality of electrolyte has been used, two forms of dielectric liquid and gaseous form can be used [5]. USM and AJM use mechanical energy for producing complex shape micro profile on hard to cut and brittle materials [6–8], But there are limited applications due to their incapability to machine ductile material. More complicated profile on most of the work material with their better surface quality can be developed by chemical energy-based machining processes but lower speed of their processing and worst dimensional accuracy is the limitation of chemical processes [9]. Now there was a requirement for a technique that can produce micro holes on different types of materials, irrespective of materials strength, hardness and conductivity. Thus, electrochemical discharge (spark generation) machining has been developed by KURA FUJI in 1968s [10]. Kura fuji used the ECDM for drilling on glass material [11]. Coteata et al. applied electrochemical discharge machining on steel material and obtained small diameter holes [12]. Author Huang and others applied electrochemical discharge (spark generation) machining by using work material as stainless steel and taking more rotational speed of tool for creating the micro holes [13]. Authors Pawar et al. observed that many authors applied electrochemical discharge (spark generation) machining on graphite for this they used tool electrode made up of tungsten carbide and for electrolysis they used NaOH electrolyte [14]. By applying ECDM drilling process on mosaic ceramic material MRR and Tool wear rate were examined and Brass is taken as cathode tool material. In drilling operation thermal cutting significance is a basic characteristic [15]. To analyze, simulate, and predict drilling cutting the finite element model is used results from input process parameters [16]. V. Jain et al. applied abrasive drilling by using electrochemically spark-assisted process in borosilicate glass using cutting tool of abrasive. This process increased machined depth and improves material removal by using abrasive cutting tools as compared to conventional cutting tool, machining performance also increase with increase in supply voltage [17]. L. Paul et al. Performed ECDM on glass of borosilicate and observed the results of process parameters. The results show that wearing rate of tool (TWR) increases by increasing the value of voltage and TWR decreases by decreasing the percentage of electrolyte in the solvent. Removal rate of material (MRR) increases by increasing the duty factor, percentage of electrolyte concentration and voltage, but Radial Overcut (ROC) also increases with an increase in voltage, percentage of electrolyte concentration and duty factor [18]. Electrochemical discharge (spark generation) machining process (ECDM) have huge scope in application for performing machining operation ceramic materials that have property of non-conductivity such as silicon nitrides, aluminium oxides, zirconium oxides, etc. Such nature of non-conductivity of engineering ceramic materials has many industrial (technical) applications in computer field parts, electronic devices, electrical and thermal insulators, bearings, artificial joints, cutting tools, aerospace components, etc (Doloi et al., 1999) [19]. There is an urgent necessity of machining ceramics in modern industries for manufacturing engineers because these advanced ceramics possess high compressive strength, high thermal shock resistance, good creep resistance, high-temperature resistance, high corrosion resistance, high compressive strength. These advanced ceramics have many engineering applications in the field of electrical, machine tool, electronics and aerospace. Application of materials with non-conductive nature such as silica, ceramics, glass and quartz, etc. in micro-electro-mechanical-system (MEMS) packaging [20], Integrated circuit (ICs) packaging [21] and through glass via interconnects related application are growing continuously due to their property of higher thermal stability, chemically non-reactive and electrically non-conductive nature. Micro machining is needed for these materials for any industrial application in cost-effective manner. Various techniques are available for machining micro features in these non-conducting materials such as abrasive jet machining (AJM), laser drilling, ultrasonic machining (USM) and plasma etching but these techniques have some limitations. Brittleness and hardness property of these non-conductive materials are a problem to machine with high efficiency, high accuracy and high reliability. To overcome these limitations and for machining of micro level holes and micro channels in brittle materials with non-conductive nature electrochemical discharge machining (ECDM) is a best alternative micromachining method [22]. In electrical discharge (spark generation) machining (EDM) spark is induced in conductive metals for removing process of material, between the work material and cathode tool electrode spark erosion frequently occurs [23,24], but some unexpected surface defects occur such as heat affected layers, micro cracks and recast layers because electro discharge machining (EDM) is an electrothermal process [25]. Various hard and difficult to machine materials can be frequently used in electrochemical machining, electro-chemical dissolution reactions occur in ECM for material removal so no surface defects and no tool wear occurs [26] but electrochemical machining has lower machining efficiency than electro discharge machining. Recently the combined effect of electro discharge (spark generation) machining (EDM) and electro-chemical machining (ECM) called electrochemical discharge machining (ECDM) has attracted research interest.

8.2   ECDM MECHANISM

In electrochemical discharge (spark generation) machining there are two electrodes one is tool electrode which is cathode and other is auxiliary electrode called anode both electrodes immersed in the solution of electrolyte like NaoH or KoH. The workpiece is dipped in the electrolyte below the tool electrode. Tip of the tool (cathode) electrode is dipped in the electrolytic solution few millimetres and auxiliary (anode) electrode which is kept a few centimeters far away from cathode tool electrode. A voltage supply or pulsed form of voltage is used between tool electrode (cathode) and auxiliary tool (anode) electrode which is anode this combination formed an electrochemical circuit (cell) (ECC). Electrolysis process starts due to applied voltage (potential difference lower than critical voltage) in the electrochemical cell across the cathode tool electrode and auxiliary electrode and consequently gas bubbles of hydrogen are generated at the tool electrode (cathode) and bubbles of oxygen are generated at auxiliary (anode) electrode. By increasing the terminal voltage, gas bubbles formation increases and mean radius of the gas bubbles also increases because of increase in current density. As a result of this layer of gas, bubble comes together and grows vicinage of tool electrode (cathode). As the applied voltage exceeds the critical limit voltage hydrogen gas bubbles generation rate increases near the tool electrode, then around the cathode tool electrode hydrogen gas film is formed by coalesce of hydrogen gas bubbles with each other [27]. This hydrogen film between the electrolyte and tool electrode acts as a dielectric medium and behaves like an insulation body around the cathode tool electrode and an adequate resistance generates between the electrodes to create high potential difference, then electrical discharges occur between electrolytic solution and tool (cathode) electrode as the given voltage (potential difference) for a given electrolyte-cathode tool electrode arrangement becomes high enough. On the workpiece surface, a large number of electrons are bombarded by the cathode tool electrode in the discharge zone which is kept closer to the cathode tool electrode and then because of bombardment of electrons temperature of the workpiece rises. Material removal starts by placing the workpiece in the sparking zone by melting, vaporization and due to thermal erosion by the spark generation effect and heating process. The typical process illustrated graphically in Figure 8.1 presenting each step marked from (a) to (e) for better understanding. Thus, it is significant to study the gas film formation dynamics to improve potential of machines.

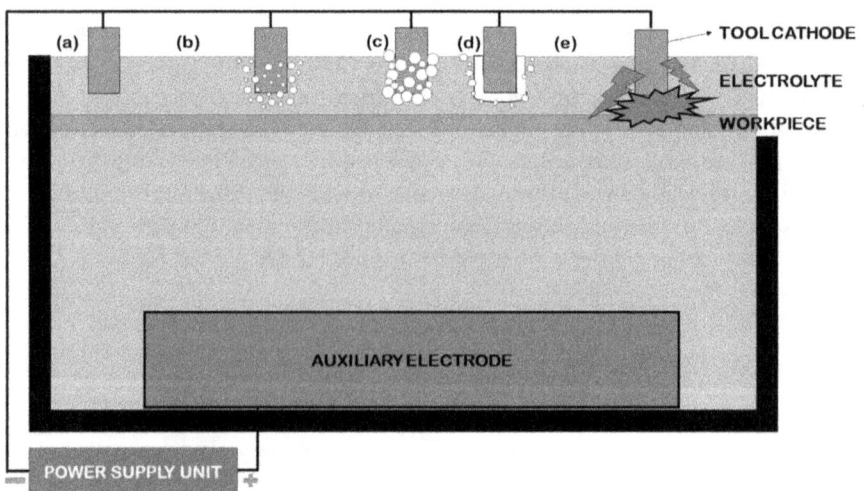

FIGURE 8.1 Technical stages of forming spark.

8.3   FORMATION OF GAS FILM LAYER

In electrochemical discharge (spark generation) machining (ECDM) process mechanism of material removal mostly because of creation of hydrogen gas bubble layer. Formation of gas layer (film) illustrated systematically in Figure 8.1.

There are many working conditions on which viscosity of the hydrogen gas layer (film) depends such as percentage of electrolyte, quality of electrolyte, electrolyte temperature, shape of the cathode tool electrode, etc. Visual observation of the gas film formation is quite difficult but with the help of high-speed cameras, visual observation of size and thickness of the gas film can be obtained. In an electrochemical discharge machining process, there are several parameters which influence the gas film quality such as tool immersion depth, electrolyte resistivity, thermo capillary flow between electrolyte and cathode tool electrode because of temperature gradients, current density, cathode tool radius and force of electrostatic attraction between cathode tool electrode and hydrogen gas bubbles.

Authors Han et al. observed that by increasing the diameter of the tool (cathode) electrode between 50 to 300 in dimension micrometre as well as depth of tooltip inside electrolyte is 8 to 303 μm, gas film thickness changes from 3 to 45 μm [28]. Guelcher observed that formation of hydrogen gas layer (film) originates because of movement of hydrogen bubbles towards each other which is caused by thermos capillary flow [29].

The gas film formed in electrochemical discharge (spark generation) machining process fluctuates constantly and is not stable that decreases the machining reproducibility. Stable gas film shows an important role in micromachining, but stability of gas film is very difficult. Fluctuation of hydrogen gas layer (film) can be reduced by minimizing the thickness of hydrogen gas layer (film). There are many ways to control the thickness of gas film, one is applying pulsed voltage, by applying pulsed voltage there is decrement in thermal disturbance of created micro dimension hole and groove machining overcutting also reduces this is experimentally proved [30]. Heating zone area reduces in this way because cooling occurs during pulse-off time. Thickness of the gas film can also be reduced in another way which is by increasing the concentration of electrolyte so that electrolysis rate can be enhanced so greater quantity of (OH) ions is produced that reduces the surface tension and then thinner gas film and better machining accuracy is achieved [31].

An electrolyte is a chemical compound or substance when dissolved in a solvent like water an electrically conductive solution is produced. Electrolyte is separated into cations which are positive and anions which are negative. Initially the electrolytic solution is electrically neutral, when potential difference is applied to the solution electrode that has excess of electrons draws cations of the solution, and electrode that has lack of electrons draws anions of the solution. Motion of the cations and anions within the solution are opposite to each other. Strong acids are strong electrolytes like HCL, HNO_3, H_2SO_4 and strong alkaline like NaoH, KoH are strong electrolytes and strong salts like Nacl, K_2SO_4 are strong electrolytes. When salt (a solid) is mixed with water electrolytic solutions are formed and dissociation reaction occurs.

$$Nacl_{(s)} = Na^+_{(aq)} + Cl^-_{(aq)S} \qquad (8.1)$$

If an electrolytic solution is highly concentrated that means high proportion of solute dissolved in water, and if electrolyte is weak low proportion of solute has mixed in the water. In electrolysis process cathode electrode and auxiliary electrode are connected by the power connection with negative terminal and positive terminal respectively and both electrodes are placed in an electrolyte. When potential difference is applied electrolyte will allow passing the electricity. A reaction occurs at the cathode (if sodium chloride, Nacl is mixed in water) so hydrogen gas will be generated.

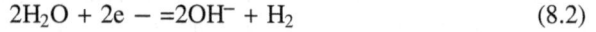

$$2H_2O + 2e - = 2OH^- + H_2 \qquad (8.2)$$

And reaction at the anode is

$$2Nacl = 2Na^+ + Cl_2 + 2e - . \qquad (8.3)$$

So, chlorine gas will be generated.

8.4   TECHNICAL PARAMETERS FOR EXPERIMENTATION

Machining performance can be identified by its rate of material removal, quality of surface it produces and machined part tolerance. Figure. 8.2 represents the possible influential parameter which contributes eminently to increase efficacy of ECDM. Therefore, the necessary combination was taken for improvisation.

In electrochemical spark generation (discharge) machining process there are two tool (electrodes) one is tool electrode (cathode) that is made up of bronze material shown in Figure 8.3 and other is auxiliary electrode made of stainless steel, these two electrodes are put into electrolytic solution of the NaOH and this electrolyte is filled in a container which is made of acrylic sheet, acrylic sheet has been used because of its several physical properties. Acrylic has the property of more resistant to impact than glass, it can withstand the change in temperature, salty water, cold weather (it has better weather resistant property), and other critical conditions. It is formed by rigid thermoplastic material which is lightweight. Work material is kept

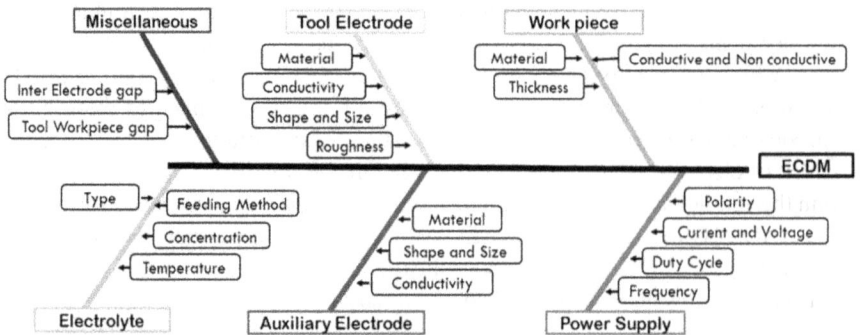

FIGURE 8.2   Cause and effect diagram of ECDM.

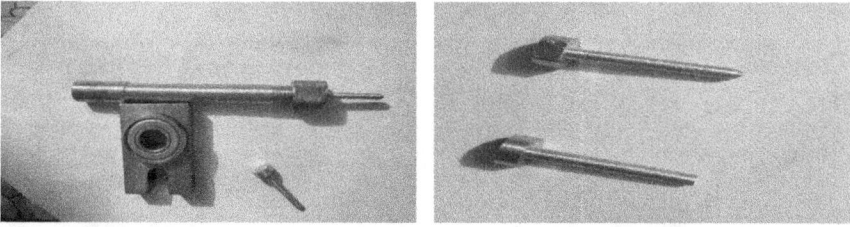

FIGURE 8.3 Left picture tool holder assembly and right picture illustrate bronze tool indigenously fabricated.

below the cathode tool electrode and stable on the holding fixture and this holding fixture is kept at the bottom of the electrolyte container of the Electrochemical discharge (spark generation) machining process. In this set up gravity feed mechanism has been applied by which when initially material is removed from the work-piece then fixture on which workpiece is placed moves upward because of gravity mechanism because counterweight has been provided on the side of holding fixture. A compound slide has been employed in this set up to control the X-Y movement of the workpiece which is fixed on the table. Cathode tool electrode is attached with compound slide for controlling the motion of cathode tool electrode. Shaft of the stepper motor gets attached with cathode tool electrode and stepper motor is coupled with the micro-step drive and the micro-step drive also requires power to rotate the stepper motor. Speed of the stepper motor is monitored by Arduino Uno board which is also connected with micro-step drive. In Arduino board program is uploaded from the computer with the help of USB cable and computer also provides power to the Arduino board to become activate. To control the revolution per minute (RPM) we have used potentiometer in this set up of ECDM. Negative connection of the power supply is attached with the cathode tool electrode and positive connection of the power supply is attached with auxiliary tool (anode) electrode. In this machining operation, NaOH and KoH have been used as electrolytes and entire machining has performed into the container, there is a drainage system at the lower portion of the container to remove the electrolyte after one experiment. It has been decided that for experiment value of voltage will be studied in three-stage 60 V, 70 V and 90 V.

8.5 RESULT AND DISCUSSION

In this work, a trial experiment has been done in the electrochemical discharge machining (ECDM) process. In this trial experiment, significant parameter is voltage. All researchers have observed that discharge is responsible for material removal and generation of discharge is depending on different-different voltage values. This trial phase of experiment focused on three values of the voltage and sparking activity on that voltage.

    a. Voltage is 60 V.
    b. Voltage is 70 V.
    c. Voltage is 90 V.

(a)

(b)

(c)

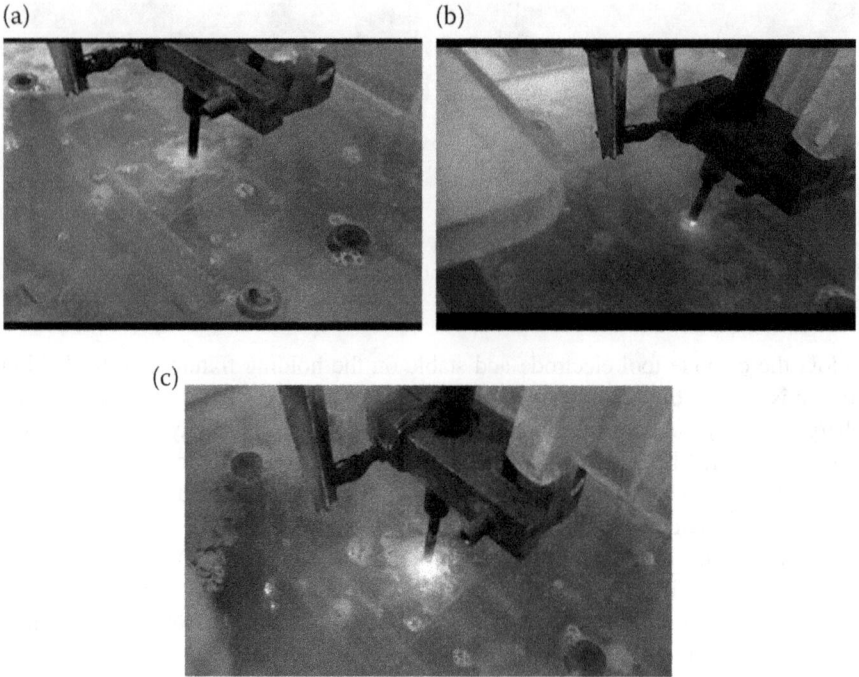

FIGURE 8.4 (a) Bubbles formation at less than 70 V, (b) discharge generation when value of voltage is 70 V, (c) discharge behaviour when value of voltage is 90 V.

Figure 8.4 shows the formation of hydrogen gas bubbles but no discharge generation when value of voltage is less than 70 V. At the initial stage when setup is ready cathode tool electrode and auxiliary electrode are connected with power here negative terminal of power supply is attached with cathode tool electrode and positive terminal of power is attached with auxiliary electrode. These two electrodes are kept in a working container in which electrolyte is filled and below the cathode tool electrode work material is placed. In this setup auxiliary tool (anode) electrode is entirely dipped in the electrolytic solution but only certain tip of tool (cathode) electrode is immersed in the electrolytic solution, complete cathode electrode does not fill in the electrolyte this is the necessary condition for spark generation. If cathode electrode is dipped more than certain limit then spark is not produced at the tip of cathode electrode but side sparking is produced this is the problem of radial overcut. When voltage is supplied then electrolysis process starts and value of voltage is less than 70 V. Electrolyte is completely dissolved in water so when voltage is supplied due to electrolysis process hydrogen ions are attracted to the cathode and negative hydroxide ions are attracted to auxiliary electrode these hydrogen ions gain electrons from cathode tool electrode and generate hydrogen gas. These hydrogen gas bubbles collected at the cathode tool electrode and when value of voltage is increased rate of hydrogen gas bubbles are increased and these bubbles combined together and form hydrogen gas film this gas film is responsible for spark generation at this stage value of voltage is less than 70 V so no spark generates because voltage value less than 70 V is not the critical voltage for this electrolyte solution. Critical voltage is that voltage on which hydrogen gas film breaks its dielectric strength and generates spark so below 70 V there is no discharge so no material removal occurs.

This means when voltage is less than 70 V this voltage is not sufficient for discharge generation so at this voltage there is no material removal. In Figure 8.4(b), discharge activity has been shown when the value of voltage is 70 V at this voltage material is heated due to sparking effect and very less material is removed. When value of voltage is 70 V spark generation has been observed means this voltage is critical voltage, at this voltage hydrogen gas film break its dielectric strength and spark is produced. Due to spark generation material gets heated since at this voltage small amount of spark generation occurs so only heating effect of material has been observed but there is no material removal that occurs. Due to this sparking effect tip of cathode tool electrode was also heated and some amount of wear has been observed. In this work, NaOH has been used as electrolyte which is completely dissolved in the water. Rotation (rpm) of tool electrode (cathode) is constant which is attached with the shaft of the stepper motor and with the help of programming fed in the Arduino Uno board this stepper motor is brought in rotating stage. Potentiometer has been for controlling the speed of stepper motor. So, it has been observed that by increasing the rotating speed of tool electrode removal rate of material can be improved. It has been observed that for spark generation electrolysis process should be fast. So, for increasing the rate of electrolysis percentage of electrolyte should be more.

Figure 8.4(c) shows more spark generation when value of voltage is 90 V. At this voltage due to more sparking some amount of material removal has been obtained but at this stage when more spark has been obtained wear rate of tool has also increased. When value of voltage is increased then rate of spark generation is increased. Due to this spark generation work material is heated and material is removed. Due to this spark, some overcut occurs because electrolytic solution does not reach at the tip of tool electrode and side sparking occurs, for producing the deep holes electrolyte should reach at the tip of tool electrode so that more material removal should occur. For this purpose, we can provide some planetary motion of cathode tool electrode so that electrolytic solution can get very less but enough space to reach at the tip of tool electrode. By doing this we can produce deep holes and macro-level machining can be done.

8.6   CONCLUSION

In this research, three sets of voltage parameters were taken into consideration for the study of discharge behaviour by incorporating a novel bronze tool. The experiment reveals generation of hydrogen gas bubbles only, below 70 V due to electrochemical reaction and dominance of buoyance force unable to coat the electrode. Once the voltage boundary cross 70 V barrier, a light orange colour discharge was observed ensuring breakdown of thin film coat around electrode. At 90 V a continuous spontaneous discharge observed due to rapid occurrence of gas bubble formation and coating over exposed surface area of tool electrode. Thus, by keeping the workpiece near to the discharge zone the material could be removed out at higher discharge energy. ECDM could be a promising choice to produce nano-manufacturing for the potential application.

REFERENCES

1. Pawar, Pravin, Raj Ballav, and Amaresh Kumar. "Measurement analysis in electro-chemical discharge machining (ECDM) process: A literature review." *Journal of Chemistry and Chemical Engineering* 9 (2015): 140–144.
2. Pawar, Pravin, Raj Ballav, and Amaresh Kumar. "Recent status of spark assisted chemical engraving: A review." *Journal of Chemical and Pharmaceutical Sciences* 10, no. 2 (2017): 1–6.
3. Shirk, M. D., and P. A. Molian. "A review of ultrashort pulsed laser ablation of materials." *Journal of Laser Applications* 10 (1998): 14.
4. Brinksmeier, Ekkard, Ralf Gläbe, Oltmann Riemer, and Sven Twardy. "Potentials of precision machining processes for the manufacture of micro forming molds." *Microsystem Technologies* 14, no. 12 (2008): 1983.
5. Piljek, Petar, Zdenka Keran, and Miljenko Math. "Micromachining–review of literature from 1980 to 2010." *Interdisciplinary Description of Complex Systems: INDECS* 12, no. 1 (2014): 1–27.
6. Liu, Qingyu, Qinhe Zhang, Guang Zhu, Kan Wang, Jianhua Zhang, and Chunjie Dong. "Effect of electrode size on the performances of micro-EDM." *Materials and Manufacturing Processes* 31, no. 4 (2016): 391–396.
7. Cheema, Manjot S., Akshay Dvivedi, and Apurbba K. Sharma. "Tool wear studies in fabrication of microchannels in ultrasonic micromachining." *Ultrasonics* 57 (2015): 57–64.
8. Yuvaraj, N., and M. Pradeep Kumar. "Multiresponse optimization of abrasive water jet cutting process parameters using TOPSIS approach." *Materials and Manufacturing Processes* 30, no. 7 (2015): 882–889.
9. Çakir, O., A. Yardimeden, and T. Ozben. "Chemical machining." *Archives of Materials Science and Engineering* 28, no. 8 (2007): 499–502.
10. Allesu, K., A. Ghosh, and M. K. Muju. "A preliminary qualitative approach of a proposed mechanism of material removal in electrical machining of glass." *European journal of mechanical engineering* 36, no. 3 (1991): 201–207.
11. Kurafuji, H. "Electrical discharge drilling of glass-I." *Annals of the CIRP* 16 (1968): 415.
12. Coteaţă, Margareta, Hans-Peter Schulze, and Laurenţiu Slătineanu. "Drilling of difficult-to-cut steel by electrochemical discharge machining." *Materials and Manufacturing Processes* 26, no. 12 (2011): 1466–1472.
13. Huang, S. F., Y. Liu, J. Li, H. X. Hu, and L. Y. Sun. "Electrochemical discharge machining micro-hole in stainless steel with tool electrode high-speed rotating." *Materials and Manufacturing Processes* 29, no. 5 (2014): 634–637.
14. Pawar, Pravin, Raj Ballav, and Amaresh Kumar. "Revolutionary developments in ECDM process: An overview." *Materials Today: Proceedings* 2, no. 4–5 (2015): 3188–3195.
15. Mourad, Abdelkrim, Brioua Mourad, Belloufi Abderrahim, and Brabie Gheorghe. "Numerical study of cutting temperature during drilling process of the C45 steel." *Optimization* 7, no. 8 (2016): 12.
16. Pawar, Pravin, Raj Ballav, and Amaresh Kumar. "Finite element method broach tool drilling analysis using explicit dynamics ansys." *International journal of Modern Manufacturing Technologies* 8, no. 2 (2016): 54–60.
17. Jain, V. K., S. K. Choudhury, and K. M. Ramesh. "On the machining of alumina and glass." *International Journal of Machine Tools and Manufacture* 42, no. 11 (2002): 1269–1276.
18. Paul, Lijo, and Somashekhar S. Hiremath. "Response surface modelling of micro holes in electrochemical discharge machining process." *Procedia Engineering* 64 (2013): 1395–1404.

19. Doloi, B., B. Bhattacharyya, and S. K. Sorkhel. "Electrochemical discharge machining of non-conducting ceramics." *Defence Science Journal* 49, no. 4 (1999): 331.
20. Sukumaran, Vijay, Tapobrata Bandyopadhyay, Venky Sundaram, and Rao Tummala. "Low-cost thin glass interposers as a superior alternative to silicon and organic interposers for packaging of 3-D ICs." *IEEE Transactions on Components, Packaging and Manufacturing Technology* 2, no. 9 (2012): 1426–1433.
21. Lee, Ju-Yong, Sung-Woo Lee, Seung-Ki Lee, and Jae-Hyoung Park. "Through-glass copper via using the glass reflow and seedless electroplating processes for wafer-level RF MEMS packaging." *Journal of Micromechanics and Microengineering* 23, no. 8 (2013): 085012.
22. Wüthrich, Rolf, and Valia Fascio. "Machining of non-conducting materials using electrochemical discharge phenomenon—An overview." *International Journal of Machine Tools and Manufacture* 45, no. 9 (2005): 1095–1108.
23. Ho, K. H., and S. T. Newman. "State of the art electrical discharge machining (EDM)." *International Journal of Machine Tools and Manufacture* 43, no. 13 (2003): 1287–1300.
24. Nguyen, Minh Dang, Mustafizur Rahman, and Yoke San Wong. "Simultaneous micro-EDM and micro-ECM in low-resistivity deionized water." *International Journal of Machine Tools and Manufacture* 54 (2012): 55–65.
25. Chen, Sung-Long, Ming-Hong Lin, Guo-Xin Huang, and Chia-Ching Wang. "Research of the recast layer on implant surface modified by micro-current electrical discharge machining using deionized water mixed with titanium powder as dielectric solvent." *Applied Surface Science* 311 (2014): 47–53.
26. Zhu, D., W. Wang, X. L. Fang, N. S. Qu, and Z. Y. Xu. "Electrochemical drilling of multiple holes with electrolyte-extraction." *CIRP Annals* 59, no. 1 (2010): 239–242.
27. Bhattacharyya, B., B. N. Doloi, and S. K. Sorkhel. "Experimental investigations into electrochemical discharge machining (ECDM) of non-conductive ceramic materials." *Journal of Materials Processing Technology* 95, no. 1–3 (1999): 145–154.
28. Han, Min-Seop, Byung-Kwon Min, and Sang Jo Lee. "Modeling gas film formation in electrochemical discharge machining processes using a side-insulated electrode." *Journal of Micromechanics and Microengineering* 18, no. 4 (2008): 045019.
29. Guelcher, Scott A., Yuri E. Solomentsev, Paul J. Sides, and John L. Anderson. "Thermocapillary phenomena and bubble coalescence during electrolytic gas evolution." *Journal of The Electrochemical Society* 145, no. 6 (1998): 1848.
30. Kim, Dae-Jin, Yoomin Ahn, Seoung-Hwan Lee, and Yong-Kweon Kim. "Voltage pulse frequency and duty ratio effects in an electrochemical discharge microdrilling process of Pyrex glass." *International Journal of Machine Tools and Manufacture* 46, no. 10 (2006): 1064–1067.
31. Cheng, Chih-Ping, Kun-Ling Wu, Chao-Chuang Mai, Cheng-Kuang Yang, Yu-Shan Hsu, and Biing-Hwa Yan. "Study of gas film quality in electrochemical discharge machining." *International Journal of Machine Tools and Manufacture* 50, no. 8 (2010): 689–697.

9 Frequency Sensitivity Performance Analysis of Single-Layer and Multi-Layer SAW-Based Sensor Using Finite Element Method

Baruna Kumar Turuk and Basudeba Behera

CONTENTS

9.1 Introduction...149
9.2 Generating Surface Acoustic Waves 151
9.3 Structure of Resonators and Delay Lines..153
9.4 Basic Configurations of SAW Devices 154
9.5 Finite Element Method (FEM) 155
9.6 Design and Simulation of the Proposed Structure....................................155
9.7 Result and Discussion 158
9.8 Measurement of Electric Potential..159
9.9 Measurement of Sensitivity 160
9.10 Conclusion 162
References..162

9.1 INTRODUCTION

COVID-19 has been declared a world health crisis and highly contagious syndrome (World Health Organization). The world health company (WHO) is operating carefully with international specialists, governments and partners to swiftly amplify clinical understanding of this unprecedented infection, to tune the unfolding and deadly nature of this disease, and to provide a recommendation to international locations and individuals on procedures to shield fitness and stop the virulence of this outbreak (https://www.un.org/en/coronavirus). Generally, the individuals who've COVID-19 disorder 2019 (COVID-19) become cured absolutely within some weeks. Nevertheless, even for individuals with minor disease disparity, some human beings reveal symptoms after their initial treatment. These individuals once in a while pronounce themselves as "lengthy haulers," The situations were mentioned as a publish-COVID-19 disorder or "long COVID-19." those fitness issues are often known as an after-COVID-19 syndrome. These are usually mentioned as aftermaths of COVID-19 that remain in the body for more than one month after you have been affected with the coronavirus. Aged persons and people with many critical clinical conditions are most likely to remain affected with the coronavirus syndrome; however, even younger, otherwise healthy human beings can remain sick for weeks to months after contamination.

The restoration and safety remedies of corona virus-inflamed patients are an entire lot crucial to keep away from the usual dispersion of this illness. Numerous unstable herbal compounds (VOCs) upward push up from the scientific respiratory device by providing respiration gases to the affected person. These organic compounds can affect the nervous system of COVID-19-infected patients [1]. The immunity process of the corona-virus-affected patient to cure this illness, a VOC detection sensor using the surface acoustic wave has been reported and realized. Smart sensors receive the signals from the environment and convert their inherent characteristics into a tested electrical signal. The data is collected from the temperature, mass, speed, pressure, or the presence of heat bodies like humans [2–4]. The electric signal is then forwarded and processed via a microprocessor to offer outputs related to a set of movements. This structure ultimately connects the output to the receivers within the desired gadgets for good working. The application of smart sensors is in the control system, motion sensors and biomedical applications like Light sensors and industry, etc. [5]. We can design several products consisting of resonators, actuators, detection of warfare agents and gas detectors. The advantage of MEMS sensors can be witnessed in the manufacturing, consequent in shallow unit costs when mass-produced, possess incredibly high sensitivity. MEMS switch and actuator can go up to very high frequencies, and MEMS devices provide insignificant power consumption [6–8]. The SAW-based sensors may be found from semiconductor process technology, which is also compatible with the CMOS technology [9]. We can detect any surface change on the piezoelectric materials if the SAW waves move through them. When volatile organic compounds are reacted with the SAW-based sensor's surface, then the electron movement in the device, as a result, changes the sensitivity of the device. In this work, polyisobutylene has been taken as an organic compound employed in many industrial uses. Polyisobutylene bio-transformation can generate carbon monoxide (CO) gas, resulting in haemoglobin in the blood cells when it enters the blood.

We measured the displacement and resonance frequency characteristic of the SAW-based sensor on changing the mass of the sensing layer during the interaction of gas molecules [7–10]. This type of sensor is preferred due to its high accuracy, low power consumption, portability, low sensitivity. The characteristic of the SAW sensors modifies with variation in pressure, temperature, humidity and quality gas concentration [11–15]. They are generating the surface acoustic wave piezoelectric materials ($LiNbO_3$, $LiTaO_3$ and quartz) are required, and the speed of the SAWs depends on the type of materials used for designing a sensor. In the proposed work, SAW sensors (Figure 9.1) can be implemented and tested for the DCM gas concentration in two ways that are single-layer and multi-layer ($SiN_4/PIB/LiNbO_3$). From the simulation, we concluded that multi-layer gets easily affected than the

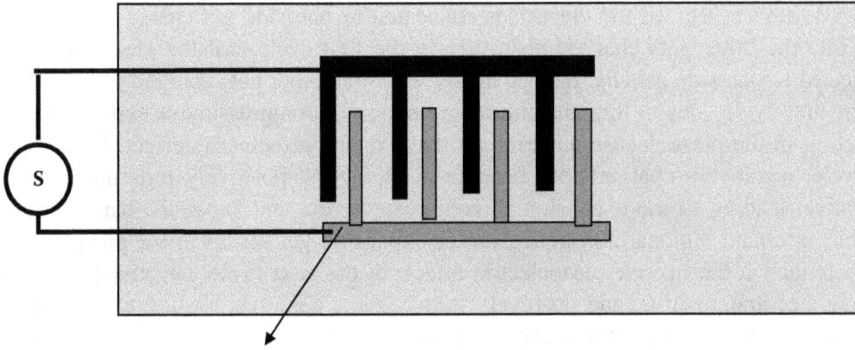

FIGURE 9.1 Shows the generation of SAW.

TABLE 9.1
Parameter List

Name	Expression	Description
P	1[atm]	The pressure of the air
T	25[°C]	The temperature of the air
C_0	100	DCM concentration in ppm
c_DCM_air	1e-6*c0*P/(Constant*T)	DCM concentration in air
M_DCM	84.93[g/mol]	Molar mass of DCM
K	10^1.4821	PIB/air partition constant for DCM
rho_DCM_PIB	K*M_DCM* c_DCM_air	Mass concentration of DCM in PIB
rho_PIB	0.918[g/cm^3]	Density of PIB
E_PIB	10[Gpa]	Young's modulus of pIB
nu_PIB	0.48	Poisson's ratio of PIB
eps_PIB	2.2	The relative permittivity of PIB
switch	0	Switch for adding DCM density
vr	3448[m/s]	velocity
width	4[μm]	Width of unit cell SAW
f0	Vr/width	Estimated SAW frequency
t_pIB	0.5[μM]	PIB thickness

single-layer sensor, which means that frequency sensitivity is higher during the variation of pressure, thickness and pressure [16,17]. Properties of the materials and factors affecting the materials have been parameterized, as shown in Table 9.1.

9.2 GENERATING SURFACE ACOUSTIC WAVES

To generate surface acoustic wave interdigital transducer (IDTs) of suitable metal needs to be fabricated on piezoelectric materials using lithography technique. On applying the electromagnetic field to the input of the interdigital transducer of the

SAW device [9]. All the electrodes connected to one side get charged negatively while the other gets charged positively in the first cycle. All the electrodes connected to one side get charged negatively while the other gets charged positively in the first cycle. Due to this, the alternating region of compression or expansion gets set up in the piezoelectric material via the inverse piezoelectric effect. In the next cycle, negatively charged rods become positive, and positively rods become negative, leading to the expansion of compressed parts and expanded parts. Due to this, alternate regions of compression or expansion get set up in the piezoelectric material via the inverse piezoelectric effect. In the next cycle, negatively charged rods become positive, and positively rods become negative, leading to the expansion of compressed parts and expanded parts. Since the polarity of the electric field is changing continuously, an acoustic wave got generated on the surface of piezoelectric material. SAW device can be used as a bandpass filter [10]. Rayleigh wave is a surface acoustic wave travelling on the solid; it can be produced in material piezoelectric transduction or localized impact. Rayleigh waves contain longitudinal and transverse motion, and the amplitude decreases exponentially as the distance from the surface [11].

Rayleigh waves can be used in various television, radio signal and radar applications. The primary importance of the Rayleigh wave is that for a particular frequency, the wavelength of the electromagnetic wave is 10^5 times greater than that of the wavelength of the Rayleigh wave (Figure 9.2) using the following relation $c = f \times \lambda$. The velocity of the electromagnetic wave is 3×10^8 m/s, whereas the velocity of the Rayleigh waves travels at an acoustic speed of a few kilometres per second [12]. If the Rayleigh wave transmits without causing serve attenuation, then the length of the transmitting device wavelength, because the electromagnetic wave of a given frequency is 105 times greater than the Rayleigh wave of the same frequency. Hence, transmit Rayleigh wave device can be made smaller than electromagnetic counterparts, which reduces the size and weight, important for application space vehicles and aircraft applications [13]. Another important property of the Rayleigh wave is non-dispersive, which means the speed of the Rayleigh is independent of frequency, hence for multi-frequency signals, a Rayleigh-wave device can be used without any change in the waveform [14].

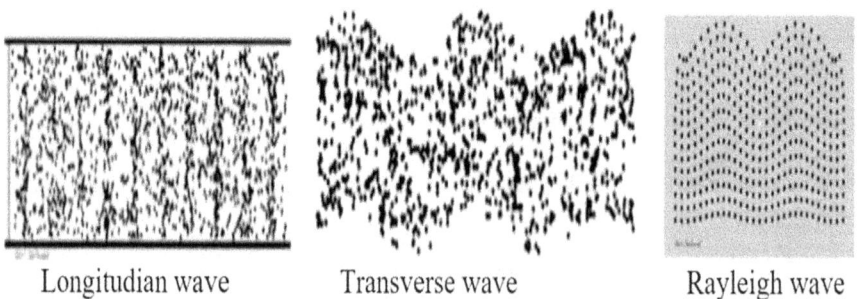

Longitudian wave   Transverse wave   Rayleigh wave

FIGURE 9.2   Generation of Rayleigh wave.

9.3 STRUCTURE OF RESONATORS AND DELAY LINES

The electrodes patterned with high conductive metals such as Aluminium, Platinum in a comb structure on the surface of a piezo substrate in a SAW device are known as an interdigital transducer (IDT). Aluminium (Al) is generally used to fabricate IDT having conducting and physical properties using photolithography. Figure 9.3(a) shows an IDT with bond pads for an electrical connection fabricated on a piezo substrate. The electrical potential applied to the IDT electrodes produces an electric field that will cut the piezo substrate, as illustrated in Figure 9.3(b), and generate stresses at the surface of the piezo substrate. Application of sinusoidal wave excitation results in a series of alternate compression and expansion regions propagating on the surface of the substrate and both sides of the IDT as SAW.

When a sinusoidal potential with period T is applied, the vibrations add positively if the centre-to-centre distance between IDT fingers p equals half of the acoustic wavelength λ for the excitation frequency as shown in Figure 9.3(a) [18]. Figure 9.3(b) illustrates the constructive generation of SAW on a piezo substrate. The stress wave generated at time t by a pair of IDT fingers travels a distance of half acoustic wavelength ($\lambda/2$) in the time interval of half period ($T/2$) with the speed of SAW phase velocity c. As shown in Figure 9.3(b), at time $t + T/2$, the generated stress wave reaches the neighbour IDT finger pair, where it adds constructively to the stress wave produced during the next half cycle of the input sinusoid [18].

Deformed waves generated by each finger pair constructively with the deformed waves generated by other finger pairs in the subsequent cycles of input excitation

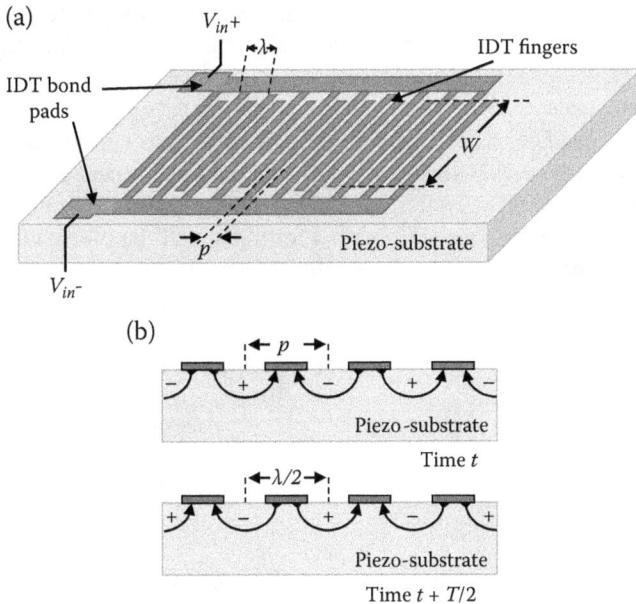

FIGURE 9.3 (a) Interdigital transducer (IDT) with bond pads on a piezo-substrate (b) Polarities of applied potentials on IDT fingers at time t and at time $t + T/2$.

resulting in resonance. The resonance frequency or the synchronous frequency f_0 is related to the pitch p of the IDT fingers and SAW phase velocity c as given in Equation (9.1) and Equation (9.2).

$$f_0 = \frac{c}{\lambda} \tag{9.1}$$

$$\lambda = 2p \tag{9.2}$$

where wavelength is denoted by λ, the number of finger pairs of the IDTs determines the operating bandwidth of the SAW device. The 3 dB bandwidth f_b of a SAW device with N number of IDT finger pairs can be estimated [19] as given in Equation (9.3).

$$f_b = \frac{0.88 f_0}{N} \tag{9.3}$$

9.4 BASIC CONFIGURATIONS OF SAW DEVICES

The SAW device can be used in two ways: a resonator and a delay line. A SAW resonator consists of IDT and different layers having different properties. SAW resonators divide into one port resonator and two-port resonators. One port SAW resonator consists of one IDT separated by two reflectors, whereas the two-port resonator consists of two IDT and reflectors placed on both sides of the IDTs. One port resonator can be made without reflectors. The reflector may be shorted metal strip or grooves. When reflector electrode periodicity is equal to half of the wavelength, then Bragg frequency is generated. In this condition reflections coefficient of all the electrodes are added coherently [20]. SAW device has a strong reflection coefficient when $N|r| > 1$, where N is the number of electrode grating and r is the reflection coefficient. Generally, the reflection coefficient is 0.02, and N is 200 or more [20]. In a two-port SAW resonator, one IDT is used for input, and the other IDT is used for output. The two-port resonator is more stable as compared to the one-port resonator. A delay line consists of two IDTs: the transmitter, the receiver, and two absorbers are attached to two sides of the IDTs, shown in Figure 9.4. Both a few wavelengths separate the IDTs. When electrical excitation is given to the input of the interdigital

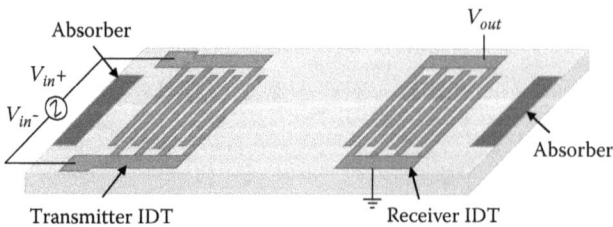

FIGURE 9.4   Schematic diagrams of the SAW delay line device [21].

Rayleigh wave propagation direction

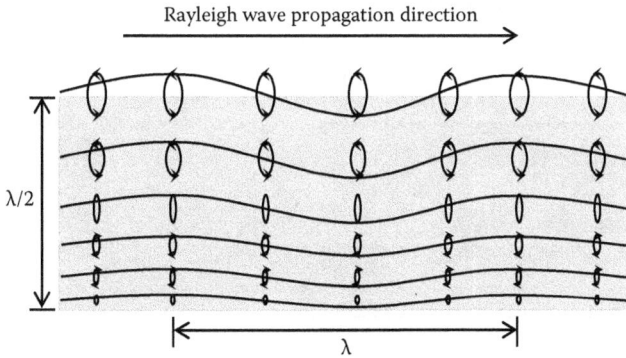

FIGURE 9.5   Schematic of motion trajectories of points in the substrate during the propagation of Rayleigh SAW [22].

transducer then a SAW wave generates that propagates towards other IDT called a receiver IDT, which converts electrical output.

Figure 9.5 exhibits the schematic trajectories motion of the points when SAW propagates on the free surface of a semi-infinite elastic body. A Rayleigh wave travelling from left to right results in anticlockwise elliptical motion of the points on the surface of the piezoelectric materials.

The amplitude of SAW exponentially decreases with increasing depth inside the substrate. The amplitude of the SAW has died after 1λ to 2λ.

9.5   FINITE ELEMENT METHOD (FEM)

In the finite element method (FEM), all the structure is divided into smaller, simpler parts called finite elements. The help of numerical technique provides rough solutions to the differential or integral equations of a complex system through a discretization process [23]. In this method formulation of a boundary value problem, the results in an algebraic equation. FEM method is the unknown function in the domain, and the simple equation of these models combined large system equation that will model the entire problem [24]. This method can be helped for solving partial differential equations in engineering and mathematical modelling such as structural analysis, heat conduction analysis, fluid flow, mass transport, solid structure and electromagnetic potential.

9.6   DESIGN AND SIMULATION OF THE PROPOSED STRUCTURE

SAW-based sensors monitor VOCs in gas paths leading to the patient. The developed structure is combined with the data reception and signal processing system. Single-layer of the proposed device consists of IDT, Lithium niobate and polyisobutylene (PIB); in this device, PIB act as a sensing layer placed over the $LiNbO_3$ substrate surface [20]. The properties of Lithium niobate and polyisobutylene (PIB) are shown in Tables 9.2, 9.3, 9.4, and 9.5, respectively. The electrode dimensions corresponding to IDT are 1 µm width and 0.2 µm height and the dimension of the sensing layer is 0.4 µm width and breadth 0.5 µm, the proposed model is shown in

TABLE 9.2
Elasticity Matrix (all the units are taken GPA)

242.39	75.19	75.19	0.0	0.0	0.0
75.19	203	57.3	0	8.49	0
75.19	57.29	203	0	−8.49	0
0.0	0.0	0.0	75.2	0.0	8.49
0.0	8.49	−8.49	0	59.49	0
0.0	0.0	0.0	8.49	0.0	59.49

TABLE 9.3
Coupling Matrix

1.34	0.24	0.24	0.0	0.0	0.0
0.0	0.0	0.0	−2.51	0	3.71
0.0	−2.51	2.51	0.0	3.71	0.0

TABLE 9.4
Relative Permittivity

28.8	0.0	0.0
0.0	85.3	0.0
0.0	0.0	85.3

TABLE 9.5
Material Contain in PIB

Property	Variable	Value
Young's modulus	E	10[GPa]
Poisson's ratio	nu	0.48
Density	rho	0.918[g/cm^3]
Relative permittivity	epsilon	2.2

Figure 9.6. Similarly, a multi-layer structure made of silicon nitride, PIB and lithium nitride, silicon nitride placed over the PIB sensing layer whose thickness is 0.5 μm and width is 4 μm, and other dimensions are the same as a single-layer structure which is shown in Figure 9.7 [18].

We designed and simulated the single- and multi-layer gas sensors using COMSOL software. Creating the SAW-based sensor Interdigital Transducer (IDT) is one of the critical components, and the electrode of the IDT is generally made of aluminium or gold. The IDT is fabricated on the surface of the piezoelectric

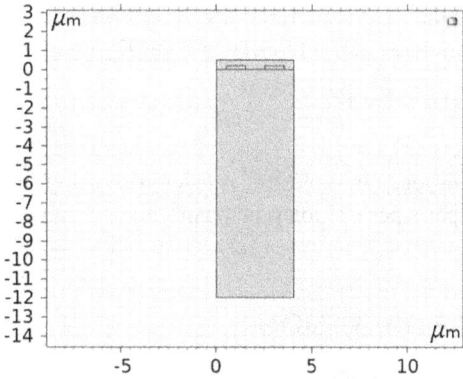

FIGURE 9.6 Models of a single-layer.

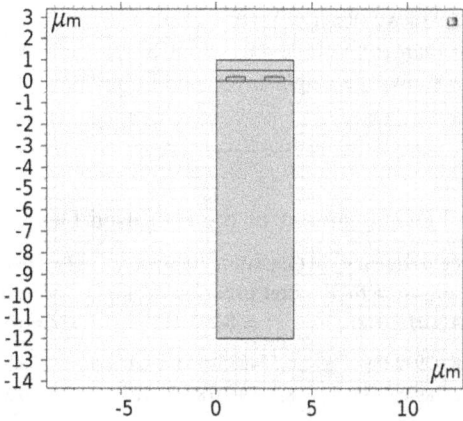

FIGURE 9.7 Model of multi-layer.

substrate using the lithography fabrication process. The length of each electrode of IDT used in the SAW sensor is 100 times more than their width. An electric field is produced when IDT is subject to a radio-frequency signal, permitting the piezo-electric coupling coefficient to travel surface acoustic waves. We can calculate resonance frequency from the sensing layer by using Equation (9.1).

When the RF input signal is applied to the electrode of the IDT in the input side and makes potential zero of the output side IDT of the SAW gas sensor, in the above expression, f, v, λ is represented resonance frequency, velocity and wavelength of the unit cell. The speed of the SAW device is one of the essential parameters for calculating the resonance frequency of the proposed structure. The total density of the sensing layer can be found using Equation (9.2) by adding the thickness of the DCM gas to a PIB (polyisobutylene) film.

$$\rho_{DCM,PIB} = KMc \qquad (9.4)$$

where $K = 10^{1.4821}$ is the air/PIB partition coefficient for DCM, m is the molar mass, and c is its air concentration can find out using the Equation (9.3)

$$c = c_0 p / RT \qquad (9.5)$$

In the DCM concentration, c in moles/m^3 is calculated using Gas Law in which c_0 is the concentration in parts per million p is the pressure, T is the temperature and R is the gas constant.

9.7 RESULT AND DISCUSSION

The simulation showed that single-layer and multi-layered resonance frequencies are 0.855 and 0.817 GHz, respectively, at standard temperature and pressure. Table 9.6 shows the frequency shift of single-layer and multi-layer SAW sensors before and after exposing 100 ppm gas concentration of DCM at standard pressure and temperature. From Table 9.6, we observed that a multi-layer device structure gives more frequency sensitivity than a single-layer for DCM gas and provides the high displacement. Figure 9.8 shows the single-layer, and Figure 9.9 shows that an additional

TABLE 9.6
Frequency Shift and Displacement of the Proposed Device

Structure of proposed device	Operating frequency (before exposure to gas) = A(MHz) Hz)	Operating frequency (after exposure to gas) = B(MHz)	Frequency shift in Hz = (A-B)	Displacement In μm
PIB/LiNbO$_3$	8.554837939722893	8.554835704396436	223.5	473
Si$_3$N$_4$/PIB/ LiNbO$_3$	8.298963189450538	8.298960783753319	240.5	783

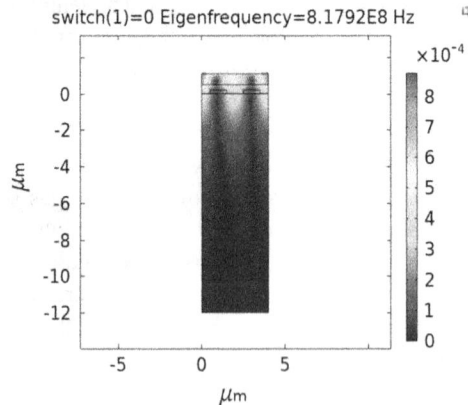

FIGURE 9.8   Displacement distribution of a single-layer.

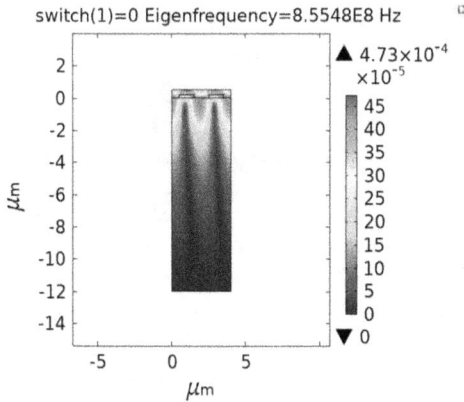

FIGURE 9.9 Displacement distribution of a multi-layer Device after exposure of DCM gas.

layer to the single-layer improves the device's displacement and sensitivity. It leads to the propagation of acoustic waves on the surface of the piezoelectric materials. An additional layer also guards the IDT electrode and the piezoelectric substrate.

The sensing layer contains a thin polymer layer with a high-affinity corresponding to Volatile organic compounds; in this device, we have taken DCM. When the PIB layer absorbs DCM gas from the air, then the mass of the sensing layer is increased. Due to increasing the mass of the PIB layer [19,21,25], a shift in the resonance remains slightly lower than resonance frequency for the same surface acoustic wave. In this model, a silicon nitride layer separates aluminium, electrode and piezoelectric substrate. Velocity of the surface acoustic wave change by increasing the mass of PIB due to the absorption of DCM gas molecules related to the surface.

9.8 MEASUREMENT OF ELECTRIC POTENTIAL

The simulation Figure 9.10 (a and b) exhibits the electric potential distribution of single-layer and multi-layer that exposes DCM gas from the air to the PIB layer. In

(a)   (b)

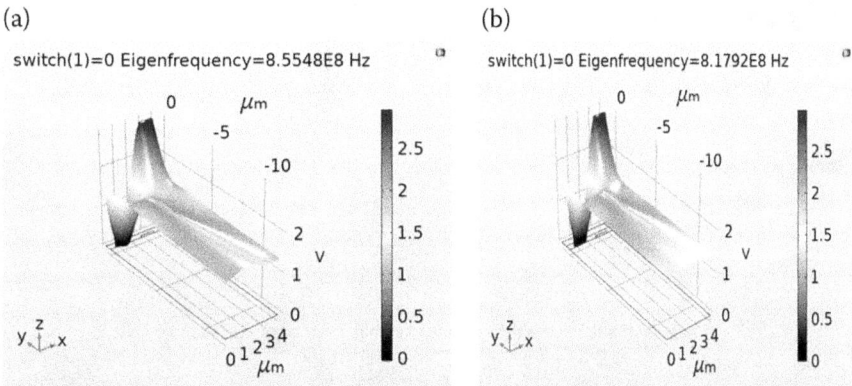

FIGURE 9.10 Potential curves for (a) single-layer and (b) multi-layer.

Figures 9.10 (a and b), we observed that generated voltage in multi-layer is higher than that of single-layer and also spreading of electric potential is symmetric along the electrodes.

9.9 MEASUREMENT OF SENSITIVITY

Figure 9.11 exhibits the variation in frequency shift with increasing sensing layer width of both single-layer and multi-layer of the proposed devices. From the graph, it is observed that the recovery time of a multi-layer is significantly less as compared to a single-layer. In a single-layer, frequency shift increases with increasing the piezoelectric layer. Still, in multi-layer, the frequency shift dies down above 0.65 μm sensing layer width, which reduces wave velocity [22–24]. From Equation (9.1), we can conclude that frequency shift is also reduced. The graph shows that the sensitivity of DCM gas detection is higher in multi-layer compared to the single-layer model for the value of extra thickness less than 0.65 μm. In Figure 9.11, the displacement of the multi-layer SAW sensor exponentially increases with an increasing sensing layer of the multi-layer SAW sensor and gives the maximum displacement at 0.8 μm of the sensing layer width after this displacement gradually decreases [26].

Figure 9.12 shows the temperature variation on the frequency shift of single-layer and multi-layer devices. From this graph, it can be observed that in multi-layer increasing the temperature above the ambient, DCM's gas molecules get evaporated and move away from the surface of the sensing layer. Consequently, a meagre amount of molecules of the gas communicate and occupy the surface of a multi-layer structure that causes decreasing the frequency shift in the proposed devices. We can conclude that frequency shift is very sensitive to the temperature of multi-layer devices.

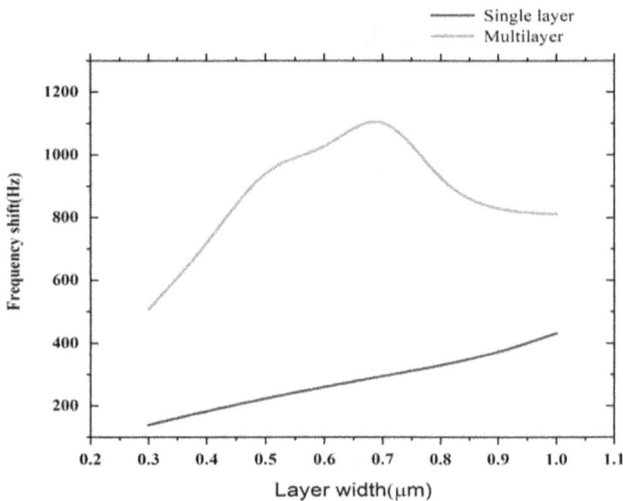

FIGURE 9.11 Frequency shift curve single-layer and multi-layer variation of sensing layer.

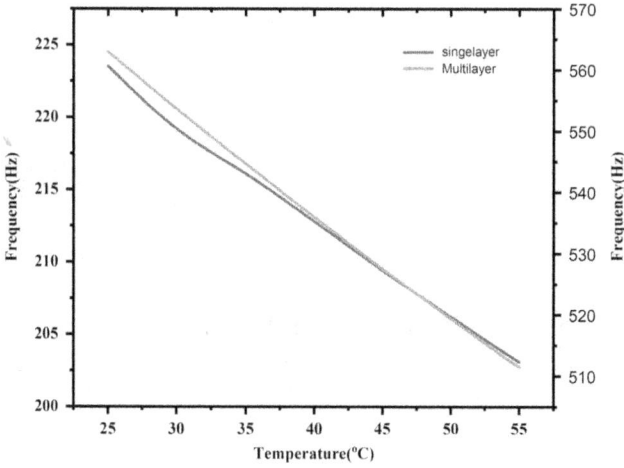

FIGURE 9.12 Frequency shift curve at different temperature ranges for single-layer and multi-layer.

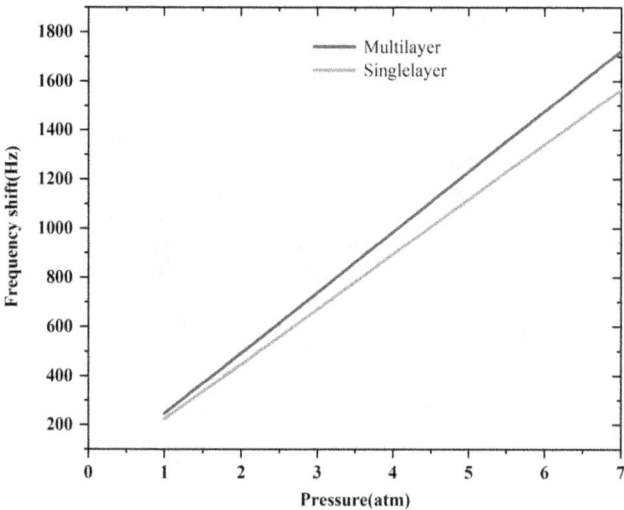

FIGURE 9.13 Displacement curve vs. pressure.

Figure 9.13 shows the graph plotted for pressure for frequency shift of single-layer and multi-layer gas sensors. In the plot, pressure effects on single-layer and multi-layer are studied. Increasing the pressure, the concentration of gas molecules on the surface of the $Si_3N_4/PIB/LiNbO_3$ SAW sensor increase [27]. More potent gas molecules result in significant modification of the sensor surface. When the pressure of the gas increases, molecules of the gas are compressed in the smaller region. The frequency shift increases with increasing pressure, as shown in Figure 9.13.

9.10 CONCLUSION

In this work, we designed and analyzed a single- and multi-layer SAW-based sensor by using COMSOL 5.4. Here the sensor has been designed to detect DCM gas produced in a gas pathway device located in the hospital. The recovery of the Corona patient can be ensured by avoiding the other infections by detecting DCM gas. We also investigated frequency shift variation of the width of the piezoelectric layer, temperature, pressure of the proposed devices. The above result shows that the multi-layer sensor exhibits more sensitivity to DCM gas than the single-layer at 100 ppm gas concentration. The single-layer structure gives less shift frequency and displacement than the multi-layer structure. The above graph shows higher sensitivity when the piezoelectric layer is thinner than 0.6 μm. The frequency of the multi-layer SAW sensor was improved with increasing pressure.

REFERENCES

1. Aichner VS (2018) How to sample and analyze VOCs of respiratory medical devices according to the new ISO 18562-3. *Pharm Ind* 80(10):1423–1427.
2. Gupta A, Kumar P, Pandey S (2017) Analysis of multi-layered SAW-based gas sensor. In: International conference on trends in electronics and informatics (ICEI). IEEE, New York, pp. 239–242.
3. Behera B (2019) Design and investigation of a dual friction-drive-based LiNbO$_3$ piezoelectric actuator employing a cylindrical shaft as slider. *IEEE Sensors Journal*, 19(24):11980–11987, 15 December. 10.1109/JSEN.2019.2938246
4. Chen HH (2005) Application of Taguchi robust design method to SAW mass sensing device. *IEEE Trans Ultrason Ferroelectr Freq Control* 52(12):2403–2410.
5. Ahmadi S, Hassani F, Korman C, Rahaman M, Zaghloul M (2004) Characterization of multi- and single-layer structure SAW sensor [gas sensor]. *Sensors*3:1129–1132.
6. Mohanan A, Islam M, Ali S, Parthiban R, Ramakrishnan N (2013) Investigation into mass loading sensitivity of sezawa wave mode- based surface acoustic wave sensors. *Sensors* 13(2):2164–2175.
7. Fechete A, Ippolito S, Wlodarski W, Kalantar-Zadeh K, Holland A, Wisistsora-at A (2005) Layered InOx/Si$_3$N$_4$/36YX LiTaO$_3$ surface acoustic wave-based hydrogen sensor. *East-West J Math* 32:465–470.
8. Johnson S, Shanmuganantham T (2014) Design and analysis of SAW-based MEMS gas sensor to detect volatile organic gases. *Carbon* 119(5):254–258.
9. Kumar M, Kumar D, Gupta AK (2015) Fe-doped TiO$_2$ thin films for CO gas sensing. *J Electron Mater* 44(1):152–157.
10. Niyat FY, Abadi MS (2018) COMSOL-based modeling and simulation of SnO$_2$/rGO gas sensor for detecting NO$_2$. *Sci Rep* 8(1):2149.
11. Behera B, Nemade HB (June 2018) Investigating translational motion of a dual friction-drive surface acoustic wave motor through modelling and finite element simulation. *SAGE Simulation: Transactions of the Society for Modelling and Simulation International* 95(2):117–125. 10.1177/0037549718778770
12. Gaur AM, Joshi R, Kumar M (2011) Deposition of doped TiO$_2$ thin film by sol-gel technique and its characterization: A review. In: Proceedings of the World Congress on Engineering, London, pp. 6–8.
13. Kumar M, Gupta AK, Kumar D (2016) Mg-doped TiO$_2$ thin films deposited by low-cost technique for CO gas monitoring. *Ceram Int* 42(1):405–410.

14. Kalantar-Zadeh K, Powell DA, Wlodarski W, Ippolito S, Galatsis K (2003) Comparison of layered based SAW sensors. *Sens Actuators B Chem* 91(1–3):303–308.
15. Ippolito SJ, Kalantar-Zadeh K, Powell DA, Wlodarski WA (2002) Finite element approach for 3-dimensional simulation of layered acoustic wave transducers. In: Conference on Optoelectronic and Microelectronic Materials and Devices. COMMAD 2002. Proceedings (Cat. No. 02EX601). IEEE, New York, pp. 541–544.
16. Hasanuddin NH, Wahid MHA, Shahimin MM, Hambali NA, Yusof NR, Nazir NS, Khairuddin NZ, Azidin MAM (2016) Metal oxide-based surface acoustic wave sensors for fruits maturity detection. In: 3rd International Conference on Electronic Design (ICED). IEEE, New York, pp. 52–55.
17. Mishra D, Singh A (2015) Sensitivity of a surface acoustic wave-based gas sensor: Design and simulation. International Conference on Soft Computing Techniques and Implementations (ICSCTI). IEEE, New York, pp. 1–5.
18. Turuk BK, Behera B (2021) Finite element simulation and characterization of one-port hetero structured surface acoustic wave resonator. *Ferroelectrics* 583:33–40.
19. Voinova MV (2009) On mass loading and dissipation measured with acoustic wave sensors: a review. *J Sens* 2009:943125. 10.1155/2009/943125
20. Kutiš V, Gálik G, Královič V, Rýger I, Mojto E, Lalinský T (2012) Modelling and simulation of SAW sensor using FEM. *Procedia Eng* 48:332–337.
21. Nazemi H, Joseph A, Park J, Emadi A (2019) Advanced micro-and nano-gas sensor technology: A review. *Sensors* 19(6):1285.
22. Zheng P, Greve DW, Oppenheim IJ (2009) Multiphysics simulation of the effect of sensing and spacer layers on SAW velocity. In: COMSOL Conference, Boston, pp. 1–7.
23. Francis LA, Friedt JM, Bartic C, Campitelli A (2004) An SU-8 liquid cell for surface acoustic wave biosensors. In: MEMS, MOEMS, and micromachining. In: Proceedings of the SPIE Photonics, vol. 5455. Europe, pp .353–363.
24. Ju BF, Bai X, Chen J (2012) Simultaneous measurement of local longitudinal and transverse wave velocities, attenuation, density, and thickness of films using point-focus ultrasonic spectroscopy. *J Appl Phys* 112(8):084910.
25. Ruther P, Colelli K, Frerichs HP, Paul O (2003) Surface conductivity of a CMOS silicon nitride layer. *Sensors* 2:920–925.
26. Nagmani AK, Turuk BK, Behera B (2021) Simulation and optimization of the geometrical structure of one-port SAW resonator using EEM. In: AIP Conference Proceedings 2341, 020043.
27. Nagmani AK, Behera B (January 2022) A Review on high-temperature piezoelectric crystal $La_3Ga_5SiO_{14}$ for sensor applications. *IEEE Transactions on Ultrasonics, Ferroelectrics, and Frequency Control* 69(3):1–15 10.1109/TUFFC.2022.3143666

10 Nanomanufacturing for Energy Conversion and Storage Devices

Shubham Srivastava, Deepti Verma,
Shreya Thusoo, Ashwani Kumar,
Varun Pratap Singh, and Rajesh Kumar

CONTENTS

10.1 Introduction...165
10.2 Nanomaterials Used in Energy Conversion and Storage........................167
10.3 Application in Energy Conversion 168
    10.3.1 Solar Energy..168
    10.3.2 Hydrogen Energy 168
    10.3.3 Biomass/Biofuels...169
    10.3.4 Ocean, Geothermal and Wind Energy...169
10.4 Application in Energy Storage...170
    10.4.1 Mechanical Systems 170
    10.4.2 Thermal Systems 170
    10.4.3 Optical Systems...170
    10.4.4 Electrical Systems 171
    10.4.5 Lithium Ion Batteries 171
10.5 Conclusion 172
References..172

10.1 INTRODUCTION

Energy is used in various fields like transportation, industry, household, etc. and therefore is one of the vital resources for all human activities [1]. With advances in technology, the demand for energy has increased in the whole world. This demand is continuously growing and shall escalate in future (as shown in Figure 10.1). With the increased demand it has become crucial to generate, store and conserve energy to meet the demands. Initially, the energy generation source was heavily based on crude oil which had lower efficiency, limited availability and harmful to environment. Therefore, the use of more efficient and sustainable technologies had become equally important.

Nanotechnology is the understanding and control of matter at the nanoscale, at dimensions between approximately 1 and 100 nm, where unique phenomena

DOI: 10.1201/9781003220602-10

World Energy consumption and projection for various sectors

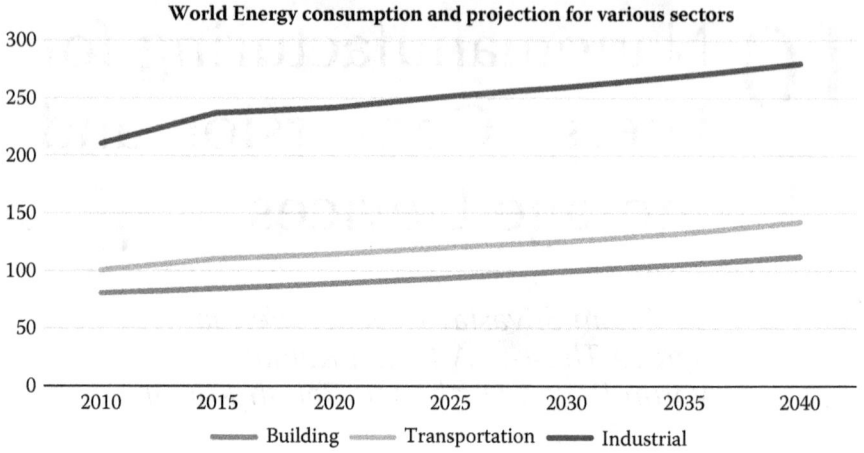

FIGURE 10.1   World energy consumption and prediction [2].

enable novel applications. It is the use of matter on an atomic, molecular, and supramolecular scale for industrial purposes. It encompasses formation and use of materials and devices at nanoscale. The energy triangle consists of energy storage, generation and transfer [3] and Nanotechnology is a potential solution to resolve this energy issue (Figure 10.2). It promises a major influence on the energy industry in terms of energy conversion and storage devices. The utilization of nanotechnology for development in this area has already begun like increased

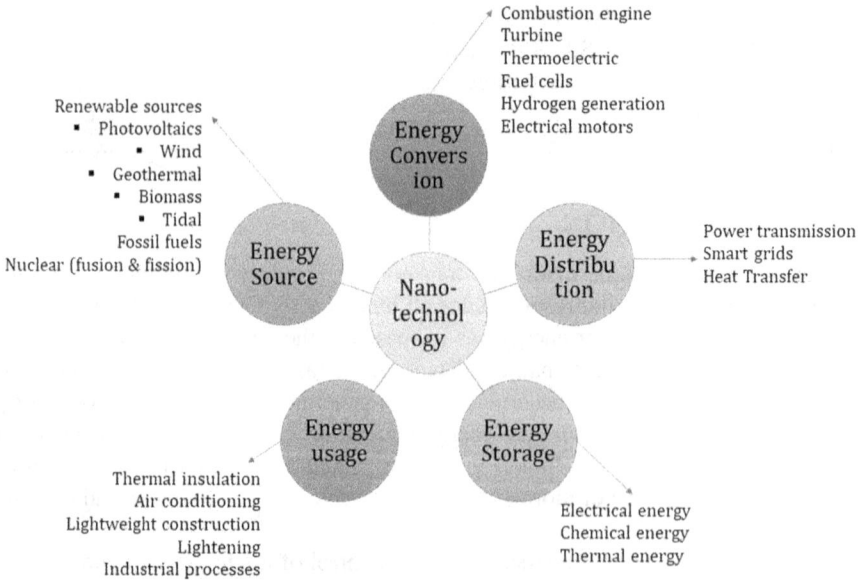

FIGURE 10.2   Applications of nanoscience in energy sector [4].

electrical storage capacity, increased efficiency in heating and lighting and reduced level of pollution in achieving the above.

10.2 NANOMATERIALS USED IN ENERGY CONVERSION AND STORAGE

The process of creating and designing devices at nanoscale or incorporating nanomaterials in any device is known as nanofabrication. Devices smaller than 100 nm have proved to be a breakthrough in storing and transferring of energy. In the nanoscale range, thermoelectric, photovoltaic, catalytic and electrochromic materials have made key contributions to various energy applications (Figure 10.3). Inorganic nanomaterials' unique properties, such as excellent electrical and thermal conductivity, large surface area and chemical stability, make them highly competitive in energy applications [5]. Nanotechnology is being widely used in portable and smart devices today as they are more durable, flexible and have longer life span [6]. Nanomaterials like Graphene is largely used and has emerged as most promising material for energy storage. There has been rapid research in use of graphene-based nanomaterials. Low weight and price have been the main driving reason for the rapid research of its use. It has relatively higher electrical conductivity and also modifies sulphur for improved electrochemical performance in batteries. Another nanomaterial widely used in solar energy is silicon-based nano-semiconductors as they absorb larger range of wavelengths. Studies have shown that such nanomaterial-based cells have 40% higher efficiency compared to regular solar cells.

In general, there are two main stages in energy processing: conversion and storage [7]. The application of nanotechnology in these two steps leads to more efficient and sustainable solution to problems in energy sector. The best part of nanotechnology is the way it encompasses all applications leading to more efficiency, flexibility and cost-effective solutions.

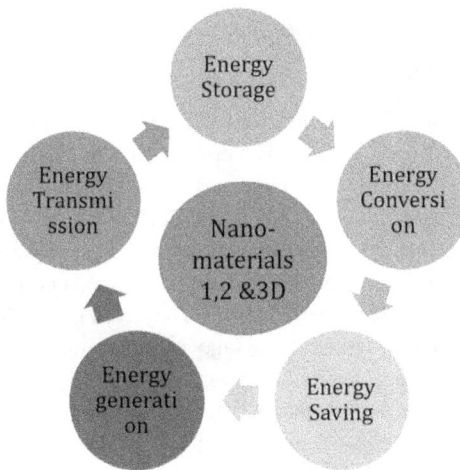

FIGURE 10.3   Nanomaterials in energy sector.

10.3   APPLICATION IN ENERGY CONVERSION

The conversion of energy from one form to another generated through different sources is the basis to various developments nowadays [8]. The application of nanotechnology in various energy conversion are discussed below:

10.3.1   SOLAR ENERGY

Limited availability and environmental issues have been major drawbacks in use of fossil fuel for energy generation. Solar energy has emerged as alternative and gained popularity in recent times. Intensive work has been carried out to convert solar energy into electrical energy [9,10]. The solar panels are put in open sunlight, the solar cells absorb the sunlight (sun's energy) and then convert it into electricity. Nanotechnology helps in boosting the absorption and retention of sunlight by the solar cells [11]. Light emitting nano-particles are widely used in improving the performance of solar cells. Also, nanotechnology has helped in development of solar cells which have a broader range of absorption. Nanofluid is mostly used in manufacturing because of its high coefficient of heat transfer in engines or heat exchangers to increase both performance and economy [12].

It is also used in devices such as solar thermal devices, flat collectors, solar water heaters or desalination plants and improve solar absorption capabilities of solar collectors. Nano coatings have a self-cleaning property and therefore help in protecting the solar cells against the sedimentary deposits on the surface of solar cells, thereby improving the efficiency. Nanophotocatalysts raise the spectrum of absorption and also provide an atmosphere free of pollutants. Nanocomposite films (di-functional) rapidly convert and store solar energy into thermal energy [13]. The photocatalyst also provides benefit of spectral selective absorption. Nanoparticles used as additives enhance the thermal conductivity of the nanocomposite used in solar water heater. Similarly, adding nanoparticles (MgO) to water minimizes the heating loss time which proves to be beneficial in solar heating systems in buildings.

10.3.2   HYDROGEN ENERGY

Obtaining electricity by chemical reaction between H^+ and O^- ions and free electrons is the basis of hydrogen energy. The combustion in diesel engines is promoted by introduction of aluminium nanoparticles [14]. Also, nanofluid helps in decreasing the concentration of smoke generated during combustion and increases the total combustion heat. Cerium oxide nanoparticles catalyse the unburnt hydrocarbon owing to their high surface area to volume ratio. Fuel cells convert chemical energy into electrical energy through redox reactions. They contain membrane which allows hydrogen ions to pass and restricts passage of other ions [15]. Nanotechnology is being used to develop longer-lasting and lighter-weight cells by developing more effective membranes using nanoparticles and by using nanocatalyst to enhance the chemical reactions (as replacement to platinum catalysts). Nanoparticles increase the performance of fuel cells especially at high temperatures. Nanofluidic cells have also been developed which have high efficiency, lower cost and are miniaturized

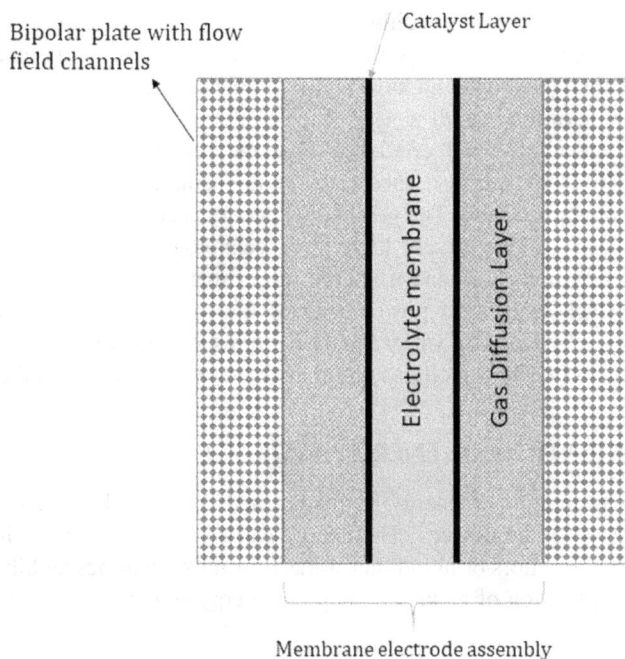

Bipolar plate with flow field channels

Catalyst Layer

Electrolyte membrane

Gas Diffusion Layer

Membrane electrode assembly

FIGURE 10.4   Polymer electrolyte membrane fuel cell [17].

power sources [16]. The carbon nanotubes and nanofibers have proved to be good catalysts for fuel cell applications.

The PEM (Polymer Electrolyte Membrane) fuel cells has emerged as example to convergence of nanotechnology to sustainable energy technologies (Figure 10.4). Photocatalytic hydrogen production is a sustainable energy system and offers reduction in emission of greenhouse gases.

10.3.3   BIOMASS/BIOFUELS

The biomass is one of the renewable resources available but its drawbacks of ash deposits, NOX and SOX emission have restricted its use on larger scale. These problems however can be overcome by use of nanotechnology by providing coating on the boilers and by adding nanomaterials as additives in order to reduce the emission of harmful pollutants. The presence of KF/CaO nanocatalyst increased the biodiesel yield and can be used to convert high acid value oil into biodiesel. Hydrophobic porous nanomaterials show greater efficiency in biphasic systems.

10.3.4   OCEAN, GEOTHERMAL AND WIND ENERGY

The wind energy as well as ocean energy is converted into utilizable form by use of turbines and supercapacitors. Nanotechnology plays an important role in these turbines and capacitor design. The nano-coatings increase the resistance of the turbines.

Nano colloidal boron nitride additives used in wind turbine gearbox form a wear protective tribofilm. Further, the turbines are made of carbon-based nanotubes materials which are lightweight and have higher strength. Nanostructured materials also help in reduction of the overall weight of the blades and turbines. They have higher durability, sustainability and efficiency than the regular ones. In wind energy, nanotube-based cables are also used to carry the same amount of electric current but are lightweight in nature. The geothermal energy is combustion-free but the cost involved in deep drilling is very high. Nanotechnology helps in overcoming this challenge introduction of a nanostructured model that acts as nano-storage particle with liquid which helps turn power turbines through evaporation. These liquids are being refined by blending various nanostructures to improve the efficiency of the geothermal plants [4,18]. Nanofluids absorb higher energy than normal thermal fluids.

10.4 APPLICATION IN ENERGY STORAGE

The huge consumption of energy in human activities can be catered only with efficient energy storage devices. The energy can be stored either in direct storage form like supercapacitors or in indirect storage form like batteries and flywheels etc. The various application of nanotechnology in energy storage devices is as follows:

10.4.1 MECHANICAL SYSTEMS

Flywheels are one of the most important mechanical systems for energy storage. Although very few researches have been made on these but nanotechnology has proved to improve its efficiency. Nanomaterials, nanocomposites and nanofillers when used in the rotors of the flywheels enhance the efficiency and increase the speed of the flywheels [19].

10.4.2 THERMAL SYSTEMS

In these systems, energy can be stored using thermochemical and thermos-physical processes. The efficiency of such systems depends on the design aspect and properties of the utilized material. The nanostructured materials and grids change the thermos-physical property of the bulk material which leads to enhancement of their performance. Encapsulated phase change materials are used for air conditioning and nanoporous materials like zeolites for reversible heat storage in buildings.

10.4.3 OPTICAL SYSTEMS

Light-matter interaction at nanoscale has provided alternative to store energy through optical systems. Nanooptics have also emerged as a result of research work in past decades which has shown promising results in enhancing the efficiency of optical systems. The application of nanostructures in the design of optical systems permits high efficient harvesting within scale limit. The asymmetric shape of the nanoparticles and the designed mechanism enhances the absorbance and optimizes the light scattering in cavities.

10.4.4 ELECTRICAL SYSTEMS

Supercapacitors store energy at lower voltage limits and can be classified as double-layer capacitor, pseudo capacitor and hybrid capacitor. Nanostructured devices are widely used in recent times. The nanostructured carbon films provide a larger surface area and therefore help achieve high specific power [20]. In energy storage devices it is important to have larger interfaces which can be achieved by nanoporous structures of uniform and thin carbon films. Nanoporous adsorption storage includes metallo organic structures for gas storage. Nanocrystalline metal hybrids form the basis of hydrogen storage. Li-ion batteries have high energy density, minimized maintenance merits and nanotechnology helps provide enhancement in efficiency of these devices. Nanostructured array batteries overcome the problems of conventional Li metal batteries (Figure 10.5). They control thermal stability, volume shrinkage, volume fluctuations. Therefore, minimum handling and explosion hazards are achieved by the use of nanotechnology.

10.4.5 LITHIUM ION BATTERIES

Optimization of lithium-ion batteries is achieved by using nanocomposites for electrode materials and nanostructured design (Figure 10.6). Carbon nanotube (CNT)

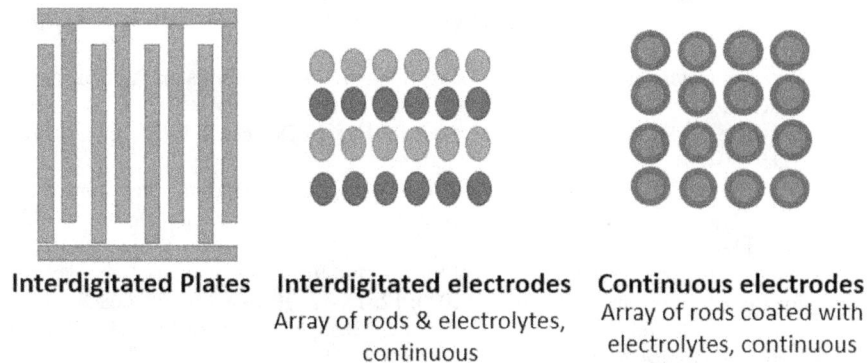

Interdigitated Plates   **Interdigitated electrodes**   **Continuous electrodes**
Array of rods & electrolytes, continuous   Array of rods coated with electrolytes, continuous

FIGURE 10.5 Geometrical design of 3D batteries in energy storing architecture [4].

Coverage approx.83%

$Li_{4.4}Sn$

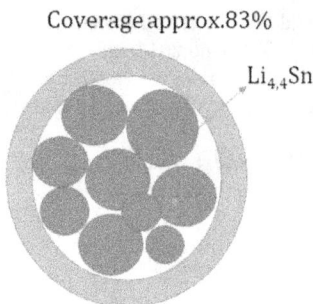

FIGURE 10.6 Schematic illustration of encapsulation process of lithium-ion nanoparticles in tin hollow carbon balls [1].

based electrodes have higher charge capacities for supercapacitors. With the application of nanotechnology in Li batteries, a reduction in chances of burning of battery is achieved by developing higher heat resistant electrodes. The time of charging reduces and the power increases by coating the surface of electrode by nanoparticles. The surface area increases, efficiency is improved and weight is reduced. A study utilized tin-based nanoparticles to improve the efficiency of the battery.

10.5   CONCLUSION

With the increase in human population, the energy demands have increased exponentially and have various drawbacks in its conventional sector. The demands will remain unfulfilled due to limited fossil fuel and pollution created therein. Therefore, it needed to shift to a cleaner and sustainable energy technology. Nanotechnology provides one of the best alternatives for the same. It not only finds its application in various fields but also enhances the efficiency, reduces pollution and is cost-effective. It works on nanoscale providing finer and better alternatives to overcome limitations and drawbacks in conventional methods of energy storage and conversion.

REFERENCES

1. Ferric Christian E, Selly, DA, and Antonius I, Application of Nanotechnologies in the Energy Sector: A Brief and Short Review Front. *Energy* 2013, 7(1): 6–18. 10.1007/S11708-012-0219-5
2. Data from International Energy Outlook 2017, U.S. Energy Information Administration, September 2017.
3. Donahoe FJ, Hierarchical Nanostructures for Energy Devices. *J Franklin Inst.* 2014; 271: 230–231. 10.1039/9781849737500
4. Abdalla AM, et al. Nanotechnology Utilization in Energy Conversion, Storage and Efficiency: A Perspective Review. *Advanced Energy Conversion Materials* 2020; 1(1):30–54
5. Wang H, Liang X, Wang J, Jiao S, and Xue D, Multifunctional Inorganic Nanomaterials for Energy Applications. *Nanoscale* 2020; 12, 14–42, 10.1039/C9NR07008G
6. Tarascon JM, and Armand M, Issues and Challenges Facing Rechargeable Lithium Batteries. *Nature* 2001; 414: 359–367. 10.1038/35104644
7. Alden SZ, and Taylor KD, *Renewable and Alternate Energy Resources*; ABO-CLIO; 2008;ISBN 978-1-59884-089-6
8. Fan S, Photovoltaics: An Alternative "Sun" for Solar Cells. *Nat Nanotechnol.* 2014; 9: 92–93. 10.1038/nnano.2014.9
9. Beard MC, Luther JM, and Nozik AJ, The Promise and Challenge of Nanostructured Solar Cells. *Nat Nanotechnol.* 2014; 9: 951–954. 10.1038/nnano.2014.292
10. Ghasemzadeh F, and Shayan ME,Nanotechnology in the Service of Solar Energy Systems., In: Sen, M., Ed., Nanotechnology and the Environment, IntechOpen, London, 2021, 1–15, https://doi.org/10.5772/intechopen.93014
11. Sen M, editor. *Nanotechnology and the Environment [Internet]*. London: IntechOpen; 2020. https://www.intechopen.com/chapters/73145. 10.5772/intechopen.93014
12. Nanometer TiO_2 Photocatalyst Titanium Dioxide (jianghutio2.com)
13. Kao M, Ting C, Lin B, et al. Aqueous Aluminum Nanofluid Combustion in Diesel Fuel. *Journal of Testing and Evaluation* 2008; 36(2): 186–190.
14. http://www.understandingnano.com

15. Lee JW, and Kjeang E, Nanofluidic Fuel Cell. *J Power Sources* 2013; 242: 472–477. 10.1016/j.jpowsour.2013.05.129
16. Patel V, and Mahajan YR, *Nanotechnology for Energy Sustainability*,Techno-Commercial Opportunities of Nanotechnology in Wind Energy,Wiley – VCH, 2017; 1079–1106. Doi: 10.1002/9783527696109.ch43
17. Kannan A.M., and Munukutla L.V., Application of Nano-Technology for Energy Conversion and Storage, 2007 Annual Conference & Exposition, Honolulu, Hawaii, American Society for Engineering Education 2007, June 24–27, Pages 12.246.1–12.246.8.
18. Schulz R, Huot J, Liang G, et al. Recent Developments in the Applications of Nanocrystalline Materials to Hydrogen Technologies. *Mater Sci Eng A Struct.* 1999; 267: 240–245.
19. Diederich L, Barborini E, Piseri P, et al. Supercapacitors Based on Nanostructured Carbon Electrodes Grown by Cluster-Beam Deposition. *Appl Phys Lett.* 1999; 75: 2662–2664. 10.1063/1.125111
20. Badwal SPS, Giddey SS, Munnings C, et al. Emerging Electrochemical Energy Conversion and Storage Technologies. *Front Chem.* 2014; 2: 1–28. 10.3389/fchem.2014.00079

11 Nanofabrication Techniques for Solar Photovoltaic Applications

Girija Nandan Arka, S.B. Prasad, and Subhash Singh

CONTENTS

11.1 Introduction 175
11.2 Nanomaterials for Semiconductive Film and Its Synthesis Process 178
    11.2.1 Synthesis Route to Develop Mesoporous TiO_2 178
    11.2.2 Deposition of Nanostructured TiO_2 180
        11.2.2.1 Spin Coating 180
        11.2.2.2 Dip Coating 181
        11.2.2.3 Doctor Blading 182
        11.2.2.4 Screen Printing 182
        11.2.2.5 Ink Jet Printing 182
        11.2.2.6 Pad Printing 183
11.3 Nano Deposition for Flexible Solar Cell 183
    11.3.1 Electrophoretic Deposition 183
    11.3.2 Chemical Sintering 183
    11.3.3 Mechanical Compression 183
11.4 Conclusion 184
References 184

11.1 INTRODUCTION

Cumulative energy necessity of the society propelling huge burden to bureaucrats of the countries. Use of conventional fossil fuels emits greenhouse gases that increase the earth's temperature and causing global warming responsible for the climate change. Moreover, burning of fuels not only damage the human race but also disfigure the whole ecosystem. Therefore, to make an adorable ecosystem we must move towards renewable energy sources to satisfy the need of the society. Since Sun as a prime source of energy enthusiastically influence a lot for the survival of ecosystem thus has prodigious potential [1]. Moreover, solar energy is respected as sustainable energy source as it is environmentally friendly,

DOI: 10.1201/9781003220602-11

abundantly available and doesn't produce CO_2 gas or solid or liquid waste derivatives as fossil fuels do [2–4].

Solar photovoltaic (PV) cells are conversion devices potentially gratified for direct translation of solar energy into electrical energy. Moreover, solar photovoltaic cells are power production systems that can be installed to receive power from micro to mega range [5]. Furthermore, these cells can be predominantly installed in remote areas where power grid service technically and economically not feasible to serve [6,7]. Further, the solar photovoltaic cells are classified into three generations depending open the technology [8]. The most frequently used PV cells are based silicon wafers classified in 1st-generation solar cells. However, manufacturing of these PV cells charges a high capital cost. Further, fabrication of crystalline silicon solar cell comprising silicon wafer or poly-silicon contributes for raising the cost [9]. Moreover, technological revolution urging potential application of nanomaterials and nano fabrications to get flexible PV solar cells. Nevertheless, silicon-based solar cells are failed to have flexibility for emerging electronic gadgets [10]. Thus, to make economical PV cell, thin film-based solar cells and emerging thin film-based solar cells are acknowledged worldwide. Thin film-based solar cells such as amorphous silicon, cadmium tellurium, copper indium gallium selenide, etc. are stacked in 2nd-generation photovoltaic cells [11]. Whereas emerging thin film-based solar cells include Perovskite solar cell, dye-sensitized solar cell (DSSC), heterojunction solar cell, organic solar cell, quantum dot cell, etc. are stacked in 3rd-generation photovoltaic cells [12–14]. This generally uses organic light-sensitive materials as perovskite or photosensitizer or donor-accepter polymer for the generation of electrons by significant photo participation [15]. Moreover, this involves a simple fabrication process that relatively reduces the cost of fabrication. Although the efficiency of these thin film-based photovoltaic cells is far below than conventional photovoltaic cells based on silicon (15%–20%) but found significant research interest in their structural and optical development [16].

Thus, fabrication of these thin film-based solar cells and emerging thin film-based solar cells involve synthesis of nanoparticles and nanoparticles coating to produce a thin film. Therefore, to enforce nanofabrication, one must understand the role and relevance of the solar cell components. For example, DSSC consists of transparent substrate, transparent conductive oxide film, semiconductive oxide film, dye, electrolyte and counter electrode respectively from which transparent substrate has the function of transmitting light into the system, transparent conductive oxide film has the function of transmitting light and collect electron from semiconductive oxide film and semiconductive oxide film has function of catching electrons from LUMO of Dye and transport electrons to transparent conductive oxide film by inhibiting recombination of electrons with oxidized dye or electrolyte respectively [17]. Table 11.1 demonstrating physical architecture of different photovoltaic solar cells confined to 2nd- and 3rd-generation solar cells with graphical contrast to comprehend the essential elements of photovoltaic cells that need nanofabrication.

Since 2nd-generation solar cell uses scarce and toxic material, thus significant researches are reported for the development of 3rd-generation solar cell. Thus, the prime objective of this book chapter is to identify the synthesis of nanoparticles and nanofabrication to obtain adorable solar PV cells. Furthermore, numerous techniques

TABLE 11.1

Indispensable Components of Different Photovoltaic Cells and Materials

PV cell	Architecture and indispensable components	Important materials
Quantum dot PV cell	GLASS SUBSTRATE / TRANSPARENT CONDUCTIVE OXIDE FILM / QUANTUM DOT FILM / METAL ELECTRODE	TCO: ITO/FTO QDs: Antimonide (Sb) based QDs such as InSb and GaSb
cupper indium gallium selenide (CIGS) PV cell	ANTIREFLECTION COATING / TRANSPARENT CONDUCTIVE OXIDE FILM / N-TYPE (CdS) / P-TYPE (CIGS) / Mo (METALLIC BACK CONTACT) / GLASS/FOIL/PLASTIC	TCO: ITO/FTO Semiconductor: CdS, CIGS
DSSC PV cell	TRANSPARENT SUBSTRATE / TRANSPARENT CONDUCTIVE OXIDE LAYER / ELECTROLYTE / PLATINUM LAYER / TRANSPARENT CONDUCTIVE OXIDE LAYER / TRANSPARENT SUBSTRATE / ● DYE ◯ SEMI CONDUCTIVE OXIDE ↓ RECOBINATION	TCO: ITO/FTO Semiconductor: TiO_2, ZnO_2
Perovskite PV cell	SUBSTRATE / TRANSPARENT CONDUCTIVE OXIDE LAYER / ELECTRON TRANSPORT FILM / PEROVSKITE / HOLE TRANSPORT FILM / METAL CATHODE	TCO: ITO/FTO Semiconductor: TiO_2, ZnO_2
Cadmium tellurium PV cell	GLASS SUBSTRATE / TRANSPARENT CONDUCTIVE OXIDE LAYER / N-LAYER FILM (CdS) / P-LAYER FILM (CdTe) / METAL CATHODE	TCO: ITO/FTO Semiconductor: CdS, CdTe

are addressed to making thin films for potential applications. Challenges and solutions for achieving flexible solar cells based on PET and PEN also have been addressed for flexible confinement.

11.2 NANOMATERIALS FOR SEMICONDUCTIVE FILM AND ITS SYNTHESIS PROCESS

Semiconductive oxide layer portraying the vibrant protagonist to cache significant electrons and effectively transport the electrons to transparent conductive oxide layer for external use. Moreover, the semiconductive oxide layer should have high surface area to offer significant adsorption of light-sensitive material [18,19]. Thus, the semiconductive oxide layer must have mesoporous structure to concoct larger specific surface area. To address above configuration, TiO_2 nanostructured material has received significant potential of interest for solar cell application [19]. Moreover, TiO_2 has favourable energy band gap and aligned electronic structure that thermodynamically trigger the electrons to hop from least unoccupied molecular orbit (LUMO) of light-sensitive materials [17]. Research revealed that several parameters that used to determine the overall performance of solar cells are particle size, crystal phase, crystal facet, pore diameter, thickness, morphology and surface area respectively. Nevertheless, small nanoparticles contribute for large surface area and larger nanoparticles contribute for small surface area.

The TiO_2 has three fundamental crystallographic phases and are anatase phase, rutile phase and brookite phase, respectively. Out of all, anatase phase extensively applied for the development of solar cells due to its wide band gap 3.2 eV, indirect band gap property, high coefficient of electron diffusion, which constitutes large surface area respectively [20]. However, rutile phase confronted with recombination issue and brookite phase confronted with conductivity and low surface area. Moreover, at atomic level TiO_2 anatase with {001} exposed surface facet contributes for high surface energy and is highly reactive. The controlling of the TiO_2 anatase with {001} exposed surface facet entirely governed by the synthesis route [21]. From above discussion, the synthesis route finds the most indispensable parameter accountable for the properties of TiO_2 mesoporous structure. Thus, it is crucial to discuss the synthesis process of nanostructures.

11.2.1 SYNTHESIS ROUTE TO DEVELOP MESOPOROUS TiO_2

Sol-gel and hydrothermal synthesis routes have acknowledged worldwide for their low processing temperature, high level of purity, good composition control and low-cost processing. Sol-gel synthesis is a wet-chemical technique that follows four stages: hydrolysis, polycondensation reactions, growth of particle and gel formation to produce metal oxide nanoparticle. On the contrary, hydrothermal process involves a chemical rection in water at high pressure and temperature kept in a sealed pressure vessel. The typical process involves mixing of stoichiometric quantity of tetrabutoxide or titanium tetraisoprpoxide in ethanol or isopropanol followed by acetic acid to get colloidal solution through magnetic stirrer particles [22,23]. Then the colloidal solution calcinated at 450°C for few hours to obtain colloidal nanoparticles [22]. On the contrary, the solution kept in Teflon line autoclave device and exposed to heat at temperature 180°C–200°C for few hours to grow colloidal nanoparticles through hydrothermal process [24]. Experimentally sol-gel route respected as promising method to get high purity crystalline TiO_2 anatase phase and hydrothermal route make relatively small nanoparticles with distorted crystal structured anatase and brookite phases. However hydrothermal route acknowledged for concocting larger specific surface area for absorbing significant light-sensitive material. Therefore, hydrothermal process can produce larger surface area but it indorses recombination due to distorted crystal structure. Nanoparticle size in the range 15–20 nm was produced via hydrothermal process [25]. Research revealed that traditional hydrolysis process produces uncontrolled reaction as a result of nonhomogeneous nanoparticles created. Thus, Kathirvel et al. [24] alter the synthesis process via solvothermal alcohlysis process by introducing isopropyl alcohol and produced less surface defect TiO_2 particles. Ramakrishnan et al. produced anatase TiO_2 nanoparticles through solvothermal method and microwave method and found microwave method as a promising candidate for producing higher crystallinity with agglomerated nanoparticles that encouraged for significant surface area resulting 14.52 mA/cm^2 current density and 7.44% efficiency [26].

Even though nanoparticles reassure surface area but confronted with low electrical conductivity resulting from the interfaces between TiO_2 nanoparticles. Moreover, the mesoporous network offers undesired recombination with penetrated electrolytes. Thus, numerous processes are discovered to synthesize numerous one-dimensional, two-dimensional and three-dimensional nano shapes such as nanotubes, nanobelts, nanowires, nanorods, nanoflakes, nanoflower, nano shrub and hollow spheres, respectively.

Since one-dimensional nanostructure such as nanotubes, nanobelts, nanowires and nanorods ambitiously afford direct electron transporting path for higher charge collection at transparent conductive oxide film. Hydrothermal process recognized for the synthesis of TiO_2 nanotubes controlled by rection temperature, reaction time, concentration of NaOH, etc. However, optimum wall thickness is highly essential for efficient electron transportation, since narrow wall thickness does not support the charge layer and length of the tube essentially lesser than the electron diffusion length for higher collection of electrons. Lee et al. studied the influence of reaction temperature on TiO_2 nanotubes property and found high crystalline anatase phase property at 120°C [27]. To upsurge the performance of solar cells, tree-like hierarchical TiO_2 nanotubes also have been synthesized by one-step hydrothermal process directly grown over substrate by treating 1.77 gram of potassium titanium oxalate with 9 mL of water and 13.5 mL ethanol [28]. Electrospinning technique employed for generating hierarchical nanorod branched nanofibers by Coe et al. and found incredible conversion efficiency [29]. Similarly, Surawut et al. incorporated electrospinning and sol-gel technique to synthesized high crystalline structure and remarkable produced TiO_2 anatase nanofibers of average diameter near to 250 nm after calcination [30]. The typical process involves mixing of 10 g isopropoxide, 4-gram polyvinyl pyrrolidone, 70 mL ethanol and 20 mL acetic acid at a temperature 70°C for 4 h in a magnetic stirrer and then 10 kV applied over 10 cm collector distance and perform electrospinning for 20 min [31]. TiO_2 nanobelt synthesized by treating TiO_2 nanopowder with NaOH aqueous solution through hydrothermal process followed by washing with HCl and calcinated [32]. Liu et al. employed hydrothermal process to synthesize TiO_2 nanowire of average diameter 80 nm and applied for the potential application in solar cells [33].

Two-dimension materials such as nanoflakes are synthesized by treating 10 M NaOH with TiO_2 nanoparticles at 130°C due to contravention of Ti-O-H and Ti-O-Ti into Ti-O-Na bond. However, it further transforms to TiO_2 nanotubes [26]. Moreover, various three-dimensional nanostructures also have been employed to encourage swift transfer of electrons by avoiding recombination. TiO_2 hollow sphere structure obtained by two steps include thermal hydrolysis of titanium sulphate in a solvent comprising of water and n-propanol followed by solvothermal reaction at 180°C for 18 h [34].

Even though many alternative TiO_2 morphology exercised but still exhibit low electron mobility. Beside this, TiO_2 nanoparticles intensively absorb light and UV-degrade the materials. Further the electronic and optical properties can be tuned by doping of different good electrically conductive materials into TiO_2 lattice. Since doping of nanomaterials formed a heterojunction architecture which alter the optical bandgap (cascaded electronic structure) favourable for the efficient electron injection and increase light harvesting efficiency. Moreover, doping upraises the availability of free ions and encourages electrical conductivity. Doping material could be of metallic or nonmetallic. Both hydrothermal and solvothermal synthesis routes employed for the development of doped TiO_2 structure by varying optimized wt.%. Nevertheless, doping can be incorporated by physical mixing with TiO_2 nanoparticles or can be deposited through anodization and thermal oxidation respectively. Many novel 2D materials such as graphene, reduced graphene oxide, metallic plasmon materials (silver, copper, gold, etc.) used as dopant prepared through sol-gel route and potentially applied to TiO_2 lattice.

11.2.2 Deposition of Nanostructured TiO$_2$

Several economical methods are available to deposit TiO_2 nanoparticles on transparent conductive substrate for greater potential application interest in solar cells. Prominently liquid phase deposition technique potentially popularized for low cost and simple deposition involves process. In the liquid phase deposition process, the depositing material is either in the form of a paste or colloidal solution. This process utilizes spin coating, doctor blading, dip coating, screen printing, pad printing and jet ink printing.

11.2.2.1 Spin Coating

Spin coating is a process where substrate kept over centre of chuck and rotated at a specific rpm for a specific time depending upon thickness of coating to be fabricated. On a respective time, the as-prepared solution drip on the centre of substrate by micropipette. The solvent on rotating substrate experiences centrifugal force and it enforced to spread over the substrate which forms a thin and uniform film over the substrate surface. Broadly the process involves static or dynamic deposition based on viscosity of the solvent, spinning speed and duration and evaporation. Out of all, spinning speed, viscosity of solvent and concentration of solution are responsible for the thickness of the film [35]. Volatile nature of solvent evaporates during coating process. One report communicates that spin coating can produce about 10 nm thinner film over the substrate [36]. However, an empirical formula is existed to relate coating thickness with input parameters as follows [37].

$$Coating\ Thickness = \left(1 - \frac{\rho_v}{\rho_{v0}}\right) \times \sqrt[3]{\frac{3\vartheta m}{2\rho_{v0} w^2}} \qquad (11.1)$$

where the symbol ρ_v represents volatile liquid density, ϑ represents viscosity of colloidal solution, m represents evaporation rate and w represents angular speed respectively. Moreover, spin coating acknowledged significant interest in solar cell applications for its quick processing and high homogeneity film production. Nevertheless, initial investment cost and confined to small substrate make the spin coating unwelcome [38]. Many researchers potentially employed the spin coating techniques in solar cell applications. Bandara et al. deposited compact layer by incorporating 2 mL HNO_3 with 0.5 g TiO_2 nanoparticle to get colloidal solution and spin-coated at 1000 rpm for 2 s and the second stage at 2350 rpm for 60 s [39]. Khan et al. investigated the optoelectrical property of spin-coated TiO_2 film by introducing 0.4 grams of TiO_2 nanoparticles in 5 mL ethanol as solvent with diethylene glycol as stabilizer to produce homogeneous solution and applied at 2400 rev/s for 30 s [40]. They found the electrical resistivity inversely relate with layer thickness and form a highly optical transparency. Hang et al. had employed spin coating technique to fabricate hill alike hierarchical nanostructure by introducing a paste made up of polyvinyl alcohol as binder followed by Triton X-100 and acetylacetone capping agent which motivated for excellent dye loading 6.46×10^{-8} mol/cm^2 resulted in 11.65 mA/cm^2 J_{SC} and 4.47 PCE [41].

11.2.2.2 Dip Coating

Dip coating technique involves dipping of a substrate into a liquid solution in which a thin layer is deposited on the substrate due to concurrent action of viscous force, gravity force and capillary force respectively. Thus, the thickness of the film primarily governed by speed of substrate renunciation, and viscosity of solution [42]. It is important to note that the speed of withdrawal of the substrate is selected in such a way that the shear rates remain within the Newtonian region. Moreover, the thickness of the film can be found out by an empirical relation given below [43].

$$Dip\ coating\ thickness = 0.94 \times \frac{(\tau \times v)^{2/3}}{\gamma^{1/6} \times \sqrt{\rho \times g}} \qquad (11.2)$$

where τ represents viscosity, v represents dragging speed, γ represents liquid vapour surface tension, g represents gravity constant and ρ represents density respectively. However, multiple passes needed to get required thickness for potential application.

11.2.2.3 Doctor Blading

Doctor blading got immense popularity to fabricate thin coating for the development of electrode for the solar cell application. Moreover, doctor blading method is also known as knife coating. The doctor blading process involves very simple steps of depositing required amount of TiO$_2$ paste and distributing it uniformly by a moving blade or unrolling by a glass rod by maintaining specific contact angle and height. For making the paste α-terpineol and ethyl cellulose or triton X and polyethylene glycol are incorporated for creating mechanical and electrical bonding between the nanoparticles [44,45]. Doctor blading method popularly employed for fabrication of porous TiO$_2$ nanostructure over glass substrate favourable for significant dye adsorption in solar cell application. However, doctor blading method needs high skill for deposition since, slide hand vibration can cause nonuniform thickness film layer. Nevertheless, doctor blading is simple and low-cost processing technique for which many researchers establish doctor blading method and got high-performance solar cells.

11.2.2.4 Screen Printing

Screen printing method is a process of depositing printable nanomaterials on a substrate via application of squeezing pressure as a result a uniform regular pattern coat over substrate. The size of the opening of screen accountable for the pattern of the screen-printed surface. Moreover, viscosity of the TiO$_2$ nanoparticles paste, surface texture of substrate, distance between screen and substrate, squeezing pressure, squeezing angle, wiping speed and snap off velocity respectively are crucial parameters of screen printing [46]. Screen printing method acknowledged for creating large module solar cell in economic way. For example, an area of 6000 cm^2 photovoltaic protype module was fabricated by screen printing method [47]. Moreover, screen printing process is a fast process but the TiO$_2$ paste essential to be designed based on Newtonian fluid with low viscosity for the potential application in solar cell. However, screen printing produces high production waste for which alternative methods are explored.

11.2.2.5 Ink Jet Printing

Ink jet printing method is a promising process in which small drop of TiO$_2$ paste formed and deposited over substrate suitable for fabricating flexible substrate. In this process a drop of TiO$_2$ paste injected by drop generator translate to electrostatically charged drops governed by electric field and accelerated towards substrate technically name as drops of demand. Further, the drops of demand could be of thermal treated or piezoelectric treated based on type of energy used for pushing the drops from the nozzle to the substrate [48]. Moreover, following relation can be used to quantify the thickness of the ink jet printing film [49].

$$Thickness = N_p \times V_p \times \frac{C}{\rho_A} \qquad (11.3)$$

where N_p represents no of delivered drops per area, V_p represents volume of a single drop, C represents TiO$_2$ paste concentration and ρ_A represents density respectively. This technique was found suitable for the fabrication of flexible solar cells.

11.2.2.6 Pad Printing

Pad printing method is a process in which a covered metallic sheet with photo-polymer act as characteristic feature through which TiO_2 paste transferred to substrate by squeegeeing and compression of pad. Pad printing technology can produce 5–30 micron thinner film. One of the reports communicates that a remarkable 6.2% conversion efficiency obtained from a solar cell having active area 0.25 cm^2 fabricated through pad printing method [50].

11.3 NANO DEPOSITION FOR FLEXIBLE SOLAR CELL

Next generation solar cell propels towards development of lightweight flexible solar cell. Since polymer substrates such as PET and PEN are thermally unstable beyond 150°C and the sintering temperature needed for the coated substrate is 450°C–550°C to evaporate organic binders or residues. Therefore, alternative strategies have been explored to develop low temperature route for the potential application in flexible solar cell. Several low temperature route techniques are employed such as chemical sintering, mechanical compression, electrophoretic deposition, etc.

11.3.1 Electrophoretic Deposition

Electrophoretic deposition (EPD) inspires to fabricate thin film over polymer substrate. The process is governed by application of potential differences between substrate and electrode in a solvent comprise TiO_2 suspended ethanol or butanol or isopropanol. However, the depositing property further increased by addition of deionized water and HNO_3 [51]. Further study reported that EPD results in inhomogeneous coating. To address above issue, acetylacetone as a dispersive agent can be incorporated to get uniform coating [52]. Nevertheless, cracking of film is a prime failure of EPD process. Thus, a secondary process could make the change by compression. Moreover, this post-treatment inevitably increases interfacial adhesion of nanoparticles and substrate which respond to lowering the surface defect and suppressing the recombination [53].

11.3.2 Chemical Sintering

Chemical sintering process uses volatile solvents instead of organic solvents which can be evaporated at 120°C in few min and create strong mechanical and electrical connections between substrate and TiO_2 film. Zhang et al. used NH_4OH as chemical agent for chemical sintering and registered significant solar conversion efficiency [54]. Holliman et al. introduced hexafluorotitanic acid (H_2TiF_6) as chemical agent and reported maximum 3.2% conversion efficiency [55]. However, it is important to note that it blocks the porous structure which reduces the specific surface area.

11.3.3 Mechanical Compression

Mechanical compression method is a pseudo sintering process that uses a cold isostatic pressure to compress the TiO_2 deposited substrate for a specific time as a result interparticle contact and the contact between TiO_2 film with the substrate would enhance. Thickness of TiO_2 film governed by compression pressure, viscosity of TiO_2 paste and duration. Weerasinghe et al. [56] and Shao et al. [57] potentially employed cold isostatic process and able to conquer mechanically stable titania nanoparticle interaction for the potential application. Similarly, Peng et al. applied it as secondary process after doctor blade coating at different compression pressure 50, 100 and 200 MPa, respectively and recorded 4% conversion efficiency [58].

11.4 CONCLUSION

The processing of the thin film solar cells is completely associated with the quality of nanostructures and nanofilm attributed to nanofabrication. Here numerous nano synthesis and nano deposition techniques are highlighted for the development of high-performance solar cell application. The range and the quality of crystal structure depend on synthesis process and calcination temperature. Various morphology can be engineered to upsurge specific surface area and efficient swift of electrons by introducing 1D, 2D and 3D nanostructures into semiconductive oxide film made of up TiO_2. Since various 2D materials are developed chronologically and have huge potential to improve the performance of solar cells [59]. Further, the depositing of TiO_2 nanostructures on the substrate is accountable for the charge transportation and charge separation. Doctor blading process popularized for generating efficient porous structure whereas spin coating method appraised for generation of thin compact layer [60]. Furthermore, low temperature route is required for polymeric flexible solar cells but still many researches are endeavoured for productive solar cells. Furthermore, development of next-generation solar cells is crucial to address green energy for sustainable ecosystem [60].

REFERENCES

1. Yue C-D, Huang G-R An evaluation of domestic solar energy potential in Taiwan incorporating land use analysis. *Energy Pol* 2011;39; 7988–8002. 10.1016/J.ENPOL.2011.09.054
2. Solangi KH, Islam MR, Saidur R, Rahim NA, Fayaz H A review on global solar energy policy. *Renew Sustain Energy Rev* 2011;15; 2149–2163. 10.1016/J.RSER.2011.01.007
3. Kabir E, Kumar P, Kumar S, Adelodun AA, Kim K-H Solar energy: Potential and future prospects. *Renew. Sustain. Energy Rev.* 2018;82; 894–900. 10.1016/J.RSER.2017.09.09
4. Ahmed F, Al Amin AQ, Hasanuzzaman M, Saidur R Alternative energy resources in Bangladesh and future prospect. *Renew Sustain Energy Rev* 2013;25; 698–707. 10.1016/J.RSER.2013.05.008
5. Hosenuzzaman M, Rahim NA, Selvaraj J, Hasanuzzaman M, Malek ABMA, Nahar A Global prospects, progress, policies, and environmental impact of solar photovoltaic power generation. *Renew Sustain Energy Rev* 2015;41; 284–297. 10.1016/J.RSER.2014.08.046
6. Schiermeier Q, Tollefson J, Scully T, Witze A, Morton O Energy alternatives: Electricity without carbon. *Nature* 2008;454; 816–823. 10.1038/454816a

7. Lewis NS Toward cost-effective solar energy use. *Science (80)* 2007;315; 798–801. 10.1126/science.1137014

8. Jean J, Brown PR, Jaffe RL, Buonassisi T, Bulović V Pathways for solar photovoltaics. *Energy Environ. Sci.* 2015;8; 1200–1219. 10.1039/C4EE04073B

9. Powell DM, et al. The capital intensity of photovoltaics manufacturing: Barrier to scale and opportunity for innovation. *Energy Environ Sci* 2015;8(12); 3395–3408.

10. Karim NA, Mehmood U, Zahid HF, Asif T Nanostructured photoanode and counter electrode materials for efficient Dye-Sensitized Solar Cells (DSSCs). *Sol Energy* 2019;185; 165–188. 10.1016/J.SOLENER.2019.04.057

11. Green MA Third generation photovoltaics: Solar cells for 2020 and beyond. *Phys. E Low-dimens. Syst. and Nanostructures* 2002;14; 65–70 S1386-9477(02), 00361-2.

12. Green MA Third generation photovoltaics: Ultra-high conversion efficiency at low cost. *Prog Photovolt: Res Appl* 2001;9(2); 123–135.

13. Ludin NA, et al. Review on the development of natural dye photosensitizer for dyesensitized solar cells. *Renew Sustain Energy Rev* 2014;31(0); 386–396.

14. Moon S-J Solid-state sensitized heterojunction solar cells – Effect of sensitizing systems on performance and stability, *EPFL* 2011.

15. Yan J, Saunders BR Third-generation solar cells: A review and comparison of polymer: Fullerene, hybrid polymer and perovskite solar cells. *RSC Adv.* 2014;4; 43286–43314. 10.1039/C4RA07064J

16. Płaczek-Popko E Top PV market solar cells 2016. *Opto-Electron. Rev.* 2017;25; 55–64. 10.1016/J.OPELRE.2017.03.002

17. Babar F, Mehmood U, Asghar H, Mehdi MH, Ul A, Khan H, Khalid H, Huda N, Fatima Z Nanostructured photoanode materials and their deposition methods for efficient and economical third generation dye-sensitized solar cells: A comprehensive review. *Renew. Sustain. Energy Rev.* 2020;129; 109919. 10.1016/j.rser.2020.109919

18. Djurišić AB, Liu F, Ng AMC, Dong Q, Wong MK, Ng A, Surya C Stability issues of the next generation solar cells. *Phys. Status Solidi – Rapid Res. Lett.* 2016;10; 281–299. 10.1002/pssr.201600012

19. Kumara NTRN, Lim A, Ming C, Iskandar M Recent progress and utilization of natural pigments in dye sensitized solar cells: A review. *Renew. Sustain. Energy Rev.* 2017;78; 301–317. 10.1016/j.rser.2017.04.075

20. Kazmi SA, Hameed S, Ahmed AS, Arshad M, Azam A Electrical and optical properties of graphene-TiO_2 nanocomposite and its applications in dye sensitized solar cells (DSSC). *J. Alloys Compd.* 2017;691; 659–665. 10.1016/j.jallcom.2016.08.319

21. Ong W-J, Tan L-L, Chai S-P, Yong ST, Mohamed, AR Highly reactive {001} facets of TiO2-based composites: Synthesis, formation mechanism and characterizations. *Nanoscale* 2014;6; 1946–2008. 10.1039/C3NR04655A

22. Aydin F, Ok S, Unal, Topal S, Cellat, Fatih Ş Synthesis, characterizeation, and application of transition metals (Ni, Zr, and Fe) doped TiO_2 photoelectrodes for dye-sensitized solar cells. *J. Mol. Liq.* 2020;299; 112177. 10.1016/j.molliq.2019.112177

23. Khan MI, Farooq WA, Saleem M, Bhatti KA, Atif M, Hanif A Phase change, band gap energy and electrical resistivity of Mg doped TiO_2 multilayer thin films for dye sensitized solar cells applications. *Ceram. Int.* 2019;45; 21436–21439. 10.1016/j.ceramint.2019.07.133

24. Kathirvel S, Sireesha P, Su C, Chen B, Li W Morphological control of TiO_2 nanocrystals by solvothermal synthesis for dye-sensitized solar cell applications. *Appl. Surf. Sci.* 2020;519; 146082. 10.1016/j.apsusc.2020.146082

25. Muniz EC, Goes MS, Silva JJ, Varela JA, Joanni E, Parra R, Bueno PR Synthesis and characterization of mesoporous TiO_2 nanostructured films prepared by a modified sol–gel method for application in dye solar cells. *Ceram Int* 2011;37; 1017–1024. 10.1016/J.CERAMINT.2010.11.014

26. Ramakrishnan VM, Muthukumarasamy N, Pitchaiya S, Agilan S, Pugazhendhi A, Velauthapillai D UV-aided graphene oxide reduction by TiO_2 towards TiO_2/reduced graphene oxide composites for dye-sensitized solar cells. *Int J Energy Res.* 2020;45; 17220–17232. 10.1002/er.5806.

27. Lee CH, Kim KH, Jang KU, Park SJ, Choi HW Synthesis of TiO_2 nanotube by hydrothermal method and application for dye-sensitized solar cell. *Mol Cryst Liq Cryst* 2011;539; 125/[465]–132/[472]. 10.1080/15421406.2011.566078

28. Chen, H, Li, N, Wu, Y, Shi, J, Lei, B, Sun, Z A novel cheap, one-step and facile synthesis of hierarchical TiO_2 nanotubes as fast electron transport channels for highly efficient dye-sensitized solar cells. *Adv. Power Technol.* 2020;31; 1556–1563

29. Cao Y, Dong Y-J, Feng H-L, Chen H-Y, Kuang D-B Electrospun TiO_2 nanofiber based hierarchical photoanode for efficient dye-sensitized solar cells. *Electrochim Acta* 2016;189; 259–264. 10.1016/J.ELECTACTA.2015.12.073

30. Chuangchote S, Sagawa T, Yoshikawa S Efficient dye-sensitized solar cells using electrospun TiO_2 nanofibers as a light harvesting layer. *Appl Phys Lett* 2008;93; 033310. 10.1063/1.2958347

31. Sun J, Yang X, Zhao L, Dong B, Wang S Ag-decorated TiO_2 nanofibers for highly efficient dye sensitized solar cell. *Mater. Lett.* 2020;260; 126882. 10.1016/j.matlet.2019. 126882

32. Zhao F, Ma R, Jiang Y Strong efficiency improvement in dye-sensitized solar cells by novel multi-dimensional TiO_2 photoelectrode. *Appl. Surf. Sci.* 2018;434; 11–15. 10. 1016/j.apsusc.2017.10.131

33. Liu W, Lu H, Zhang M, Guo M Controllable preparation of TiO_2 nanowire arrays on titanium mesh for flexible dye-sensitized solar cells. *Appl Surf Sci* 2015;347; 214–223. 10.1016/J.APSUSC.2015.04.090

34. Rui Y, Wang L, Zhao J, Wang H, Li Y, Zhang Q, Xu J Template-free synthesis of hierarchical TiO_2 hollow microspheres as scattering layer for dye-sensitized solar cells. *Appl. Surf. Sci.* 2016;369; 170–177. 10.1016/j.apsusc.2016.02.049

35. Hall DB, Underhill P, Torkelson JM Spin coating of thin and ultrathin polymer films. *Polym Eng Sci* 1998;38; 2039–2045. 10.1002/pen.10373

36. Scriven LE Physics and applications of DIP coating and spin coating. *MRS Proc* 1988;121; 717. 10.1557/PROC-121-717

37. Ahmadi S, Asim N, Alghoul MA, Hammadi FY, Saeedfar K, Ludin NA, Zaidi SH, Sopian K The role of physical techniques on the preparation of photoanodes for dye sensitized solar cells. *Int J Photoenergy* 2014;2014; 1–19. 10.1155/2014/198734

38. Luurtsema GA, Spanos J, Advisor R Spin coating for rectangular substrates. 1997. http://www.davidlu.net/gluurtsMS.pdf?G¼736&ln¼jp. [Accessed 14 August 2019].

39. Bandara TMWJ, Jayasundara WJMJSR, Fernado HDNS, Dissanayake MAKL, Silva LAA, De Albinsson I, Furlani M Efficiency of 10% for quasi-solid state dye-sensitized solar cells under low light irradiance. *J Appl Electrochem* 201;45; 289–298. 10.1007/ s10800-015-0788-1

40. Khan MI, Bhatti KA, Qindeel R, Althobaiti HS, Alonizan N Structural, electrical and optical properties of multilayer TiO_2 thin films deposited by sol-gel spin coating. *Results Phys.* 2017;7; 1437–1439. 10.1016/j.rinp.2017.03.023

41. Huang W-Y, Hsieh T-L Dyes amount and light scattering influence on the photo-current enhancement of Titanium dioxide hierarchically structured photoanodes for dye-sensitized solar cells. *Coatings* 2020;10; 13.

42. Low FW, Lai CW Recent developments of graphene-TiO_2 composite nanomaterials as efficient photoelectrodes in dye-sensitized solar cells: a review. *Renew Sustain Energy* Rev 2018;82; 103–125. 10.1016/J.RSER.2017.09.024

43. Landau L, Levich B Dragging of a liquid by a moving plate. *Dyn Curved Front* 1988; 141–153. 10.1016/B978-0-08-092523-3.50016-2

44. Jahantigh F, Ghorashi SMB, Bayat A Hybrid dye sensitized solar cell based on single layer graphene quantum dots. *Dye. Pigment.* 2020;175; 108118. 10.1016/j.dyepig.2019.108118

45. Aiswarya KM, Raguram T, Rajni KS Synthesis and characterisation of nickel cobalt sulfide nanoparticles by the solvothermal method for dye-sensitized solar cell applications. *Polyhedron* 2020;176; 114267. 10.1016/j.poly.2019.114267

46. Riemer DE The theoretical fundamentals of the screen printing process. *Microelectron Int* 1989;6; 8–17. 10.1108/eb044350

47. Hinsch A, Veurman W, Brandt H, Loayza Aguirre R, Bialecka K, Flarup Jensen K Worldwide first fully up-scaled fabrication of 60 × 100 cm 2 dye solar module prototypes. *Prog Photovoltaics Res Appl* 2012;20; 698–710. 10.1002/pip.1213

48. Hinsch A, Behrens S, Berginc M, B€onnemann H, Brandt H, Drewitz A, Einsele F, Faßler D, Gerhard D, Gores H, Haag R, Herzig T, Himmler S, Khelashvili G, Koch D, Nazmutdinova G, Opara-Krasovec U, Putyra P, Rau U, Sastrawan R, Schauer T, Schreiner C, Sensfuss S, Siegers C, Skupien K, Wachter P, Walter J, Wasserscheid P, Würfel U, Zistler M Material development for dye solar modules: results from an integrated approach. *Prog Photovoltaics Res Appl* 2008;16; 489–501. 10.1002/pip.832

49. Krebs FC Fabrication and processing of polymer solar cells: A review of printing and coating techniques. *Sol Energy Mater Sol Cells* 2009;93; 394–412. 10.1016/j.solmat.2008.10.004

50. Song MY, Kim SH, Jang YJ, Kim WJ, Kim YG, Song IW, Lee KC, Non-planar/curved dyesensitized solar cell and a method manufacturing the same, US 2012/ 0118367 A1, 2012.

51. Xue Z, Jiang C, Wang L, Liu W, Liu B Fabrication of flexible plastic solid-state dye-sensitized solar cells using low temperature techniques. *J. Phys. Chem. C* 2013;118; 16352–16357. 10.1021/jp408663d

52. Kocaoglu BC, Ozenbas M Production of flexible polymeric photoanodes using binder-free electrophoretic deposition and compression for dye-sensitized solar cells. *Phys. status solidi* 2015;1250; 1246–1250. 10.1002/pssc.201510090

53. Kocaoglu BC, Icli KC, Ozenbas M Optimization of selective electrophoretic deposition and isostatic compression of titania nanoparticles for flexible dye-sensitized solar cells. *Electrochim. Acta* 2016;196; 535–546. 10.1016/j.electacta.2016.02.198

54. Zhang P, Wu C, Han Y, Jin T, Chi B, Pu J, Jian L Low-temperature preparation of hierarchical structure TiO_2 for flexible dye-sensitized solar Cell. *J. Am. Ceram. Soc.* 2011;95; 1372–1377. 10.1111/j.1551-2916.2011.04984.x

55. Holliman PJ, Connell A, Davies M, Carnie M, Bryant D, Jones EW Low temperature sintering of aqueous TiO_2 colloids for flexible, co-sensitized dye-sensitized solar cells. *Mater. Lett.* 2019;236; 289–291. 10.1016/j.matlet.2018.10.118

56. Weerasinghe HC, Sirimanne PM, Simon GP, Cheng YB Cold isostatic pressing technique for producing highly efficient flexible dye-sensitised solar cells on plastic substrates, *Prog. Photovolt: Res. Appl.* 2012;20; 321–332.

57. Shao J, Liu F, Dong W, Tao R, Deng Z, Fang X, Dai S Low temperature preparation of TiO_2 films by cold isostatic pressing for flexible dye-sensitized solar cells, *Materials Letters* 2012;68; 493–496.

58. Peng Y, Liu JZ, Wang K, Cheng YB Influence of parameters of cold isostatic pressing on TiO$_2$ films for flexible dye-sensitized solar cells, *International Journal of Photoenergy* 2011; 2011;410352.
59. Singh S, Arka GN, Gupta S, Prasad SB Insights on a new family of 2D material mxene: A review, *AIP Conference Proceedings* 2021;2341(1); 040017. 10.1063/5.0049984
60. Arka GN, Prasad SB, Singh S Comprehensive study on dye sensitized solar cell in subsystem level to excel performance potential: A review, *Solar Energy* 2021;226; 192–213. 10.1016/j.solener.2021.08.037

12 Emerging Nanomanufacturing Techniques with 2D Materials

Mamta Kumari, Ashok Kumar Jha, and Subhash Singh

CONTENTS

12.1 Introduction...189
12.2 Application Area of Nanomanufacturing 191
12.3 2D Materials and Its Feature Characteristics...192
12.4 Techniques and Processes Involve in Nanomanufacturing......................194
    12.4.1 Factors for Impacting Industrial Advancement through Nano Materials 195
    12.4.2 Top-Down Manufacturing Approach..196
    12.4.3 Bottom-Down Manufacturing Approach 198
12.5 Future Work Perspectives 200
12.6 Conclusion 201
References...203

12.1 INTRODUCTION

Nanomanufacturing has emerged its participation towards rapid development in the field of fabrication has generated wide range of expectations for commercial and industrial enlargement and expansion because of its uniqueness in the features for instance structure, shape, size composition and implication of function with various examples such as nanotubes, nanoparticles, nanowires, nanofilms and quantum dots including with many utilization in coating, ceramics fabrication, field of bio-technology cosmetics and skin care products, thin films, semiconductors, etc. In near future, the development and implementation in terms of utilization have been predicted to increase by observing its demand day by day. Processing and manufacturing of products at nanoscale is known as nanomanufacturing and is also described as the control of dimensions between 1 and 100 nm where it acquires solitary function at the molecular and supermolecular extent to generate and engage the elemental structure of material and system with enhanced features and applications having the mechanical, physical, chemical and biological properties at

DOI: 10.1201/9781003220602-12

controlled nanoscale range [1,2]. Surface area increases due to some certain changes occur in the properties of the material at molecular level and a large surface area to volume ratio. Some evidence from various literature study and analysis have represented that copper becomes transparent however it acts as a cloudy at macro scale while platinum performs as a catalyst which is originally inert in its natural form. Distinguishably stable property of aluminium is however combustible acts as an insulator at molecular state in addition to this silicon proceed as a conductor at nanoscale range [2]. Comparing with melting points the properties of gold are remarkable the reason behind this is it can change its colour from yellow to blue, pink, red and violet which can depend on the size of fine particles has built the capability at this stage to function as a catalyst. A nanoscale measure defines the material properties like electrical, optical, thermal, magnetic properties and expected to be less cost-effective due to less quantity of materials utilization [2].

Human need and imagination have emerged the circumstances to grow new technology in the field of nanoworld. Earlier it was not in exposure due to various reasons like lack of technical development to facilitate the fabrication process but drastically it has grabbed the attraction in the field of manufacturing due to rapid growth and technical advancement worldwide. However, the concept of nanotechnology is not new because some literature has clearly shown and presented the evidence of using the word "nano." The term "nano metre" was first introduced by Richard Zsigmondy, a Novel prize acclaimed in the year 1925 who was the first to measure the nanoparticle size using a microscope. The Golden age of nanotechnology has begun when a lecture was delivered by Richard Feynman in American Physical Society in the year of 1959 titled "There is Plenty of Room at the Bottom" which became very famous later had a great impact on new field of Physics to manipulate and analyze the elements which are in nanoscale range of parameters and various fundamental problems of small scales and contemplated as the father of modern nanotechnology. Fifteen years later of this incident Norio Taniguchi who was a Japanese scientist raised the topic of nanotechnology when his work was going on semiconductors and their processes in molecular deformation of materials [3]. Later, various scientists had given their inputs to verify its importance, occurrence, and application in the field of modern science and technology. Advancement in methodology and techniques has made it possible to remove material at micro and macro level, enhanced the development scenario, leading to commercialization of nano products world widely. Current situation focuses on application-based products from thin films to wire bonding, packaging, and assembly enables the multistate nanomanufacturing system in active and drivers' mode of function. From this point of view, it is possible to tailor material structure in order to full fill the requirements at an exceedingly small range of scale without disturbing the specific properties; it can acquire more strength, extensively lighter in weight, more durable, better in electrical conduction capacity and many more in commercial products like nano components, useful systems, devices, etc. for value-adding benefits and securing greater position in industrial growth as well as in research and development area. Additionally, being an interdisciplinary specialization because it enables the production and fabrication of new types of materials and products in the field of nanoelectronics, nanomaterials and nanobiology which focuses on material removal processes, medical devices, lithography,

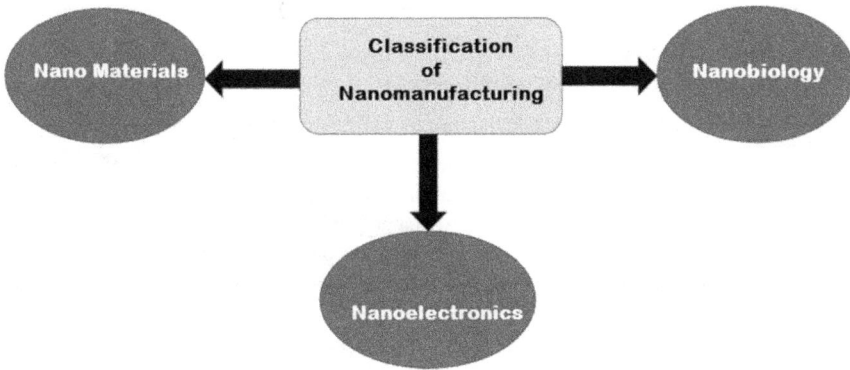

FIGURE 12.1 Classification of nanomanufacturing.

electrostatic coating, agriculture, health care devices, information technology, construction, robotics, energy, security, biomedical, etc. [4]. Detailed information about classification of nanomanufacturing is show in Figure 12.1.

12.2 APPLICATION AREA OF NANOMANUFACTURING

Nelson et.al. [2] and Nicholas R. et al. [5] have given the compiled form of applications on nanomanufacturing process presented below in Table 12.1 for better clarification towards the understanding of wide range utilization of interdisciplinary branch of N.M.

TABLE 12.1
Application of Nanomanufacturing

Serial no.	Nanomanufacturing	Application
1	**Nano Materials** 1D materials	a. Waterproof fabrics, thin films for electronics devices, fuel cell surfaces and catalysts in some devices. b. In energy storage devices and catalysis for inorganic nanotubes e.g., Molybdenum disulphide.
	2 D materials	a. Silicon nanowires for data storage, optoelectronic devices, photodetector, energy storage devices and batteries. b. Antistatic packaging works, sensors used with carbon nanotubes, devices for electric current transmission. c. DNA molecules in Biopolymers used in hydrogen fuel cells for confinement of nanotubes.
	3 D materials	a. In cosmetics industries for preparing beauty products. b. In textile industries for making related products, preparation of paints, lubrication agents, polymeric molecules dendrimers used in coating and covering and delivering of drugs in electronic circuits.

(Continued)

TABLE 12.1 (Continued)
Application of Nanomanufacturing

Serial no.	Nanomanufacturing	Application
2	**Nano Biology** Bio-nanosensors	a. For implanting silicon and combined form of enzymes in human or animal body for monitoring health issues.
		b. Delivering of drugs, diseases diagnosis, imaging of molecules and biomimetic structure.
	Tissue engineering	a. Gene correction therapies, new techniques for drug insertion.
		b. Regeneration and repairing of damaged tissues using nano material-based scaffolds.
3	**Nano Electronics** Optoelectronic devices	a. Used in cameras and personal computers in the form of nanowires and quantum dots, storage devices as batteries.
		b. Semi-conductors, optical devices, to check the quality of soil fertility made of nanomaterial devices.
		c. Used in food industries and in many more products.

12.3   2D MATERIALS AND ITS FEATURE CHARACTERISTICS

Interdisciplinary branch of technology illustrated in above table mainly focuses on utilization of nanomaterials in industrial, medical, and other fields however the major aim of this presentation is to give information about nanomaterial i.e., now has no limited use, become versatile in every field as a complete system can be generated through 2D materials.

In this regard, 2D materials have secured sufficient attraction towards the development of novel electronic appliances and optoelectronics with high grade of adaptability and feasibility as unique in electrical, high in mechanical, physical, optical properties, great thermal conductive properties with high tensile strength. Any material can be possible to make 2D material, thin it down until the thickness of the material become only at a few atomic levels.

Graphene came from successful exfoliation of 2D material i.e., graphite [6] in 2014, was first discovered and designed nearby 2004 [7] acquires the high range of properties specifically treated as versatile nanomaterial as this is a one-atom-thick plane sheet of paper bonded carbon atoms firmly filled up into a honeycomb crystal lattice form in plane σ c-c bonds and out -of -plane π bonds which enhances the above-mentioned properties including with outstanding characteristics features as high electron mobility capacity, large surface area and high rate of transmittance. After containing several layers of the same material, it becomes stable in the air including some changes in polarity by graphene oxide doping. Some other attractive features include, graphene extending 10^6 cm^2 V^{-1} s^{-1} at 1.8 K and 10^5 cm^2 V^{-1} s^{-1} at the room temperature and absorbs 2.3% of exploiting white light and opaqueness increases by 2.3% after the addition of some layers on it, evidence came from theoretical study of lattice structure and monolayer graphene with a 2D Young's modulus 340 N/m, and third elastic stiffness -690 N/m in nonlinear condition [8].

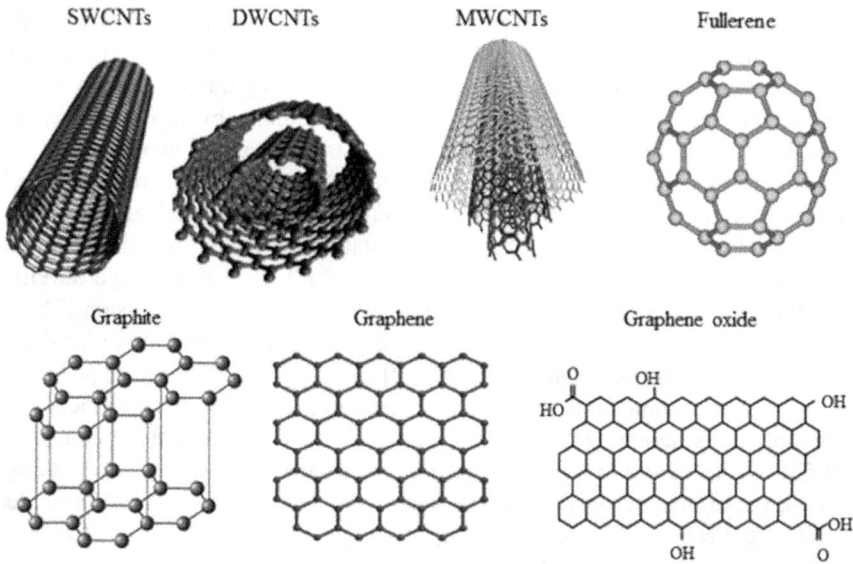

FIGURE 12.2   Different structures of carbon nanoparticles of graphene [10].

Carbon nanotubes can be produced by rolling the graphene into a tube which is a one-dimensional 1D carbon material made by one or more layers of graphene and converted into seamless cylinder which indicates outstanding performance and properties like graphene [7–9] (Figure 12.2).

Unlike graphene there are other various forms of 2D materials which have great and excellent impact on both application and properties Transition Metal Dichalcogenides (TMDs), Bucky Tape, Black Phosphorus or phosphene, hexagonal boron nitride (h-BN) [11]. The TMDs possess sandwich-like structure of two layers of chalcogens with transition layer of metals whereas graphene possesses only carbon layer of atom thickness [12,13]. TMDs accompany with MX_2 structure (Mo, W, Nb, Re, etc) where M acts as a transition metal and X is a chalcogen atom (S, Se., or Te) combining effect of M and X can vary from semiconductive to metallic and even superconductive behaviour complimenting with insulating and semi-metallic layer material. Besides this, Phosphene, silicene, germanene, borophene, mono-chalcogenides i.e., GaSe, SnSe and MXene [14] where M is Ti, Nb, V, Ta, etc. and X is C and N carrying high range in electronics and photonic use and exercise also acquiring high band gap (1–2.5 eV) [7] near to solar spectrum, dominate deeper level of defects for papery thin field effect transistors (FETs) and p-n junction usable in harvesting, sensitive photodetection, low threshold lashing and can withstand high amount of strain without mortifying its properties. GO thin sheet film or membrane possesses lighter weight due to the presence of carbon, oxygen and hydrogen which are broadly used in ceramics, formation of alloys and producing graphene-based materials [6,15].

Addition to above mentioned information regarding 2D material black phosphorous can also be considered as a specific material as far as properties and utilization is concerned revealing its wrinkled structure is an allotrope of phosphorous known as most stable element at atmospheric pressure band gap of 0.3–2 eV developing an adhesive residue on black phosphorous surface while working on it, remedial solution has found to be applied after modifying the preparation techniques as this material continuously forcing to get thinner by tape and pressing method to a polydimethylsiloxane (PDMS) substrate to obtain two layers of nanosheets, as this process is clean and occurring high quality of BP which is much required constituents of getting excellent results in studying, analyzing the basic characteristics of devices to get over the limitation of having thicker nanolayers comparing with others which is difficult to control till date research and development level restricting wide range of study and industrialization expansion [7].

Carbon Nitride (BCN) compared with other composite material considered as a new material having controllable band gaps (0–5.5 eV), several advanced properties such as electrochemical action, cathodoluminescence, optical sponginess and structural configuration varies with the change in bonds between three atoms i.e., and C used in various fields like capacitors, energy storage devices, gas storage devices, photocatalytic operations, etc. Many stimulation processes are presented for increased stability and atoms at lowest energy level of B-C, C-N, B-B, N-N materialize in OD, 1D or 2D forms [8].

In this context it will be considered half discussion without inclusion of hexagonal boron nitride (hBN) and its advance features, is contemplated as one of most stable and non-reactive, remain stable in air at a rising temperature of 1500°C with other present chemicals in the materials. As hBN acquires excellent dielectric properties and characteristics a single layer of this used in the high-performance devices of cantered on graphene and can be considered as supreme coating material for oxidation and oxidation-resistant coating on Ni surfaces for high utilization for oxygen quantities. Further, surface modification like electrochemical polishing, using of filler instead, against corrosion can be improved and enhanced. It has some additional utilization in the preparation of mainly thin films, ceramic composites (in the form of fibres) and electronics [16].

12.4 TECHNIQUES AND PROCESSES INVOLVE IN NANOMANUFACTURING

This section presents the general and common methods of preparing nanomaterials i.e.,

    i. Top-down manufacturing approach (To start from bulk or thicker material to thinner).
    ii. Bottom-down manufacturing approach (atomic ingredients and joining together [17].

Bottom-up method is preferable than top-down method as so far research study and practise is concerned, top-down method is failing down having more waste producing at the time of working so bottom-up tools will be ultimate method for producing sustainable metals in coming days. Figure 12.3(a) & (b) [17,18] is showing the complete methods of producing 2D materials by nanotechnology.

(a)

(b)

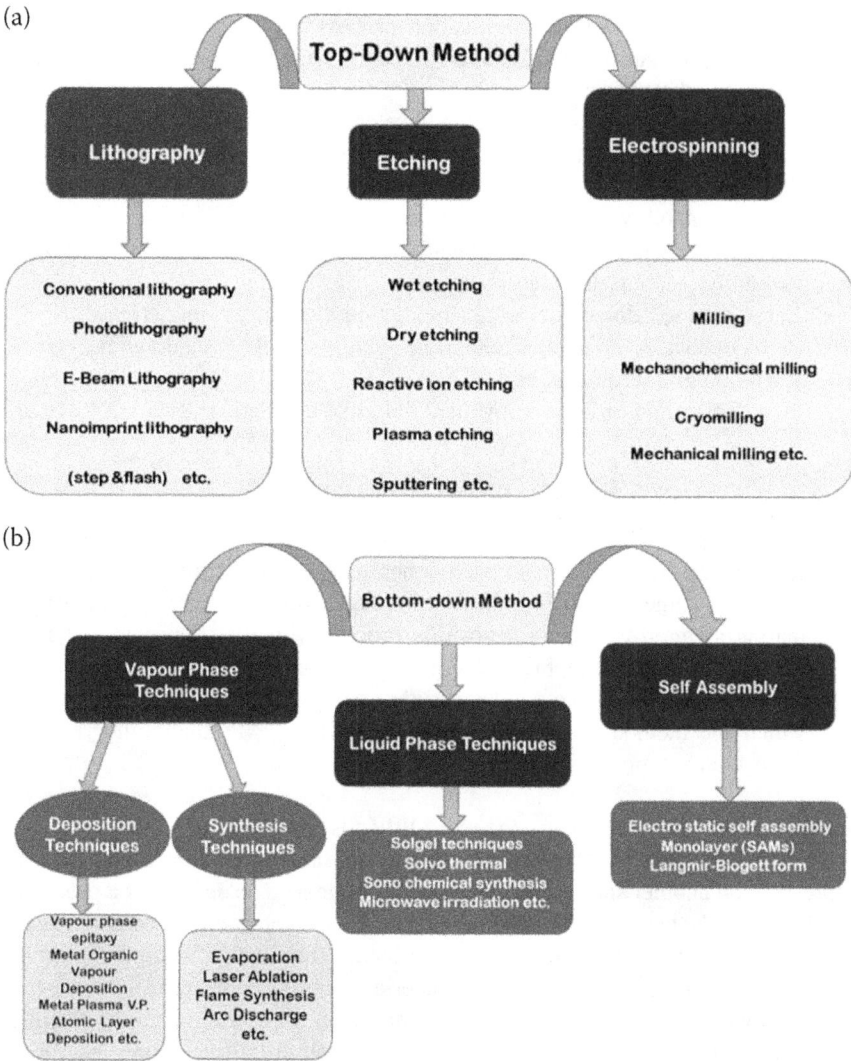

FIGURE 12.3 (a) Methods of producing 2D materials by top-up approach (b) methods of producing 2D materials by bottom-down approach.

12.4.1 FACTORS FOR IMPACTING INDUSTRIAL ADVANCEMENT THROUGH NANO MATERIALS

- The influencing factors behind the industrial enhancement and development in almost every sector of fabrication are very crucial part of discussion so far as nano word is concerned, reason has unlocked the door of increased dependency on nano products and its functionality.

- Raw material used in the fabrication process can be controlled and manipulated at nanoscale to convert it into finished nanoparticles and nanostructural systems.
- More than 200 processes have been developed in the field of nano designing and development industries so far which come under top-down mainly lithography and bottom-up known as assembly pattern making it easy because of the availability of wide of nanoscale materials and processes.
- Continuous roll-to-roll, top-down/bottom-up processes: printing, self-assembly, coating, lamination, etc.
- Large area top-down/bottom-up: lithography, direct writing, etc.
- Semiconductors or parallel chemical /fluid/thermal: fibre drawing, electrospinning, microfluids, etc.
- Nanofabrication of DNA in medical field like templating using DNA [19].

12.4.2 TOP-DOWN MANUFACTURING APPROACH

i. **Lithography** – The development of top-down approach has initialized and recommended various major techniques to facilitate in the field of electronic instruments and industries to enhance the quality in global nano market as improvement in deposition process which is connected and accompany with lithographic techniques for instance ion beam lithography, X-ray lithography, electron beam lithography. Some other examples in which the methods are used to prepare different NPs photolithography, atom lasers, coprecipitation, etc.

According to Oxford dictionary, the word **lithos** is derived from Greek word stone is basically drawing or writing down on yellow salty limestone to perform spot impression on another surface. In other words, it can also be defined as a process of printing initially based on non-miscible of oil and water and from a stone i.e., lithographic stone or a metal plate with a smooth surface or can be used to print any text onto paper or on the appropriate material counting with huge advantages in various micro and nanofabrication processes due its useful features and characteristics which gives high resolution having the ability of patterning, number of molecular and biomolecular inks with direct write techniques on semiconductors, metals, monolayer surfaces.

In this chapter writing work many studies and analysis have been so far completed in review of organizing this context about lithography and a very keen and crucial factor regarding this come in the existence associated with advancement in high-resolution process in recent years like 193 nm immersion lithography, extreme ultraviolet lithography (EUV), 3D two-photon lithography, molecular glass photoresists and nanofabrication with the help of block co-polymers. Recent publications have been focused on various aspects of research area in current advancement, however step-and-flash imprint lithography is one of the convenient techniques which are included in non-conventional lithography after being adopted by industry [20]. Nano imprint lithography is a process of an atomic force microscope dipped into chemical fluid and used to write and generate features on nanoscale by stamping and printing on

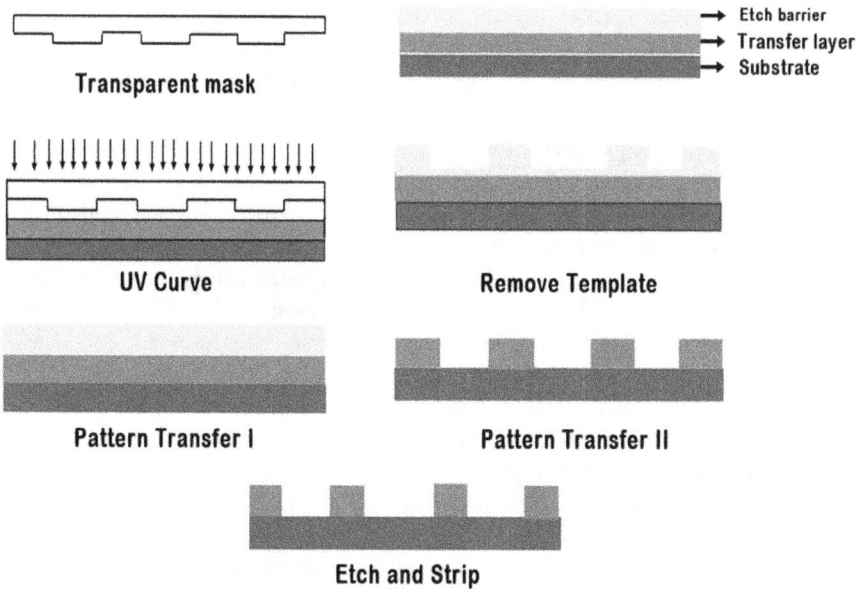

FIGURE 12.4 Step-and-flash imprints lithography process.

any metal surface or substrate. Step-and-flash imprint is considered one of the most convenient and suitable lithographic process as shown in Figure 12.4. in which the stepwise procedure has been elaborated. In this process layer of organic polymer, transfers on to a spin-coated silicon surface. Organosilicon monomer and crosslinker generally known as etch barrier are spread in a single stepper field with low viscous mixture of photopolymerizable and a UV transparent mould having some required forms of pattern are situated and compressed to withstand within it, followed by some exposure by UV radiation to protect the pattern form. Mould is then removed, and a short fluorocarbon reactive ion etch step is executed at room temperature with low pressure to remove the residual organosilicon polymer called "scum layer." A mask of pattern will be formed on material surface and oxygen plasma etch will be employed on transfer layer. Major factors are considered before designing the step-and-flash imprint lithographic process etch barrier and its adhesion properties, wettability, kinetics of photopolymerization, shrinkage capacity and etch selectivity. Each is very crucial parameter to perform this process of lithography. To achieve elevated accuracy in performance, etch barrier must contain more than 15 wt % silicon and high oxygen plasma etch selectivity should be in prime consideration over the primary organic transfer layer and organic monomer is selected to give adequate solubility for good adhesion properties as well as initiator to bottom transfer layer for several antireflection film and coating [20,21]. There are some other lithography processes and methods also available for preparing 2D materials which have different fields of application and utilization like known traditional and conventional lithography, E-beam lithography, photolithography which is used for mask generation only and focused ion-beam lithography, Next-generation, X-ray lithography and extreme ultraviolet lithography, soft lithography, low wavelength lithography.

ii. **Etching** – Etching is a method of removing unwanted materials from working surface by using chemical or physical processes which consists of plasma dry etching, chemical wet etching and other chemical and physical processes [22]. Dry etching can be used for manufacturing and assembly of GaN-based electronic devices. Chemical stability is responsible for the prevention of group-III nitrides used in wet etching but in contrary surfaces are affected by negatively charged ions and guided by high energy rate of plasma irradiation and bombardment of ions. Disorder in atomic bond arrangement and nitrogen-vacancy can be influenced by interface stage and high surface density which can show some functional stability issues like collapsing of current, gate leakage and voltage instability in threshold. Annealing treatment can be performed to recuperate affected surfaces by etching process. Previously the wet etching was famous in formation of semiconductors in order to facilitate micro and nanostructure because of highly sensitive rate, same etching speed in both longitudinal and lateral directions for isotropic etching with high yielding rate along with low cost. Having some disorders of using wet etching techniques, is gradually replaced by dry etching acquiring various advantages in high precision works. Some very crucial transformation has been monitored using dry etching that ion sputtering like laser etching, plasma etching, reactive gas etching and some other examples are popular which do not contain wet and moisture treatment. Etching techniques pay great attention towards the formation of 2D material with high impact on optoelectronic devices because of unique electrical and optical properties. These techniques are required in order to attain shape of 2D materials and controlled number of layers which is very difficult to achieve [22,23].

iii. **Electrospinning** – To facilitate and fabricate 1D material of nanofibers and some other interconnected materials, electrospinning techniques is the most famous for synthesis processes and getting the best possible results regarding inorganic 2D nanostructure. Controlled parameters, safe, low production rate including some crucial designing techniques are very important to get large aspect ratio of TDNBs with minimum thickness of 50 nm with some micrometres of length. Consideration of facts are always there to get feasible result in synthesis process because it is quite difficult on 2D materials and to get single layer of atomic structure of particular material naturally so, electrospinning techniques are used in preparation and formation of less thickness material i.e., less than 100 nm [24].

12.4.3   Bottom-Down Manufacturing Approach

i. **Liquid phase and vapour phase deposition process** – These synthesis techniques are mainly used for generating and manufacturing nanoparticles nano structured formation of nanomaterials and compound nano configuration including with solvothermal method is used to generate self-assembly of GO nanosheets into 3D interconnected network of lightweight graphene hydrogels by consistently heating the solution of GO aqueous at 180°C in an autoclave.

ii. **Sol-gel polymerization** – It is another process of preparing graphene-based aerogels of mesoscopic size in which covalent bond between assembling unit cell is initiated and control the aerogel properties, dense pore diameter, large specific surface area of super elastic graphene of 3D porous structure and carbon monolith of ultra-density are described with this synthesis process. GO nanosheets synthesis process can be controlled at the molecular and nanoscale level and spherical solid shelled bubbles at the micro-scale with micro fluids into centimetre-scale of assembled structure of the material with extraordinary density and Young's Modulus through this process [19].

iii. **Chemical Vapour deposition process** – For preparing high-quality of semiconductor film crystal of large area, chemical vapour deposition is a regular method to obtained this with redox reaction between the reactants to produce solid expedite film comparing with other technique CVD is more convenient and flexible process. Formation of heterojunction, doping and alloying synthesis has emerged the diversity in synthesis of W and Mo-based dichalcogenides. Large scale monolayer Mos2 membrane and thin films reported useful and very small amount of MoO3 i.e., 0.01 mg through CVD process.

Reaction occurred in chemical vapour deposition process including with polycrystalline growth and development of plasma enhanced chemical vapour deposition reaction used to produce pure and high performance of membrane and thin films hence, it highly recommended method by researchers for fabrication of single to more layers of nanosheets in either crystalline or amorphous structure and configuration which is little difficult for formation of thicker layers. High precision with controlled and crucial parameters such as temperature, pressure, flowrate is required for producing high-quality and defect-free graphene nanosheets which are basically useful in anticorrosion resistance in various highly sensitive sensing devices, electronics transistors, electrodes and some other devices.

iv. **Epitaxial growth** – One of the most efficient method to produce graphene sheet is epitaxial growth on silica carbide (SiC), which is exhibited at an extreme temperature around 1200 to 1500°C in vacuum giving silicon ions which results in carbon atomic layers on the working surface. This process is also feasible for self-assembly of hexagonal sheets of carbon layers which further transfer into some substrate layers including with some exceptions of having band gap of 260 meV in epitaxial graphene which offers advanced electronic analytical mechanism for B-doped diamonds and distinct biosensors [11].

v. **Molecular beam epitaxy and atomic layer epitaxy** – Other remarkable deposition method in this context are molecular beam epitaxy and atomic layer epitaxy which are one of the highly controlled methods for making thin films and depositing one atomic thick layer on the substrate respectively. Ultra-thin plastic membrane is forming through high-volume process of roll-to-roll on nanoscale.

vi. **Self-assemble techniques** –As the name suggested from the headline, this technique is described as a group of different components approach and proceed together to form an ordered and sequencing structure without providing any outside direction.

12.5 FUTURE WORK PERSPECTIVES

There is always an opportunity in the field of engineering research and development work for enhancing the techniques and technologies, methodologies, processes and assembly for better result and output with less effort, cost and input. Some statements are presented below to display future aspects of nanomanufacturing of 2D materials.

i. Various scopes are there in future as hexagonal HbN, TMDs and BP reveal high spectrum of properties like bone and teeth fixing and promotion of some other therapies of stem cells and osteoblast into fibroblast including with implantation of tissues linking [25].

ii. Chemical synthesis for example colloidal synthesis by chalcogenide of 2D material by nano formation with TiO2 nanotube layer [26] required form and design are nowadays increasing over many complementary conventional gas-phase, bound synthesis, exfoliation and platform to evaluate the single to -few-layers of TMD materials where it accommodate 2D nanosheet growth and controlling the growth by overcoming propensity situation for nano formation versus multi extended portion results in nanosheet growth and development needs deeper understanding of chemical reaction and nucleation processes.

iii. Heterostructure of nanosheet and diverse group of TMD system enables it to form up to dated two-dimensional materials that can be formed with controlled size and high yielding capacity [6].

iv. Doping process is one of the best and versatile future solution for forming and designing the band structure and for multi-functional 2D electronics and optoelectronics in doping by electromagnetic fields. Having some advantages and limitations, electromagnetic properties in this regard is yet to study and analysis, synthesis process and furthermore development can provide novel solution to overcome the drawbacks like increment and decrement in alloying doping level and types of inherent defects in 2D lattice structure is still under cover. Alloy formation and long-range ordered superstructure and heterojunction with total phase separation can be further studied, analyzed and modified to full all the required parameters for experimental functionality and characterization guideline to get best possible result.

v. Several opportunities are there to enhance new functionality in graphene-based mechanism by controlling the coupling to other surface materials. Some examples are given for better understanding in using the substrate which is coupled with spin-orbit can be able to give information about quantum spin hall state in graphene without magnetic fields. Another point is there to discuss related with development of measurement techniques to provide unique state of material. Heterostructure can also be created by combining two different 2D materials considering any crucial properties of CVD-synthesis 2D TMDs having some challenges and changes for better 2D interaction and device performance output.

vi. Producing next-generation and designing of nanoscale optoelectronics devices are there to analyze and develop some optical properties of STMDs and STMDs with 2D materials which have the capability to react the direct light-matter interaction at some critical scales and nano-optical imaging excitonic properties that have the chances for both utilization and improvement in synthesis techniques in controlled nanoscale level.

vii. Till now development in nano optics reaches only on nanoscale excitation phenomenon of electron energy loss spectroscopy which provides valuable understanding towards excitons in 2D transition metal semiconductors for nanoscale electronic properties of graphene.

viii. Design and fabrication process of TMDs and heterostructure devices required explanation of high ON-current and low subthreshold swings. Remarkable functionalization of TMDs is still in the process of gaining high-quality high dielectric layer to modify the properties of the material compared with isolated state [6].

ix. 2D materials are appealing member for nanomanufacturing as this process is quite feasible for producing power and energy storage devices famously known as batteries, focusing on volumetric storage electronic capacities having the qualities of lightweight energy devices with high-quality performance for next generation. Major considered factors can be raised in this context as Li-ion battery has two electrodes (anode and cathode) with a partition between them which serves as a medium for diffusion agent as most of the Li-ion batteries are made with graphite and Li-metal oxide e.g., lithium cobalt oxide. Anode can be replaced in Li-ion battery and Li-sulphur and Na-ion having low cost and with high potential, acquiring high specific capacities in near future research [5].

12.6 CONCLUSION

This work has presented so far, the importance and utilization of 2D materials which has great and excellent impact on both application and properties in nanomanufacturing.

i. TMDs, Bucky Tape, Black Phosphorus or phosphene, hexagonal boron nitride (h-BN) play a pivotal role in developing nano products. The TMDs possess sandwich-like structure of two layers of chalcogens with transition layer of metals whereas graphene possesses only carbon layer of atom thickness [12,13] which accompanies with MX_2 structure (Mo, W, Nb, Re, etc.) where M acts as a transition metal and X is a chalcogen atom (S, Se., or Te) combining effect of M and X can vary from semiconductive to metallic and even superconductive behaviour complimenting with insulating and semi-metallic layer material.

ii. Phosphene and other materials like silicene, germanene, borophene, mono-chalcogenides, MXene, SnSe and GaSe where M is Ti, V, Ta, Nb, etc. and X is C and N carrying high range in electronics and photonic use and exercise acquiring high band gap (1–2.5 eV) near to solar spectrum, dominate deeper level of defects for papery thin field effect transistors (FETs) and p-n junction usable in harvesting, sensitive photodetection, low threshold lashing and can withstand high amount of strain without mortifying its properties.

iii. So far present scenario has considered in optoelectronics of TMDs, BP and graphene the mixture and fusion of presented materials have great influence on direct synthesis and direct grown lateral junction mechanism used in optoelectronics devices.

iv. Automated assembly through nanomanufacturing process is more feasible and reliable process than nanomachining techniques.

v. It is showing the huge gap between fundamental study, analysis and industrial commercial utilization for future opportunity simultaneously. In order to fulfil this, some convenient and research strategy should be adopted, applied and explored to get best feasible result in quality and also exploring some new properties of 2D materials. Furthermore, so far optical and electrical properties only have got attention in development of nanomanufacturing.

vi. Mechanical and magnetic properties have still some opportunities to grow and develop including with some physical phenomenon. Many techniques are still under progress to get highly desired van der Waals heterogeneous structure along with environmental stability and encapsulating 2D material with hBN fabrication of device with effective methods.

vii. When the surfaces are affected by negatively charged ions and disorder in atomic bond arrangement and nitrogen-vacancy influenced by interface stage and high surface density showing some stability issues like collapsing of current, gate leakage and voltage instability in threshold wet etching and annealing treatment give great solution to recuperate affected surface.

viii. Designing techniques are used to get large aspect ratio in case of using TDNBs with minimum thickness of 50 nm with some micrometres of length with controlled parameters, safe, low production rate because it is difficult to get single layer and atomic structure of 2D materials so, electrospinning techniques overcome this problem including with those having difficulty in synthesis process of less than 100 nm thick materials.

ix. Mass production in electronics components is a recent challenge in practical and experimental approach towards the integration of 2D materials which comprises with photodetector and high-quality performance devices.

x. Fabrication process can be controlled and manipulated at nanoscale to convert raw material into finished nanoparticles and nanostructural systems/devices.

xi. Aluminium is combustible and acts as an insulator at molecular state in addition to silicon proceed as a conductor at nanoscale range and copper become transparent however it acts as a cloudy at macro scale while platinum performs as a catalyst which is originally inert in its natural form.

xii. Chemical etching is used for synthesis of Mxenes and exfoliated and terminated the surfaces with some another group of etchants available i.e., from MAX phase and it depends on delamination and etching process for best outcome in terms of properties and utilization [27].

REFERENCES

1. JE Hulla, SC Sahu, AW Hayes, Nanotechnology: History and future, human and experimental toxicology, *SAGE* 2015, 34(12), 1318–1321.
2. NA Ochekpe, PO Olorunfemi, NC Ngwuluka, Review article nanotechnology and drug delivery Part 1: Background and applications, *Tropical Journal of Pharmaceutical Research*, June 2009, 8(3), 265–274.
3. L Leon, EJ Chung, C Rinaldi, A brief history of nanotechnology and introduction to nanoparticles for biomedical applications, Eun Ji Chung, Lorraine, Leon, and Carlos, Rinaldi In *Micro and Nano Technologies, Nanoparticles for Biomedical Applications*, Elsevier, 2020. 10.1016/B978-0-12-816662-8.00001-1
4. A Ivanov, K Cheng, *Non-traditional and hybrid processes for micro and nano manufacturing*, Springer-Verlag London Ltd., November 2019.
5. NR Glavin, R Rao et al., Emerging applications of elemental 2D materials, advance materials, *Int. J. Adv. Manuf. Technol,* 2019, 105, 4481–4482. 10.1007/s00170-019-04711-0
6. Z Lin, A McCreary, N Briggs et al., 2D materials advances: From large scale synthesis and controlled heterostructures to improved characterization techniques, defects, and applications, *2D Mater.* 2016, 3, 042001.
7. J Cheng, C Wang, X Zou, Lei Liao, Recent advances in optoelectronic devices based on 2D materials and their heterostructures, *Adv. Optical Mater.* 2019, 7, 1800441. 10.1002/adom.201800441
8. SVS Prasad, RK Mishra et al., Introduction, History, and Origin of Two Dimensional (2D) Materials. In: Singh, S., Verma, K., Prakash, C . (eds) *Advanced Applications of 2D Nanostructures. Materials Horizons: From Nature to Nanomaterials.* Springer. 10.1007/978-981-16-3322-5_1
9. X Wu, F Mu, H Zhao, Recent progress in the synthesis of graphene/CNT composites and the energy-related applications, *Journal of Materials Science & Technology*, 2020, 55, 16–34. 10.1016/j.jmst.2019.05.063
10. C Liu, X Huang, Y-Y Wu et al., Advance on the dispersion treatment of graphene oxide and the graphene oxide modified cement-based materials, *Nanotechnology Reviews* 2021, 10(1), 34–49. 10.1515/ntrev-2021-0003
11. M Garg, N Vishwakarma, AL Sharma, S Singh, et al., Chapter 2 different types and intense classification of 2D materials. In: Singh, S., Verma, K., Prakash, C. (eds) *Advanced Applications of 2D Nanostructures. Materials Horizons: From Nature to Nanomaterials.* Springer. https://doi.org/10.1007/978-981-16-3322-5_2

12. D Akinwande, CJ Brennan, JS Bunch, P Egberts, et.at., A review on mechanics and mechanical properties of 2D materials, *Graphene and beyond, Letters* 2017, May, 42–77.

13. R Lv, et al., Transition metal dichalcogenides and beyond: Synthesis, properties, and applications of single- and few-layer nanosheets, *Accounts of Chemical Research* 2015, 10.1021/ar5002846

14. H Zhang, 2D materials chemistry, *American Chemical Society Rev.* 2018, 118, 6089–6090. July 11, 2018, 10.1021/acs.chemrev.8b00278 Chem

15. S Singh, K Rathi, K Pal, Synthesis, characterization of graphene oxide wrapped silicon carbide for excellent mechanical and damping performance for aerospace application, *Journal of Alloys and Compounds*, 2018, 740, 436–445. 10.1016/j.jallcom. 2017.12.069

16. M Nasrollahzadeh, SM Sajadi, M Sajjadi et al., An introduction to nanotechnology, Nasrollahzadeh, M., Mohammad Sajadi, S., Sajjadi, M., Issaabadi, Z., Atarod, M, *Interface Science and Technology* 2019, 28. Elsevier. 10.1016/B978-0-12-813586-0. 00001-8

17. K Cooper, Scalable nanomanufacturing—A review, *Micromachines* 2017, 8, 20. 10.3390/mi8010020

18. SVS Prasad, SB Prasad, S Singh, Nanostructured 2D materials as nano coatings and thin films. In: Singh, S., Verma, K., Prakash, C. (eds) *Advanced Applications of 2D Nanostructures. Materials Horizons: From Nature to Nanomaterials*. Springer. 10.1007/978-981-16-3322-5_5

19. Y Wang, W Zhou, K Cao, X Hu, L Gao, Y Lu, Architecture graphene and its composites: Manufacturing and structural applications, *Composites Part A: Applied Science and Manufacturing* 2020, 140, 106177. 10.1016/j.compositesa.2020.106177

20. H Sengul, TL Theis, S Ghosh, Toward sustainable nanoproducts an overview of nanomanufacturing methods, *Journal of Industrial Ecology* 2008, 12(3), 329–359. 10.1111/j.1530-9290.2008.00046.x

21. D Bratton, D Yang, J Dai et al., Review Recent progress in high resolution lithography, *Polym. Adv. Technol.* 2006, 17, 94–103. 17 January 2006, Wiley Inter Science (www. interscience.wiley.com). 10.1002/pat.662

22. T He, Z Wang, F Zhong et al., Etching techniques in 2D materials, *Adv. Mater. Technol.* 2019, 1900064. 10.1002/admt.201900064

23. S Matsumoto et al., Effects of a photo-assisted electrochemical etching process removing dry-etching damage in GaN, *Jpn. J. Appl. Phys.* 2018, 57, 121001.

24. W Ji, G Zhao, C Guo et al., A novel method to fabricate two-dimensional nanomaterial based on electrospinning, Journal homepage: www.elsevier.com/locate/compositesa, *Manufacturing* April 2021, 143, 106275.

25. Md M Iqbal, A Kumar, S Singh, 2D Nanomaterials Based Advanced Bio-composites. In Singh, S., Verma, K., Prakash, C. (eds), *Advanced Applications of 2D Nanostructures. Materials Horizons: From Nature to Nanomaterials*. Springer. 10.1007/978-981-16-3322-5_12

26. GN Arka, SB Prasad, S Singh, Comprehensive study on dye sensitized solar cell in subsystem level to excel performance potential: A review, *Energy* September 2021, 15, 192–213. 10.1016/j.solener.2021.08.037

27. S Singh, GN Arkab, S Gupta et al., Insights on a New Family of 2D Material Mxene: A Review, Materials, Mechanics & Modeling (NCMMM-2020) AIP Conf. Proc. 2341, 040017-1–040017-7. 10.1063/5.0049984

13 Biodegradable and Biocompatible Polymeric Nanocomposites for Tissue Engineering Applications

Sahana S. Sringari, Gourhari Chakraborty, and Arbind Prasad

CONTENTS

13.1 Introduction..205
13.2 Market of Nanocomposite in Tissue Engineering....................................208
13.3 Biodegradable and Biocompatible Polymeric Materials...........................208
    13.3.1 Chitosan...208
    13.3.2 Alginates..210
    13.3.3 Starches...211
    13.3.4 Cellulose..211
    13.3.5 Gelatin...212
13.4 Biopolymer Nanocomposite Hydrogels for Tissue Engineering
    Applications...213
13.5 Latest Trends in Nanocomposites in Tissue Engineering..........................213
13.6 Bioactivity and Biodegradation of Nanocomposites in Tissue
    Engineering..215
13.7 Challenges...216
13.8 Applications and Future Scope...217
    13.8.1 Natural Nanocomposite Scaffolds for Tissue Engineering
        Applications...218
13.9 Conclusion..221
Acknowledgement 222
References..222

13.1 INTRODUCTION

Defects that currently exist in human body may be caused due to various reasons. There are several methods which are outlined to rectify problems associated with

DOI: 10.1201/9781003220602-13

few early motion therapies. The following five ways could be used to achieve the healing of tissues, namely: (a) spontaneous healing, (b) autologous tissue transplantation, (c) cell-free biomaterial implantation, (d) cell therapy, and the last one being (e) tissue engineering approach. The right approach can be selected by properly examining tissues involved, defective region, and the body's healing capacity which varies with age. These selected methods help in solving the existing problems of tissue defects. Regeneration of constructs designed by artificial means which consists of scaffolds in addition to live cells is referred to as tissue engineering. The generalized methods which are combined from both the material sciences and life sciences are utilized for this regeneration [1,2]. Recently investigations of tissue engineering as a promising approach towards regeneration of tissues have been performed [3]. Chemistry, material sciences, engineering, and medicine are applied for tissue engineering as it's a multidisciplinary field whose goal is repairing and replacing tissues and organs [1]. Tissue engineering helps in overcoming the problems associated with certain therapeutic issues, like loss or failure of tissues or organs. In different developing and developed countries, tissue engineering has been used for improving quality of healthcare. Hence, it has attained a special care [2]. In 1988, National Science Foundation workshop described the term tissue engineering. It defined it as the application of principles and methods of engineering and life sciences towards fundamental understanding of structural and functional relationships in both normal and mammalian tissues and the development of biological substitutes in order to restore and maintain tissue functions." Various biomaterials like bio-polymers, bio-ceramics, and other bio-inorganics, bioactive molecules, are used in order to develop this newly developed bioengineering area. They may also be combined in order to stimulate signals into various transplant configurations and helps in enhancing the proliferation towards regenerating tissues in desired region which includes diseased or damaged areas along with organs of the body. Crucial knowledge of structural and functional connections is included in a typical pathological tissue and the advancements of organic substitutes which are involved in re-establishment, sustainment, or enhancement of the function of tissues [2]. For the purpose of growth of the cells, their attachment and proliferation, biomaterials are required [3]. Producing living tissues in-vivo involves the development of cell cultures on substrates which undergo degradation bio actively. This gives both physical and synthetic signs which help in managing their separation and get together into 3D structures [2].

Strategies of tissue engineering are categorized into three different constituents. They are namely, scaffold, cells, and biological signalling molecules like growth factors. Natural polymers, synthetic polymers and purely biological molecules can be used to produce scaffolds which are three-dimensional structures. The scaffold should have the capacity of mimicking structural and biological function of natural ECM. It should be able to mimic both in terms of its structural properties along with its chemical composition. Extracellular matrix basically comprises of proteoglycans, proteins, and signalling molecules. The main role of the extracellular matrix is to provide support structurally to cells, aid as a location for migration of cells, show favourable surface properties which help in adhesion, proliferation and differentiation of cells. Therefore, they are greatly applied in tissue engineering research [1]. Designing an appropriate, porous, biodegradable, non-toxic, non-mutagenic and non- immunogenic and biocompatible scaffold are an important challenge. Bioactivity of scaffold is affected by the type of material, its microstructure and its mechanical properties. Scaffolds must therefore meet some characteristics such as, it should have more regular and re- producible three-dimensional structure, specific pore size, sufficient porosity, high inter- connectivity needed for uniform seeding of cells, their distribution, function, and tissue regeneration [4]. Two approaches are followed by tissue engineering. These approaches will help in differentiating from the cell therapies. Initially during the in-vitro conditions, cells will start communicating and interacting with one another to synthesize the extracellular matrix. However, during in-vivo conditions, cells are seeded onto a scaffold material directly before implanting or when defective region acts as primary centre in which the cell suspension is implanted directly prior to implantation. Biodegradability remains the most important factor that is taken into consideration (Figure 13.1). This consists of restoring cells in addition to physiological deterioration of bio-materials that are utilized as scaffolds in tissue engineering. This will result in newly formed tissue that is healthy which completely fulfills the need and requirements of detective region. Naturally occurring tissues are highly adjusted to local condition. But, adaption to body is needed for the artificially constructed biomaterials. Hence it is very much necessary to eliminate the synthetic part for smooth biological tissue formation and re-modelling. Biomaterial selection is very difficult as it has the presence of complex and sensitive host system [2].

Polymeric scaffolds have shown great promise in tissue engineering research field [5,6]. Polymers that undergo biodegradation are greatly utilized for making scaffolds as they undergo degradation during new tissue formation and finally would not leave anything that is foreign to body. The scaffolds should also be easily sterilizable, on the surface and in the bulk for preventing infections [7]. Hence, designing and preparing multi-component polymeric system represents a good strategy for developing innovative multifunctional biomaterials [6].

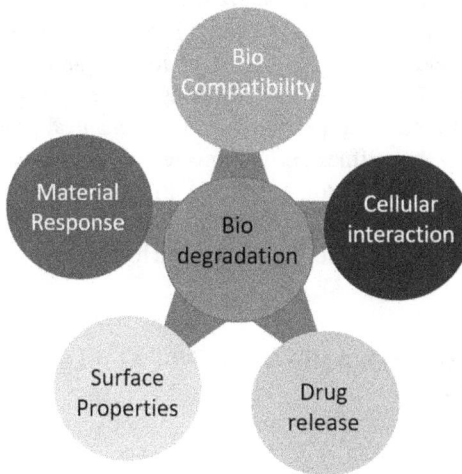

FIGURE 13.1   Importance of biodegradability in biomedical application.

13.2 MARKET OF NANOCOMPOSITE IN TISSUE ENGINEERING

Large particles which have size of about 100 nm are less unique when compared to nanoparticles which have novel characteristics. "Nano" word comes from Greek word which means "dwarf" which is a spatial unit of measurement. Its value is given by 1.0×10^{-9} m. Both the physical and chemical properties of particles are changed when size is reduced from micrometre to nanometre scale. At present, extensive research has been carried out on different kinds of nanomaterials in various technological disciplines. Some of the latest advancements include nanofibers, nanofilms, nano-ceramics, nanogels, nanotubes, nanocomposites, etc. [8,9]. Nanotechnology includes materials which have one or more dimensions in nanometre length [2]. They are materials which consist of two phases which are not miscible. One includes a polymer and another one includes filler (nano reinforcement) having nanoscale characteristics in at least 1D [4]. There has been development, characterization and evaluation of various nanocomposites like organic-organic, inorganic-inorganic, and organic-inorganic during the past few years. These were many carried out for their application in a variety of biomedical applications. Mainly, research has been centred on tissue engineering applications during the past few decades. All the basic functional cells, their subunits are precisely described at nanoscale. Hence it is necessary to understand nanotechnology, nanobiology, and nanomaterials. This helps in characterizing new frontline in the tissue engineering research [2].

On comparing with individual components, nanocomposites have shown tremendous balance between both strength and toughness. There is an overall improvement in their characteristics. One component which majorly affects the property of nanocomposites is adhesion between the nanoparticle and the polymeric matrixes. Several methods had been employed in the past for increasing interfacial strength between two phases. Preparation of nanocomposites can be done by the addition of nanoparticles or nano fibres into various polymeric matrices. Filler particle size plays an important role [10,11]. Those fillers which are nanosized, have larger surface area on comparing with traditionally used fillers, like having large interfacial area per unit volume of the particles, large number density of particles per particle volume, particle-particle co-relation arising at low volume fraction, etc. [12,13]. These fillers can therefore form tight interfaces along with composite polymeric matrixes. This leads to increased performance in mechanical properties. Mechanical strength in addition to the stiffness of composites is also increased due to the contribution of intrinsic properties of nanosized fillers towards various interactions among the fillers and polymeric matrixes [10]. Furthermore, research has been focused on different nanocomposite materials which are made up of biodegradable polymers [2].

13.3 BIODEGRADABLE AND BIOCOMPATIBLE POLYMERIC MATERIALS

13.3.1 CHITOSAN

Chitin is a primary constituent of crustacean's outer skeleton and fungi cell wall. Chitosan (β-1,4-linked N-glucosamine) is obtained naturally from this. Both of them are not homo-polymers. It is a cationic aminopolysaccharide. It has anti-bacterial, biodegradable, and biocompatibility has its intrinsic properties. Food and Drug Administration has notified that Chitosan is "generally recognized as safe" (GRAS) material. Very minimal foreign body reaction is evoked by chitosan-based systems. Hence, they are under investigation to be applied in tissue engineering. Moreover, they are easily moldable into various different forms of porous structures. This can be done by using different techniques like freeze-drying, rapid prototyping and internal bubbling process. Both the mechanism and rate of depolymerization of chitosan will affect viscosity, solubility and biological activity in aqueous solutions. Chitosan has played a major role in bone tissue engineering since many decades as they do not possess toxic reactions. Chitosan along with its derivatives have been found useful in osteogenic bone substitution at defective/diseased bone region. [2,14]

A group of researchers carried out preparation and characterization of intercalated nanocomposite material of chitosan and hydroxyapatite (HAp). Sol-gel method was utilized to synthesize HAp in nano size. The synthesized HAp nanoparticles had the ability to self- assemble. This property helped during drying of solvent casting films leading to development of a homogeneous chitosan hydroxyapatite nanocomposite. These composites when used in bone tissue engineering will help in promoting the development of apatite similar to that of bone on its surface. This enhances attachment, proliferating and differentiating capacity of cells similar to that of osteoblasts. In situ preparation method was recently adapted for the preparation of a homogeneous chitosan-Polylactic acid/Hap nanocomposites material (Figure 13.2). Inorganic particles of rod shape having a composition of randomly directed subparticles of around ten-nanometre diameters were suspended in this polymeric matrix. Incorporating HAp into matrix increased elastic modulus and compressive strength. In addition to this, polylactic acid also played a significant part in controlling structure of inorganic particles and enhancing mechanical characteristics [2,15].

A group of researchers used in-situ hybridization method to prepare and characterize chitosan-nanohydroxyapatite (nHAp). Characterization was done using

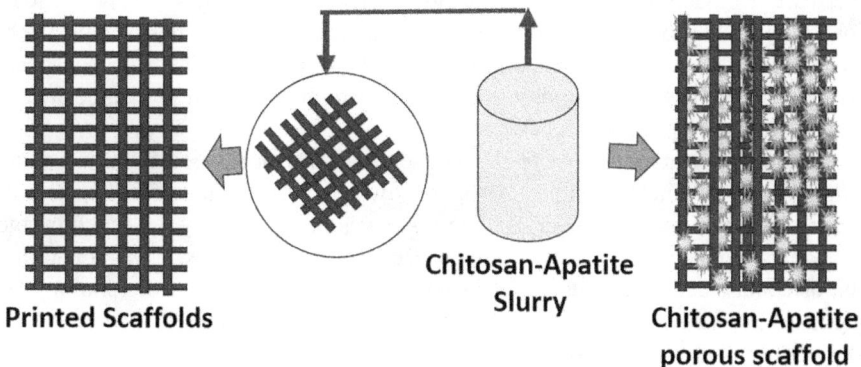

Printed Scaffolds

Chitosan-Apatite Slurry

Chitosan-Apatite porous scaffold

FIGURE 13.2 Schematic for preparation of 3D printed PLA/HA-CS composite hydrogels through freeze gelation technique.

scanning electron microscopy, atomic force microscopy; Fourier transformed infrared, and X-Ray diffraction. On conducting mechanical compressive test, satisfactory mechanical performance was observed. Therefore, it could be used in re-placement of bone tissues. Chitosan-based nanocomposite materials were synthesized by Keller et al. for repairing bone defects. Chitosan having different molecular weights and concentrations was used for preparing these scaffolds. Silica nanoparticles were reinforced as nanofillers into the scaffold. The concentration of chitosan utilized in chitosan-silica nanocomposite scaffold did not affect the pore size as it was noticed that the scaffold had similar pore size. Usage of silica nanoparticle filler as reinforcement enhanced the mechanical compression resistance of the prepared scaffold by 30%. Alamar blue assay was used to check for biocompatibility of the 3D scaffold suggesting the in-vitro biocompatibility of chitosan-silica nanocomposite materials. Thereby, signifying excellent prospective in regenerating bone tissues [2].

13.3.2 ALGINATES

Alginate is a marine bio polysaccharide. Their extraction is from brown algae. They are anionic linear and unbranched in nature. They are constituted by changing the quantity of (1,4)- linked β-D-mannuronic acid and α-L-guluronic acid. Along the length of the polymeric chain, its composition and sequence will change. Alginate physical properties mainly depend on its molecular weight and composition. Purifying alginate is required for achieving control of degradation of material. Since last few decades, there has been a wide research on alginates as raw material in tissue engineering applications. Most of the nanocomposites prepared from alginate have shown increased biocompatibility, cell proliferation, mechanical strength, porosity, cell adhesion, differentiation, and excellent mineralization. This showed its effective use in tissue engineering of bones. A group of researchers prepared scaffold using alginate/halloysite nanotubes. Solution-mixing and freeze-drying techniques were used for this purpose. Mechanical behaviour of scaffolds was enhanced by reinforcing halloysite nanotubes into alginate matrix. On comparing the scaffolds made from pure alginate, alginate/halloysite nanotube composite scaffolds showed improvement in compressive modulus and strength. It was noticed that these scaffolds had improved stability against enzymatic degradation in phosphate buffer solution. Cytocompatibility test also showed improved cell growth behaviour and cell adhesion property [2,14].

Alginate/HAp nanocomposite fibrous scaffold was developed by Chae et al. using electrospinning technique. Also, HAp which mimics mineralized bone collagen fibrils was made. Nanocrystalline HAp was homogenously deposited on the collagen nanofibers by using this method. During in-vitro studies, it was found that the osteoblast cells were increased in length into spindle-shaped morphology on alginate/HAp nanocomposite fibrous scaffold. On the other hand, it showed round-shaped morphology. On combining nanofibrous topography alginate/HAp nanocomposite scaffold with hybridization of alginate and nHAp proved as a helpful tool for regeneration of bones [2].

13.3.3 STARCHES

Starch is a major plant-derived biodegradable polysaccharide. This polymer has a very good biocompatible nature. They are known as the storage organs in plants. It constitutes both a linear poly (1,4-α-D-glucopyranose) (amylose) and branched poly (1,4-α-D-glucopyranose) with branches of (1,6-α-D-glucopyranose) (amylopectin) which occurs for every twenty-five glucosidic residues. It is produced in semi-crystalline granule form naturally. However, its size and composition vary according to its source of origin. Many techniques can be used to process them into three-dimensional scaffolds as they are biodegradable and less expensive. Hence, they find great potential in tissue engineering. In recent times, composites based on starch are being developed to be used in tissue engineering. Investigations made on composites based on starch have already begun fabricating scaffolds for tissue engineering of bones. This will further facilitate good porous structure for the migrations of cells. Recently, fabrication of composites based on starch has been carried out by re-inforcing biopolymers for improving the properties of scaffold [2,14].

Synthesis and characterization of nHAp-starch biocomposites via a biomimetic process were performed by Meskinfam et al. Morphological studies were carried out using SEM and TEM. Rod-shaped nHAp formation was observed within the starch matrix. MTT assay was done to check for biocompatibility of these biocomposites which suggested that nHAp had its effect on the proliferation of cells. Also, no adverse effects were seen on morphology of cells, its viability and proliferation. A similar nanocomposite was synthesized by another group of researchers for application in bone tissue engineering. This was carried out in-situ following biomimetic steps of bone-like nHAp development in presence of wheat starch. The shape and morphology of nHAp was influenced when wheat starch was used as template. Rod-shaped nHAp was formed which is very much similar to the HAp present in inorganic component of natural bones [2,16].

13.3.4 CELLULOSE

Cellulose being an abundant renewable resource is found globally. Preliminary sources are lignocellulosic material found in fores`t, algae, bacteria biosynthesis, and chemo-synthesis. It has a straight polysaccharide chain containing 100-10,000 β – (1– 4) – connected d-glucose units. Due to presence of greater molecular weight, intramolecular and intermolecular hydrogen bonds, cellulose has good mechanical strength and chemical stability. They are also biodegradable and biocompatible in nature. Dissolving it in frequently used solvents is quite difficult because of its chemical properties. Therefore, its application in tissue engineering is complex. By using cellulose derivatives, this problem can be overcome. These derivatives include carboxymethyl and bacterial cellulose for precipitating scaffold materials. Bacterium *Acetobacter xylinum* extracellularly synthesizes a nanomaterial known as bacterial cellulose (BC). It is highly pure, nanofibrous structure, having more crystallanity, tensile strength and good biocompatibility. It is not a branched polymer. It has β -1, 4-linked glucopyranose units. Plant cellulose has chemical identity with BC. However, it differs in its macro-molecular structure and properties [2,14].

A group of researchers had developed HAp/BC nanocomposites for enhancing osteoconductivity and mechanical properties. They also had the aim of increasing the small pore size of BC. Phosphorylated BC had been first soaked in SBF. Later, there was a uniform formation of HAp crystals containing carbonate on BC fibres. On comparing with pure BC, human bone marrow MSCs when incorporated on nanocomposite materials showed improved adhesion and activity. May be due to the improvement in the pore size and existence of inorganic components, proliferation of cells on nanocomposites were much faster. Along with this, it was observed that there was enhancement in cell osteoblastic differentiation even in the absence of osteogenic reagents. Recently, another method was used for preparing BC- HAp nanocomposites. This method involved adding of mineral phase into bacterial culture media while cellulose fibrils were being formed. Upon usage of carboxymethyl cellulose for suspending HAp nanoparticles in culture medium when cellulose nanofibrils were formed, the average diameter of BC fibres was less when compared to BC fibres that were not modified [2].

13.3.5 GELATIN

Gelatin is a naturally occurring protein. It is a biodegradable and biocompatible polymer which possesses good plasticity. It is also non-antigenic. This biopolymer has the ability to mimic some helpful characteristics on extra cellular matrix for the purpose of cell adhesion and proliferation. Investigation has been carried out for fabricating tissue engineering scaffolds using gelatin as the raw material for regeneration of injured bone tissues. They are used as a substitution for grafting of bones. Currently, composites based on nanofibrous gelatin have been developed for mimicking the composition of bone in addition to maintaining its architectural behaviour of bone. Thermally-induced phase separation method was used in order to prepare nanofibrous gelatin matrix. These types of composite scaffolds are under investigation for tissue engineering of bones because of their large surface area in addition to enhanced mechanical power. This conclusion is made after comparing with commercially available gelatin of different grades with same features of porosity and pore size [2].

Another study conducted by Samadikuchaksaraei et al. developed osteoblast cells constrained with nHAp/gelatin composite scaffolds for application in regeneration of bone tissues. This composite scaffold was prepared using the techniques such as casting, freeze- drying, lamination, etc. Later, it was constrained by osteoblast cultures on surface and taken out by continual freeze-thaw method. Result was that mechanical and biological analysis conducted in-vitro showcased that in-vitro biological characteristics as well as mechanical profiles weren't affected by technique of osteoblasts conditioning. Motive behind the osteoblasts cell conditioned nHAp/gelatin composite scaffolds was for maintaining cell growth along with adhesion of cells. Apart from this, the in-vitro cytotoxicity of this composite scaffold was also examined with the help rat mesenchymal stem cells. The osteoblasts cell conditioned nHAp/gelatin composite scaffolds were assessed for in-vitro implantation studies. In the critical sized calvarial defect repair model in rats' bone which showed that the osteoblast cell conditioning had improved osteoinductivity of the scaffold, as well as biocompatibility. The collagen amounts were speeded up throughout the bone healing procedure because of these osteoblasts cell conditioning nanocomposites scaffold besides the in vitro results even exhibited that newly emerged bone tissues occupied approximately total bone defects within the implantation of three months [2].

13.4   BIOPOLYMER NANOCOMPOSITE HYDROGELS FOR TISSUE ENGINEERING APPLICATIONS

Hydrogels whose origin is either from natural or artificial sources are three-dimensional structures comprising of chemical or physical crosslinking hydrophilic polymeric chains. They are able to absorb a lot of water content. But they do not get dissolved. Their soft, three-dimensional environment which mimics extracellular matrix is provided by their hydrated architecture. This will enable three-dimensional regeneration of damaged tissues [17]. In present times, developing and designing hydrogels made from polymers has gained great interest in articular cartilage tissue engineering field [18]. It helps in sustainable delivery of various biological agents. This plays a vital role in the cell development process. Due to certain positives and negatives of hydrogels as bio-materials researchers have made hydrogel composites which demonstrates high performance in bio-medical field [17].

A group of researchers prepared bi- polymeric hydrogel composite by incorporating some quantity of stick-like Titanium dioxide nanostructures. This was done to enhance the adhesion of osteoblast and favour surface behaviour. On investigating, it was noticed that stick-like nanostructures of Titanium dioxide were very well inserted into PVA/PVP the prepared matrix. This was mainly because of promotion of physical cross-linking density in hydrogel networks due to long chains of PVA molecules. This hydrogel matrix showed improvement in surface topography and enhanced osteoblast attachment [18].

Shin et al., synthesized gelatin methacrylate and carbon nanotube-based gels. When multi-walled carbon nanotubes are present in polymeric matrix, it makes the gel biocompatible, non-toxic and increases its mechanical strength. These gels can be used for encapsulating cells. One more study involved the usage of carbon nanotube and gelatin methacrylate-based hydrogels were used in cardiac tissue engineering. Varying concentrations of multi-walled carbon nanotubes were used which showed good electrophysiological, biocompatible, and mechanical properties. These properties make it very useful in multifunctional cardiac and muscle tissues [19].

13.5   LATEST TRENDS IN NANOCOMPOSITES IN TISSUE ENGINEERING

The scaffolds used in bone tissue engineering are expected to be osteoconductive, osteoinductive & osseointegration. Most of the scaffolds are formulated to be similar to the characteristics of human bone [20]. SWNTs, especially ultra-short SWNTs, possess potential to be used as a reinforcing polymeric scaffold material. Single-walled carbon nanotubes are being used as reinforcing agents due to their outstanding mechanical properties. They also possess unique properties that claim them to be the best choice for orthopaedic applications. The density of SWNTs is almost same as that of graphite [21].

Nano hydroxyapatite and GO were being used as reinforcement in aliphatic polyester scaffolds that gives mechanical support to cell ingrowth and proliferation while tissue regeneration. Besides, these nanofillers are found to increase osteoblastic cell adhesion, bioactivity, expansion and osteogenic differentiation of MSCs. Nanoceramic exterior mimic the interior environments of the human bone, hence enhancing protein adsorption and initiating new bone birth. Specifically, nanocrystalline hydroxyapatite supports huge adsorption of vitronectin and BMP-2 on its surface, thereby speeding up the new bone formation effectively. Tjong and co-workers showed that the hybrid nanocomposites of PLA which has the reinforcements of both nanohydroxyapatite and GO fillers showed good compatibility than binary poly lactic acid and nanohydroxyapatite composites. It was due to the promotion of new bone emergence by controlling osteoblastic adhesion and growth, and osteogenic differentiation of mesenchymal stem cells (MSCs) on exteriors by graphene and its derivatives encourage new bone formation [20].

To obtain improved structural integrity and to improve the mechanical characteristics of biodegradable polymer once implanted underweight tolerating state. Carbon nanostructures like 0D fullerenes and 1D SWCNTs are being researched intensively as reinforcing material. 2D nanomaterials like single and multiwalled graphene oxide nanoribbons (SWGONRs, MWGONRs), graphene oxide nanoplatelets (GONPs), and molybdenum disulphide nanoplatelets (MSNPs) were used as reinforcements in PPF composite [22].

Two-dimensional carbon and inorganic nanostructures such as graphene oxide nanoplatelets, graphene oxide nanoribbons (GONRs) and molybdenum disulphide nano-platelets (MSNPs) possess unique physicochemical properties have been harnessed for a number of applications like components in energy and semiconductor electronic devices, water filtration membranes, agents for bio-imaging and drug delivery and distribution agents for processing of lubricants, liquid crystals and porous scaffolds for tissue engineering [22].

Chitin is one of the most abundant forms of polymerized carbon available in nature, which can be made use for biomedical applications like drug delivery systems due to its versatile characteristics like compatibility with living tissues, bio recyclability and eco- friendly. A group of scientists made Chitin and Chitosan dependent nanofibrous materials with good biological and mechanical properties that possess outstanding capacity as antibacterial agents in wound dressing, drug delivery, biosensor and tissue engineering. But the low sensitivities, conductivity, weak mechanical toughness and poor cross-linked architecture of chitosan are its disadvantages that stop it from being used in biomedical applications. However, these nanocomposites having lower concentration of nanofillers show distinct optical, toughness, conductivity, biocompatible and biodegradable characteristics [19]. Extracellular matrices at the nano range are being developed to control cells and facilitate tissue recreation. Such kind of organized three-dimensional scaffolds will not only serve as a physical template for growing cell and forming tissue but also gives bimolecular, mechanical, chemical, and geometrical indications to cells. The newly budding biomaterials are showing considerable advancement in tissue engineering [23,24].

13.6 BIOACTIVITY AND BIODEGRADATION OF NANOCOMPOSITES IN TISSUE ENGINEERING

The important problems of tissue engineering scaffold are that a large number of materials are not mechanically competent, biodegradable, and bioactive simultaneously, therefore combining these features, hybrid and nanocomposite material gives a good chance to make scaffolds with required biological, structural and physical characteristics [10,25]. Many types of biodegradable polymers are being made and some are already being produced in factories. In accordance with their emergence, polymers are classified into three types. Synthetic polymers like aliphatic polyester, such as poly (L-lactide), poly, etc. [21], polyesters that are formed using microbes that include varying types of poly (hydroxyalkanoate), including poly (β-hydroxybutyrate) and poly (3-hydroxybutyrate-co-3-hydroxyvalerate, etc. [20], polymers that originate from environmental sources like starch, cellulose, chitin, chitosan, lignin, etc. [26].

PLA is one of the best biodegradable materials among all the aliphatic polyesters. It is greatly biocompatibility, biodegradable and has mechanical toughness. It can also be extracted from renewable resources. Incorporation of a variety of nanoparticles in PLA matrix appreciably enhances the characteristics of this substance. Dubois and his group made poly (L, L-lactide)/organo-modified montmorillonite nanocomposites with both interjecting and exfoliated architecture by making use of in situ ring-opening polymerization technology. It was discovered that the kind of nanofiller had an important role in its final dispersing morphology [26–30].

Because of its low cost and availability, starch is seen to be one of the best polymers for making nanocomposites. It is easily made into thermoplastic materials using plasticizer alone, heat and shear forces. Manufacturing of starch nanocomposites is growing as a good choice to enhance the mechanical and barrier properties. Its good efficiency and reduced-price make the layered silicate a most preferred choice to be incorporated [26]. By adding inorganic nanoparticles into various polymeric matrixes, nanocomposite materials can be prepared. The filler particles size is main variable. Fillers of nano-size have larger surface area in contrast with conventional sized i.e., micro-sized fillers. Nano range fillers therefore will be able to make dense interface with the matrix of polymer in composite and therefore excellent performance in mechanical characteristics is achieved [10]. Suitably organically modified clay when combined with biopolymers gives environmentally friendly nanocomposites.

The introduction of nanoparticles in polymeric matrix will decrease permeability of penetrant molecules thereby leading to development of higher barrier nanocomposites. Biodegradable nanocomposites are a group of materials which are widening enormously whose environmental impacts need to be studied to be used in industrial and in automotive applications. It is also necessary to assess the total life cycle of such new materials for the full realization and utilization [25]. These days nanotechnology is being frequently employed on polymers with the aim of enhancing the overall performance. The two important methods to make polymer nanomaterials are to introduce nanoparticles into a polymeric matrix or to make polymer/nanoparticle composites or to fabricate polymers in nanoscale [26].

13.7 CHALLENGES

While nanoparticles have good capacity to be used in tissue engineering like improving mechanical, biological characteristics, bactericidal effects, gene delivery and establishing engineered tissues, there are certain problems that need good attention before they are introduced into widespread clinical applications. For instance, at first, we will need better assessment tools and methods for detecting and analyzing nanoparticle carcinogenicity, teratogenicity and toxicity. Secondly, toxicity, teratogenicity and carcinogenicity of nanoparticles are greatly reliant on dose and subjection. In most of the applications, the nanoparticles made use such that their concentration is below the threshold at which they are mostly not damaging [31]. Nano bio ceramics scaffolds help in drug release and tissue recreation. We don't know information about their prolonged use in the body [7].

There are few problems in association with the application of nanocomposites in other tissues; therefore, a better knowledge of the interaction between the inorganic and organic phases and biological response is needed [32]. In the next few years, the level of exposure to the nanoparticles like titanium oxide, zinc oxide, nickel oxide, aluminium and silver the fraction of people with exposure will grow. Some characteristics of nanoparticles in factory applications may be harmful for environment and biological systems. This has brought a perspective of nanotoxicology, which is the structured methodology of testing of possible toxic effects exhibited by the products on cells or scaffolds [7].

In the last few years, there has been tremendous success attained in using nanocomposite scaffolds in tissue engineering applications. Most of the areas of tissue engineering have seen appreciable advancements that have enabled better understanding of fundamental processes that helps to face the problems in the making novel scaffolds [32]. The bioceramics made of PCL have posed problems such as releasing crystals from bone scaffold that will pile up in the joints and trigger inflammatory responses in the place of prosthesis therefore better understanding in biological systems is crucial [7].

Even though many scaffolds have capacity for clinical application, their safety and efficacy in animal samples and clinical trials are to be assessed with precision [32]. The well-known fact is that the nanoparticles may bioaccumulate within the body for a longer time. Hence nanoparticles used in the human body will tend to accumulate over prolonged periods and reach an accumulation that leads to toxicity of cells, deteriorating consequences of reproductive systems or cancers as well as harmful effects on fetuses before their birth [31].

There are many products consisting of nanoparticles out for sale, but there is some lack of knowledge on a particular threat that could be implicit. For instance, there are no international standards available for nano precise threat evaluation, and specific data demands and evaluation methodologies [31]. Another way towards tissue restoration is growing biomimetic organs by making use of methods like organ-on-chip platforms. One more problem to be faced is storage of scaffolds for an extended time use. Anyways, it is expected that clubbing of skills in multi-faceted fields like revolutions in genetic editing like Clustered regularly interspaced short palindromic repeats, immunomodulation, engineered biomaterials and bio-fabrication will help in some more development and encourage the future of this area [32].

Advancement in fabrication methods like 3D bioprinting and software-assisted programmed self-assembly, better understanding of body's immune system interaction with biomaterials and mechanism for delivery of cytokines and growth factors and progress in development of new biomaterials has led to improved methodology for tissue engineering [32]. The assessments of risk of nanomaterials are costly and laborious. For now, manufacturers are working on assessing the safety of their nanoparticle-based products and to apply the necessary safety measures. Till date, there are no regulatory tools that are nano- specific; for example, the requirements of data for criteria for classification, notification of chemicals and labelling requirements for safety data sheets are still not available widely. Hence, we will need precautionary measures for applications of nanoparticles when there is a possibility of chronic bioaccumulation in the body [31].

13.8   APPLICATIONS AND FUTURE SCOPE

There have been various functions in tissue engineering in which nanoparticles have been employed be it from enhancement of biological mechanical and electrical properties to gene delivery, viral transduction, DNA transfection and patterning of cells, to facilitate the growth of various types of tissue to molecular detection and biosensing [31,33]. Figure 13.3 shows the application of bio nanocomposite for various biomedical applications.

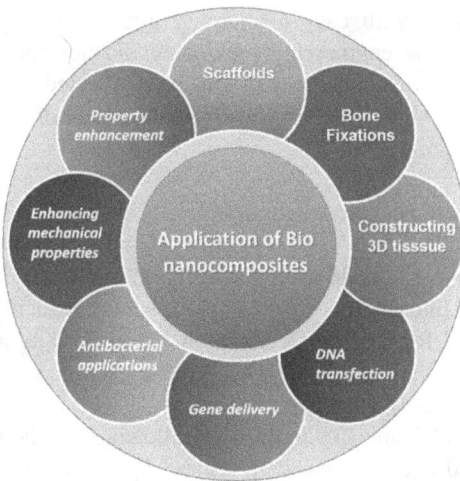

FIGURE 13.3   Application of bio nanocomposite in biomedical applications.

13.8.1   NATURAL NANOCOMPOSITE SCAFFOLDS FOR TISSUE ENGINEERING APPLICATIONS

a. *Cellulose-based nanocomposites scaffolds*

Mostly researched and versatile material is cellulose. For instance, bacterial cellulose that is produced by bacteria by fermentation is being intensively researched for the development of scaffolds because of its inherent characteristics such as high crystallanity, in situ flexibility, nanofibrous networking, good biocompatibility and comparatively simple method of development. Bacterial cellulose in vivo biocompatibility was assessed first time by implanting it subcutaneously in rats that resulted in no inflammation, chronic inflammatory reactions or fibrotic capsule formation. Nanocomposite scaffolds of bacterial cellulose are being extensively used in tissue engineering. In a study, bacterial cellulose was made with the average diameter of 50 nanometres and was mixed with PVA to form BC- PVA nanocomposite. This nanocomposite had properties that mimicked the cardiovascular tissue therefore Bacterial Cellulose – PVA nanocomposites show caliber for cardiovascular soft tissue renewal applications. In one of the latest studies celluloses were extracted out of cotton linters and was used to make semi-interpenetrating polymer network nanocomposite scaffolds [34].

b. *Collagen- and gelatin-based nanocomposite scaffolds*

Collagen is the major naturally occurring biomolecule in human body, which mostly, exists in bone, tendon, cartilage, etc. It is prepared using the recombinant DNA method in transgenic animals nowadays. Denatured and insoluble form of collagen is gelatin which is prepared by a hydrolysis process consisting of pretreatment and extraction. Gelatin has unique, superior, inherent properties when contrasted with collagen-like elasticity, reduced immunogenicity and flexibility to chemical modifications with easy availability and being economical. Scaffolds made use in bone tissue engineering previously did not encourage neovascularization, in order to face this deficit cobalt ions were injected into bioactive collagen glass/collagen glycosaminoglycan leading to enhancement of expression of vascular endothelial growth factor [34].

c. *Alginate-based nanocomposite scaffolds*

Alginate is an extensively made use polysaccharide in tissue engineering application after chitin and chitosan. It is consisting of mannuronic acid and guluronic acid. It is excellently used for scaffold preparation because of its good characteristics such as low toxicity, high biocompatibility, tunable gelation and biodegradation and it is economical and easily available.

The advantage of alginate is its moldability allowing it to be shaped in variety of structures such as hydrogels, microspheres, microcapsules foams and fibres. Chitosan alginate (CA) scaffolds are being tested to be used in various tissue engineering areas. CA polyelectrolyte complex scaffolds and 3D porous CA scaffolds are being used for rejuvenation of bone problems. These scaffolds help in expansion and acts support for MG-63 osteoblast-like cells and undifferentiated MSCs in vitro [34].

d. *Chitosan and chitin in nanocomposite scaffolds*

There is a great attention towards chitosan and chitin in recent decades because they can be used in making scaffolds as a only material or in association with other materials for it to be used in tissue engineering. Chitosan is derivative of chitin and it is derived from different fungal animal and plants. The chitosan is the mostly researched polymer following cellulose because of its outstanding flexible physicochemical characteristics. Its other characteristics are anti-haemorrhagic, analgesic, antimicrobial, muco-adhesive and antifungal natures. The polymers chitosan and chitin show a flexible deterioration rate, non-toxic residue formation and high biocompatibility. Microspheres of chitosan filled with alendronate scaffold, lead to improved osteogenic differentiation of adipose-derived stem cells, good mechanical and degradation properties and consistent drug release. This suggested that chitosan filled with alendronate scaffolds was an outstanding option for tissue engineering and controlled drug release [34].

The amalgamation of biodegradable polymers and nanomaterials leads to opening of a approach in applications of nanodevices in biomedical with tailorable morphological, heat, electrical and mechanical characteristics [6]. For fabricating of green electronics biodegradable nanocomposites are preferred and hence widely being studied as it can provide an optimal solution for the management of e-waste and protection of environment. Besides, the green electronics have great potential in biomedical applications. However, the functional nanomaterials have enhanced the whole performance of the polymers that are biodegradable but characteristics like biodegradability biocompatibility, flexibility and conductivity need to be improved silk. Silk is a non-toxic bioresorbable and biodegradable polymer. It is a suitable material for implanting in human without triggering immunological reaction and hence it is made use in embeddable electronic therapeutic devices [35–43].

In a study conducted by a group of scientists fabricated ultrathin electronic sensor array on silk, whose performance was evaluated in vivo by placing it on the exposed brain tissue. It showed that the silk material was safely resorbed and dissolved leading to formation of a conformal coating on folded brain tissue with the sensor array. There are other studies which have also shown good usage of silk as a precursor in food sensors and implantable electronics [35]. Using cellulose nano paper, a group of all solid-state ionic dielectrics was fabricated by Dai et al. These dielectrics were found to have low surface roughness, high transparency, excellent properties and good heat durability. Successful use of cellulose as substrates and dielectric materials depicts that it has a great potential in flexible, biodegradable and environmentally friendly electronic devices [35]. It is essential to minimize the thermal degradation of the main polymeric matrix during the development of biodegradable nanocomposite. The nano-composites of starch have been successfully developed with desirable mechanical particle properties, but the moisture sensitivity is too poor to keep them in the class of packaging materials especially in case of liquids [25].

Scaffolds of polymers play an important role in tissue engineering by helping in cell seeding, new tissue formation in 3D and proliferation depicting great promise in the research of engineering different tissues [7]. The new generation nanocomposites in the biomedical field are expected to be hybrid, having active components and intelligence. Polymer matrix composites have the advantage of being tailored for desired properties and being versatile [6]. HA and calcium phosphates are the bioactive ceramics that are majorly used as biomaterials for bone repairs. These classes of biomaterials showed appropriate biocompatibility and osteoconductivity because they are structurally and chemically similar to the mineral phase of native bone [7].

e. *Hydroxyapatite (HA)-based nanocomposites*

Since 65% of bone consists of HAp, it promotes the bone ingrowth and is also biocompatible. Therefore, both the synthetic abs natural HA has been rigorously studied as a desired constituent of scaffold material in tissue engineering of bone. Scientists have witnessed better osteoconductive HAp by changing composition, morphology and size. HAp in nano dimension might have other special properties because of its miniature size and bigger specific surface area. Research conducted by Webster et al it was found that there was a substantial increment in osteoblast adhesion and protein adsorption on ceramic materials of nanoscale [7]. The HAp nanocomposites are being surface modified to ease the delivery of biomolecules [12].

In tissue engineering, the present research mostly focuses on the stromal cells and biopolymer interfaces. The combination of stem cell seeding with the synthetic biopolymeric nanocomposite scaffolds with biologically active inorganic phases has been the topic of interest in the near future. The researchers had established a worthful studies related to various bioresorbable polymers along with nanohydroxyapatite for bone-related problems [44–49]. The perspective of biopolymeric nanocomposite has led to the scaffold surface mimicking complex local biological functions and might show in-vitro and in-vivo development of tissues and organs in the near future [6,7]. Table 13.1 shows the nanocomposite fabrication techniques and their applications [34].

TABLE 13.1

Various Applications of Nanocomposites (Reproduced with the Permission of Publisher)

Nanocomposites combinations	Applications	Fabrication techniques
Aspirin-loaded (graphene oxide/ chitosan- hydroxyapatite)	Bone tissue engineering	Layer-by-layer (LBL) assembly technology combined with biomimetic mineralization method
Alginate/nanoTiO$_2$ needle	Tissue engineering	Lyophilization Technique
Bacterial cellulose/ hydroxyapatite	Biomedical applications	Carboxymethyl cellulose and hydroxyapatite powders were added to medium

TABLE 13.1 (Continued)

Various Applications of Nanocomposites (Reproduced with the Permission of Publisher)

Nanocomposites combinations	Applications	Fabrication techniques
Bacterial cellulose/silk fibroin sponge	Tissue engineering	Soaking, shaking, and freeze drying
Chitinchitosan/nano TiO2	Tissue engineering	Freeze drying
Chitosan/PVA-reinforced, single-walled carbon nanotube	Neural tissue engineering	Ultra sonication and electrospinning
Chitosan Graft Poly(acrylic acid-co-acrylamide)/hydroxyapatite	Implants and drug carriers in bone tissue Engineering	Multistep freeze drying method
Dexamethasone-loaded poly (glycerolsebacate) PCL/gelatin	Soft tissue engineering	Coaxial electrospinning
Hydroxyapatite nanoparticles/ poly- hydroxybutyrate	Cell seeding and bone tissue engineering	Electrospinning
Polyethyleneoxide/chitosan/ graphene oxide	Treatment of lung cancer	Electrospinning

13.9 CONCLUSION

Nanocomposites are an emerging generation of polymers that are getting in every area of our lives. They exhibit promising application as highly performing biodegradable material that are of completely latest type and using plants, animals and natural materials. These materials are dumped in compost; they are easily broken down into water, carbon dioxide and hummus by the microbes. Many plastic items can be restored with natural biopolymeric plastics that have various applications. In order to recyclable polymer-based bio nanocomposites to cope with wide range of demands in the applications these nanocomposites characterization should be studied and researched in depth so that it can be modified to obtain better characteristics, which can also be tuned based on the consumer requirement. The nanomaterials that are presently used in commercial and industries have a distinct chance of being translated into the biomedical applications. Chitin has been used in the form of an Ag to improve the functioning of tissue-engineered constructs in recent times. Chitin possesses extraordinary characteristics like biocompatibility, biodegradability, zero toxicity and also shows excellent wound healing in humans. Besides it is also physiologically compatible with living tissues. But its mechanical property is poor hence introduction of biomaterials such as HA, bioactive glass ceramic will improve the mechanical strength as well as helps in tuning the material to match the native tissues. Synthetic silicates have currently made into the medical field as a food additive absorbent and anticaking agent and have by now shown its impact on regenerative medicine research. Tissue engineering is a hopeful way to the present-day surgery methodologies for different tissue repair work. Its aim is to give a total resurgence of the tissue that requires repair leading to recovery of its actual structure and functionalization. A huge range of biodegradable polymers and composites are being recommended for that purpose. Most of the research studies are going on determining the effect of nano inclusions on to polymeric phases. There has been inclusion of nanofillers that are inorganic into biodegradable polymers such as metal nanoparticles, HAp or carbon nanostructure to enhance the properties of the biodegradable polymer. Although this field has seen appreciable growth, the present tissue engineering approaches pose some disadvantages in factors like degradation rate, biological response and mechanical properties that are required to be solved. In the future, new methodologies, like three-dimensional printing might give an instant and trustworthy result for manufacturing tissue scaffolds. A major future step for bioinspired regenerative medicine research is concurrent spatial and temporal control over stem-cell development through polymeric and nanomaterial manipulation.

ACKNOWLEDGEMENT

The authors would like to thanks Elsevier publication for providing the license number 5272871500143 to reuse Table 13.1 mentioned in this chapter.

REFERENCES

1. Asghari F, Samiei M, Adibkia K, Akbarzadeh A, Davaran S. Biodegradable and biocompatible polymers for tissue engineering application: A review. *Artificial Cells, Nanomedicine, and Biotechnology*. 2017 Feb 17;45(2):185–192.
2. Hasnain MS, Ahmad SA, Chaudhary N, Hoda MN, Nayak AK. Biodegradable polymer matrix nanocomposites for bone tissue engineering. In *Applications of nanocomposite materials in orthopedics* 2019 Jan 1 (pp. 1–37). Woodhead Publishing.
3. Mahanta AK, Maiti P. Chitin and chitosan nanocomposites for tissue engineering. In: Dutta, P (ed) *Chitin and chitosan for regenerative medicine* 2016 (pp. 123–149). Springer.
4. Liao X, Zhang H, He T. Preparation of porous biodegradable polymer and its nanocomposites by supercritical CO2 foaming for tissue engineering. *Journal of Nanomaterials*. 2012 Jan 1;2012: 1–12. 10.1155/2012/836394
5. Xing ZC, Han SJ, Shin YS, Kang IK. Fabrication of biodegradable polyester nanocomposites by electrospinning for tissue engineering. *Journal of Nanomaterials*. 2011 Jan 1;2011.
6. Armentano I, Dottori M, Fortunati E, Mattioli S, Kenny JM. Biodegradable polymer matrix nanocomposites for tissue engineering: A review. *Polymer Degradation and Stability*. 2010 Nov 1;95(11):2126–2146.
7. 10moto M, John B. Synthetic biopolymer nanocomposites for tissue engineering scaffolds. *Progress in Polymer Science*. 2013 Oct 1;38(10-11):1487–1503. (Okamoto & John, 2013)
8. Tanaka M, Sato K, Kitakami E, Kobayashi S, Hoshiba T, Fukushima K. Design of biocompatible and biodegradable polymers based on intermediate water concept. *Polymer Journal*. 2015 Feb;47(2):114–121.
9. Zhao R, Torley P, Halley PJ. Emerging biodegradable materials: Starch-and protein-based bio-nanocomposites. *Journal of Materials Science*. 2008 May;43(9):3058–3071.
10. Allo BA. Sol-gel derived biodegradable and bioactive organic-inorganic hybrid biomaterials for bone tissue engineering.2013.Electronic Thesis and Dissertation Repository. https://ir.lib.uwo.ca/etd/1192

11. Hassan M, Dave K, Chandrawati R, Dehghani F, Gomes VG. 3D printing of biopolymer nanocomposites for tissue engineering: Nanomaterials, processing and structure-function relation. *European Polymer Journal.* 2019 Dec 1;121: 109340.

12. Ravichandran R, Sundarrajan S, Venugopal JR, Mukherjee S, Ramakrishna S. Advances in polymeric systems for tissue engineering and biomedical applications. *Macromolecular Bioscience.* 2012 Mar;12(3):286–311.

13. Puppi D, Chiellini F, Piras AM, Chiellini E. Polymeric materials for bone and cartilage repair. *Progress in Polymer Science.* 2010 Apr 1;35(4):403–440.

14. Silva M, Ferreira FN, Alves NM, Paiva MC. Biodegradable polymer nanocomposites for ligament/tendon tissue engineering. *Journal of Nanobiotechnology.* 2020 Dec; 18(1):1–33.

15. Bharadwaz A, Jayasuriya AC. Recent trends in the application of widely used natural and synthetic polymer nanocomposites in bone tissue regeneration. *Materials Science and Engineering: C.* 2020 May 1;110:110698.

16. Prakash J, Prema D, Venkataprasanna KS, Balagangadharan K, Selvamurugan N, Venkatasubbu GD. Nanocomposite chitosan film containing graphene oxide/hydroxyapatite/gold for bone tissue engineering. *International Journal of Biological Macromolecules.* 2020 Jul 1;154:62–71.

17. Papaparaskeva G, Louca M, Voutouri C, Tanasă E, Stylianopoulos T, Krasia-Christoforou T. Amalgamated fiber/hydrogel composites based on semi-interpenetrating polymer networks and electrospun nanocomposite fibrous mats. *European Polymer Journal.* 2020 Nov 5;140:110041.

18. Cao L, Wu X, Wang Q, Wang J. Biocompatible nanocomposite of TiO_2 incorporated bi-polymer for articular cartilage tissue regeneration: A facile material. *Journal of Photochemistry and Photobiology B: Biology.* 2018 Jan 1;178:440–446.

19. Kouser R, Vashist A, Zafaryab M, Rizvi MA, Ahmad S. Biocompatible and mechanically robust nanocomposite hydrogels for potential applications in tissue engineering. *Materials Science and Engineering: C.* 2018 Mar 1;84:168–179.

20. Li Y, Liao C, Tjong SC. Synthetic biodegradable aliphatic polyester nanocomposites reinforced with nanohydroxyapatite and/or graphene oxide for bone tissue engineering applications. *Nanomaterials.* 2019 Apr;9(4):590.

21. Sitharaman B, Shi X, Walboomers XF, Liao H, Cuijpers V, Wilson LJ, Mikos AG, Jansen JA. In vivo biocompatibility of ultra-short single-walled carbon nanotube/ biodegradable polymer nanocomposites for bone tissue engineering. *Bone.* 2008 Aug 1;43(2):362–370.

22. Lalwani G, Henslee AM, Farshid B, Lin L, Kasper FK, Qin YX, Mikos AG, Sitharaman B. Two-dimensional nanostructure-reinforced biodegradable polymeric nanocomposites for bone tissue engineering. *Biomacromolecules.* 2013 Mar 11; 14(3):900–909.

23. Guo B, Ma PX. Synthetic biodegradable functional polymers for tissue engineering: A brief review. *Science China Chemistry.* 2014 Apr 1;57(4):490–500.

24. Pandey JK, Kumar AP, Misra M, Mohanty AK, Drzal LT, Palsingh R. Recent advances in biodegradable nanocomposites. *Journal of Nanoscience and Nanotechnology.* 2005 Apr 1;5(4):497–526.

25. Allo BA, Costa DO, Dixon SJ, Mequanint K, Rizkalla AS. Bioactive and biodegradable nanocomposites and hybrid biomaterials for bone regeneration. *Journal of Functional Biomaterials.* 2012 Jun;3(2):432–463.

26. Yang KK, Wang XL, Wang YZ. Progress in nanocomposite of biodegradable polymer. *Journal of Industrial and Engineering Chemistry.* 2007;13(4):485–500.

27. Hasan A, Morshed M, Memic A, Hassan S, Webster TJ, Marei HE. Nanoparticles in tissue engineering: Applications, challenges and prospects. *International Journal of Nanomedicine.* 2018;13:5637.

28. Chakraborty G, Gupta A, Pugazhenthi G, Katiyar V. Facile dispersion of exfoliated graphene/PLA nanocomposites via in situ polycondensation with a melt extrusion process and its rheological studies. *Journal of Applied Polymer Science.* 2018 Sep 5;135(33):46476.

29. Chakraborty G, Dhar P, Katiyar V, Pugazhenthi G. Applicability of fe-CNC/GR/PLA composite as potential sensor for biomolecules. *Journal of Materials Science: Materials in Electronics.* 2020 Apr;31(8):5984–5999.

30. Chakraborty G, Katiyar V, Pugazhenthi G. Improvisation of polylactic acid (PLA)/ exfoliated graphene (GR) nanocomposite for detection of metal ions (Cu_{2+}). *Composites Science and Technology.* 2021 Sep 8;213:108877. doi.org/10.1016/ B978-0-12-818415-8.00006-1

31. Laurenti M, Cauda V. Biodegradable polymer nanocomposites for tissue engineering: Synthetic strategies and related applications. In: Grumezescu, Valentina and Grumezescu, Alexandru Mihai (eds). *Materials for Biomedical Engineering* 2019 Jan 1 (pp. 157–198). Elsevier.

32. Zuluaga-Vélez A, Quintero-Martinez A, Orozco LM, Sepúlveda-Arias JC. Silk fibroin nanocomposites as tissue engineering scaffolds–A systematic review. *Biomedicine & Pharmacotherapy.* 2021 Sep 1;141:111924.

33. Zhao H, Liu M, Zhang Y, Yin J, Pei R. Nanocomposite hydrogels for tissue engineering applications. *Nanoscale.* 2020;12(28):14976–14995.

34. Wahid F, Khan T, Hussain Z, Ullah H. Nanocomposite scaffolds for tissue engineering; Properties, preparation and applications. In Asiri, Abdullah M. and Mohammad, Ali (eds). *Applications of nanocomposite materials in drug delivery* 2018 Jan 1 (pp. 701–735). Woodhead Publishing.

35. Liu H, Jian R, Chen H, Tian X, Sun C, Zhu J, Yang Z, Sun J, Wang C. Application of biodegradable and biocompatible nanocomposites in electronics: Current status and future directions. *Nanomaterials.* 2019 Jul;9(7):950.

36. Carrow JK, Gaharwar AK. Bioinspired polymeric nanocomposites for regenerative medicine. *Macromolecular Chemistry and Physics.* 2015 Feb;216(3):248–264.

37. Liu S, Qin S, He M, Zhou D, Qin Q, Wang H. Current applications of poly (lactic acid) Composites in tissue engineering and drug delivery. *Composites Part B: Engineering.* 2020 Jul 28:108238.

38. Ojijo V, Ray SS. Processing strategies in bionanocomposites. *Progress in Polymer Science.* 2013 Oct 1;38(10–11):1543–1589.

39. Manikandan A, Mani MP, Jaganathan SK, Rajasekar R. Morphological, thermal, and blood-compatible properties of electrospun nanocomposites for tissue engineering application. *Polymer Composites.* 2018 Apr;39:E132–E139.

40. Bramhill J, Ross S, Ross G. Bioactive nanocomposites for tissue repair and regeneration: A review. *International Journal Of Environmental Research and Public Health.* 2017 Jan;14(1):66.

41. Pina S, Oliveira JM, Reis RL. Natural-based nanocomposites for bone tissue engineering and regenerative medicine: A review. *Advanced Materials.* 2015 Feb;27(7):1143–1169.

42. Piskin E. Biodegradable polymers as biomaterials. *Journal of Biomaterials Science, Polymer Edition.* 1995 Jan 1;6(9):775–795.

43. Campodoni E, Heggset EB, Rashad A, Ramírez-Rodríguez GB, Mustafa K, Syverud K, Tampieri A, Sandri M. Polymeric 3D scaffolds for tissue regeneration: Evaluation of biopolymer nanocomposite reinforced with cellulose nanofibrils. *Materials Science and Engineering: C.* 2019 Jan 1;94:867–878.

44. Mulchandani N., Prasad A., Katiyar V. Resorbable polymers in bone repair and regeneration. In Grumezescu, Valentina and Grumezescu, Alexandru Mihai (eds). *Materials for biomedical engineering* 2019. Elsevier Inc.

45. Prasad A. Bioabsorbable polymeric materials for biofilms and other biomedical applications: Recent and future trends. *Materials Today: Proceedings.* 2021a;44: 2447–2453.

46. Prasad A. State of art review on bioabsorbable polymeric scaffolds for bone tissue engineering. *Materials Today: Proceedings.* 2021b;44:1391–1400.

47. Prasad A., Bhasney S., Katiyar V., Ravi Sankar M. Biowastes Processed hydroxyapatite filled poly (lactic acid) bio-composite for open reduction internal fixation of small bones. *Materials Today: Proceedings.* 2017;4(9):10153–10157.

48. Prasad A., Devendar B., Sankar M. R., Robi P. S. Micro-scratch based Tribological characterization of hydroxyapatite (HAp) fabricated through fish scales. *Materials Today: Proceedings.* 2015;2(4–5):1216–1224.

49. Prasad A., Sankar M. R., Katiyar V. State of art on solvent casting particulate Leaching method for orthopedic scaffolds fabrication. *Materials Today: Proceedings.* 2017;4(2):898–907.

14 Design and Manufacturing of Nanorobots and Their Industrial Applications

Abhishek Shrivastava and Vijay Kumar Dalla

CONTENTS

14.1　Introduction...227
14.2　Design of Nanorobot...229
　　　14.2.1　Architecture of Nanorobot..231
　　　14.2.2　Estimated Model of Nanorobot..231
14.3　Manufacturing Approaches of Nanorobots...232
　　　14.3.1　Nubots...232
　　　14.3.2　3D Printing...232
　　　14.3.3　Biohybrids...233
　　　14.3.4　Surface Bound Systems...234
14.4　Nanomanipulation...234
14.5　Nanorobotic Devices..234
　　　14.5.1　Nanocoils Assembly by Nanorobots...235
14.6　Applications of Nanorobots in Industry...235
14.7　Disadvantages of Nanorobots..236
14.8　Conclusion and Discussion..236
References...237

14.1　INTRODUCTION

The process of creating, producing and manipulating materials at the nanoscale is referred to as nanotechnology [1]. Nanorobotics is the study of robots at the nanoscale, and nanotechnology is the embedded technology. Nanorobots are tiny robots that can sense, actuate, signal, process information, intelligence, or display swarm behaviour [2]. They are made up of numerous components that perform specialized activities; the components are built at the nanoscale and can range in size from 1 to 100 nm [3,4]. Nanobots (also known as nanorobots) are the present focus of a developing field of research. They are also known as nanites, nanoids, nanomachines or nano-mites [5]; nevertheless, the term nanobot is made up of two words; nano and bot (Figure 14.1). Nano means very little or minute, and bot means

DOI: 10.1201/9781003220602-14

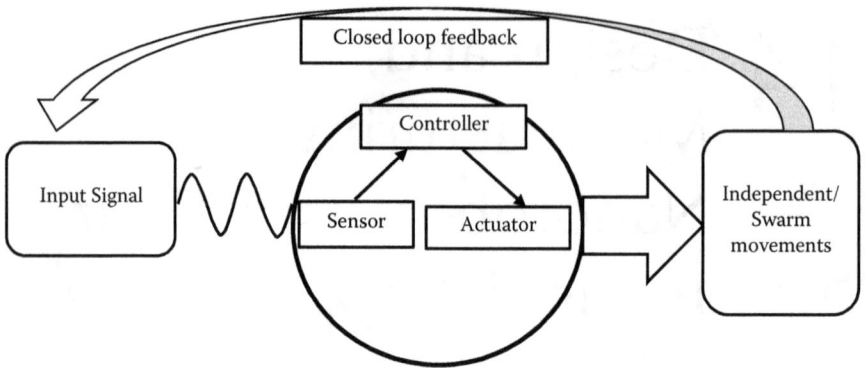

FIGURE 14.1 A schematic of nanorobots/nanites/nanoids/nanomachines/nano-mites working.

a device that can be operated by a software, i.e., a robot. Cancer therapy, surgery, precision medicine, diabetes monitoring, dentistry, blood monitoring and medication administration are all possible uses for nanorobots in the medical area [6]. Mobile phones are currently proven to be an effective feedback mechanism for data transfer for interaction, regulation and energy supply within the body. Furthermore, the advent of novel materials and their implementation in the fabrication of sensors and actuators has culminated in nanoscale devices [7]. Because nanorobots have the possibility of working independently or in swarms in a wide range of applications, including drug, environmental surveillance systems, armed services and many more innovative applications, it is critical to undertake an analysis of current state-of-the-art nanoscale systems [8].

The current applications of nanorobots in industries are shown in Table 14.1 and Figure 14.2. The terms microrobots and nanorobots are frequently used interchangeably. The information presented in this study about microrobots also applies to nanorobots. Nanorobots must be extremely effective, controlled, cost-efficient in mass manufacturing, and completely operational with minimum supervision [9]. They must, however, be tiny enough not to physically degrade live tissues upon

TABLE 14.1

Current Application and Possible Application of Nanobots in Industries

Possible medical applications	Possible industrial applications
Cancer therapy	Environmental surveillance systems
Surgery	Armed services
Precision medicine	Nanoscale travel
Diabetes monitoring	Data transfer from nanoscale environment to macroscale environment
Dentistry	Nanoscale manufacturing
Blood monitoring	Crack detection
Medication administration	

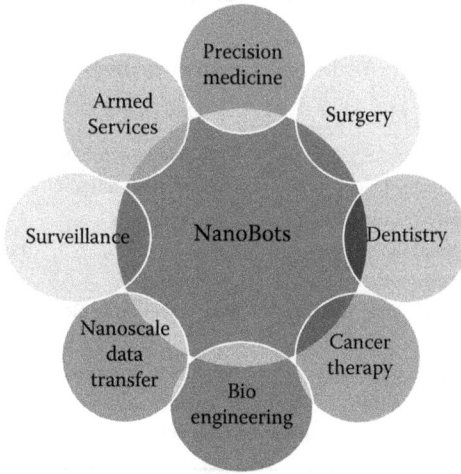

FIGURE 14.2 Possible applications of nanobots.

entering the body, and big enough to process both endogenous and external data from numerous sensory systems. Other difficulties that researchers confront include creating and constructing nanorobots with dimensions less than a nanometre, as well as controlling and managing a huge number of nanorobots (known as swarms). Furthermore, additional unique design concerns such as sensing, guidance, power transmission, mobility and constituent manipulation pose significant hurdles in the field of nanorobotics. Another difficulty in terms of nanorobotic structure is regulating matter at the molecular levels to influence the behaviour of nanorobots. Altogether, the mechanization, power and fabrication of nanorobots is a difficult and unique topic. Because of their capacity to be remotely controlled, nanorobots may perform admirably in a wide range of industrial applications, from cleaning up the environment to conveying medicinal drugs to a specific area in the body [10,11]. The nano swimmers have the potential to be next-generation nanorobots that can benefit a variety of industries. They can help with focused medication administration and successful therapy in the medical area, as well as reduce the degree of damage caused by chronic illnesses [12]. The goal of this review study is to give some helpful and classified information on the available possibilities for design of nanorobots. The information presented in this paper should assist others in accelerating their research and developing new concepts and approaches. In this work, design and manufacturing of nanorobots and their industrial applications are discussed. A comparison between macro-bot, micro-bot and nano-bot has been made in Table 14.2 which may help in selecting an appropriate bot.

14.2 DESIGN OF NANOROBOT

Because of dimensional limits, the standard robotics paradigm sensor actuator-control must be modified at the nanoscale. This means that sensors, actuators will be brought together and incorporated into nanorobot frameworks in a manner as shown

TABLE 14.2

Difference Between Macro-, Micro- and Nano-Bots

Parameter	Macrobot	Microbots	Nanobots
Control system	Open loop	Open and closed loop	Closed loop
Size	Greater than 1 mm	1 μm to 1000 μm	1 nm to 1000 nm
Position accuracy	Greater than 10 μm	1 μm to 10 μm	1 nm to 1000 nm
Fabrication material	Solid, polymers, composites.	Solid, polymers, composites	Polymers, biological materials
Driving and manipulation	Mechanical principle	Mechanical principle	Electrochemical
Dynamics	Newtonian mechanics	Newtonian mechanics	Quantum mechanics
Usage	Mechanical	Biological and mechanical	Biological and mechanical

in Figure 14.1. The usual approach of reading sensors, analyzing the data and constructing a control system to send orders to actuators in order to reduce the difference between the desired and actual values of a parameter (e.g., velocity, location, etc.) may not be practicable [13]. Mechanical nanobots are made up of a variety of materials and coatings. The machine coating or body is meant to breakdown in biological fluids, propelling the nanorobot in the event of chemical propulsion and/or releasing the relevant therapy to treat the ailment. The magnetic-propelled nanobot is by far the most common model in this categorization because of its simplicity of actuation, in which nanorobots incorporating magnetic elements are pushed using a harmless external magnetic force. Physically driven nanorobots made of synthetic and/or metallic materials can be activated here using a chemical reaction or external energy inputs like magnetic, ultrasonic and light fields [14]. These nanobots are complicated on a billionth scale, with joints and appendages that allow for agile swimming or walking. A schematic of design methodology is described in Figure 14.3.

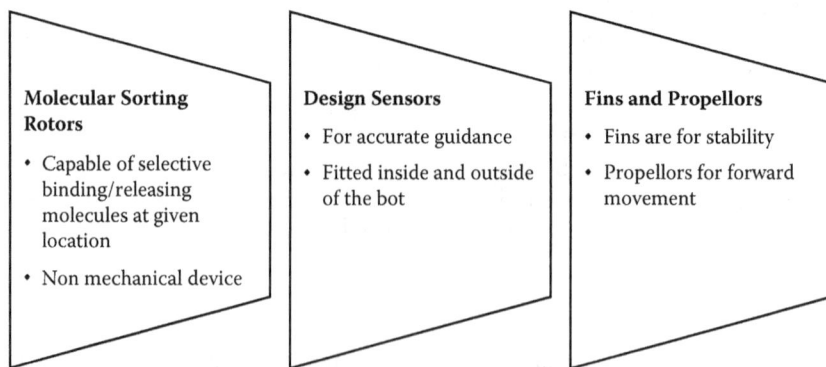

Molecular Sorting Rotors

• Capable of selective binding/releasing molecules at given location

• Non mechanical device

Design Sensors

• For accurate guidance

• Fitted inside and outside of the bot

Fins and Propellors

• Fins are for stability

• Propellors for forward movement

FIGURE 14.3 General design methodology of a nano-bot proposed by DOBRA.

14.2.1 Architecture of Nanorobot

The body of the nanorobot will be made of carbon nanotubes because of its inherent ability to absorb near-infrared light waves, which pass harmlessly through human cells. For collision avoidance, ultrasonic sensors are affixed to the nanorobot's body [15]. This is done to keep the nanorobots from colliding with one another or with other cells in the blood arteries. Nanoactuators play a significant role in the positioning of nanorobots and nanorobotic manipulators. While nanosized actuators for nanorobots are still being researched and are a long way from being implemented, MEMS-based efforts are focusing on lowering their sizes. Carbon nanotubes are among the most commonly utilized nanomaterials, owing to the material unique properties. They are already being utilized in applications where great wear resistance and break strength at a low weight are required, including such as bike frames, bulletproof jackets, industrial robot arms, sailboat hulls and spacecraft components [16]. Because of their distinct structure, these nanotubes are also useful in water filtration and medicine delivery. The carbon rings in the structure of nanotubes can filter out various chemical, biological and physical contaminants from water. Their form allows them to wrap around inner components, protecting them during medication administration. Carbon nanofibers are occasionally employed in the production of safety clothing, particularly biotextiles, where they can give a variety of extremely valuable attributes such as liquid and stain resistance, as well as antibacterial capabilities [17]. Carbon nanoparticles can also be combined with hefty, non-nano materials such as steel. These nanoparticles, when disseminated throughout steel, can boost their strength. They can eventually minimize the quantity of material required, resulting in lighter-weight devices that do not rely solely on nanoparticles.

14.2.2 Estimated Model of Nanorobot

The estimated model of nanorobot is depicted as

$$\hat{x}(t) = [Y(t)\ 1]\theta(t) \tag{14.1}$$

with

$$\hat{x}(t) = \begin{bmatrix} \hat{p}(y_1(t)) \\ \hat{p}(y_2(t)) \\ \hat{p}(y_n(t)) \end{bmatrix}, \ Y(t) = \begin{bmatrix} y_1^T(t) \\ y_2^T(t) \\ y_n^T(t) \end{bmatrix} \tag{14.2}$$

where

$y_1(t), y_2(t)....,y_n(t)$ are $k \times 1$ position vector in k-dimensional space, $\theta(t)$ is $(K+1) \times 1$ vector having entries equivalent to 1. The estimated error is depicted by the difference between estimation, $\hat{x}(t)$ and measurement, $x(t) = [\hat{p}(y_1(t)), \hat{p}(y_2(t)), \hat{p}(y_n(t))]^T$. To minimize the difference, the least squares (LS) estimator can be used. Using an LS estimator, we get the following parameter estimation.

$$\theta(t) = \begin{bmatrix} Y^T(t) & Y(t) \\ 1^T Y(t) & 1^T 1 \end{bmatrix}^{-1} \begin{bmatrix} Y^T(t) \\ 1^T \end{bmatrix} x(t) \tag{14.3}$$

$$\dot{u}_c(t) = [I \quad 0]\theta(t)v_i = - \sum_{j \in M(i)} \omega_{1ij}(y_i - y_j - y_{ui} - y_{uj}) + c_0 \hat{g}_c \tag{14.4}$$

where $\dot{u}_c(t)$ is gradient estimation at time t, I is $k \times k$ identity matrix.

The following control law for each nanorobot is based on the estimated gradient, Eq. (14.4).

$$\begin{aligned} v_i = &- \sum_{j \in M(i)} \omega_{1ij}(y_i - y_j - y_{ui} - y_{uj}) + c_0 \hat{g} \\ &- \sum_{j \in M(i)} \omega_{2ij}(u_i - u_j) - \frac{c_1}{n} \sum_{i=1}^{n} u_i - \frac{2c_2}{n} \sum_{i=1}^{n} u_i \end{aligned} \tag{14.5}$$

where y_{ui} and y_{uj} are the desired position of ith and j^{th} robot joints.

14.3 MANUFACTURING APPROACHES OF NANOROBOTS

Manufacturing nanomachines from molecular components is a difficult process. Because of the complexity, many engineers and scientists are continuing to collaborate across diverse techniques in order to reach breakthroughs in this new field of research. As a result, the significance of the following unique methodologies now used in the manufacture of nanorobots is rather obvious. A flow chart of nano-bot manufacturing is shown in Figure 14.4.

14.3.1 NUBOTS

A nucleic acid robot (nubot) is a nanoscale organic molecular machine. The structure of DNA may be used to create 2D and 3D nanomechanical systems. Small chemicals, proteins and other DNA molecules can be used to trigger DNA-based machinery [18]. Biological circuit gates based on DNA materials have been developed as molecular machines to enable in-vitro medication delivery for specific health issues.

14.3.2 3D PRINTING

3D printing is the technique of creating a three-dimensional structure using additive manufacturing technologies. Many of the same processes are used in nanoscale 3D printing, although on a much smaller scale. To create a structure in the 5–400 nm size, the 3D printing machine's accuracy must be substantially increased [19]. As an enhancement methodology, a two-step 3D printing procedure including 3D printing and laser etched plates was implemented. To achieve more precision at the nanoscale, the 3D printing technique employs a laser etching equipment, which etches the features required for the segments of nanorobots onto each plate. The plate is then fed into a 3D printer, that fills the engraved portions with the required

```
┌─────────────────────────┐
│   Identifying working   │
│      environment        │
└─────────────────────────┘
            ⇓
┌─────────────────────────┐
│  Selecting appropriate  │
│       materials         │
└─────────────────────────┘
            ⇓
┌─────────────────────────┐
│  Proper architecture for │
│    feedback control     │
└─────────────────────────┘
            ⇓
┌──────────────────────────────────┐
│ Selecting proper sensor, actuator │
│    and controlling device         │
└──────────────────────────────────┘
```

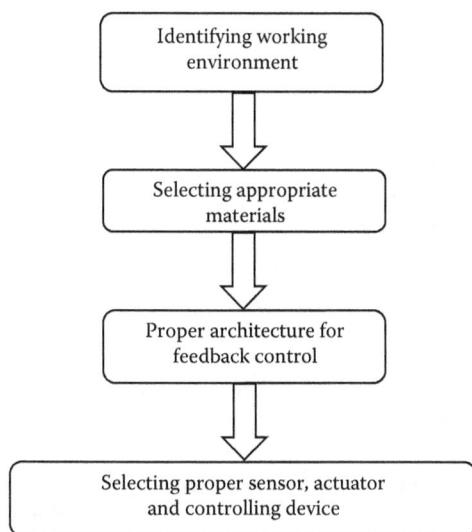

FIGURE 14.4   Flow chart of nanobot manufacturing.

nanoparticles. The 3D printing procedure is continued until the nanorobot is constructed from the ground up. There are several advantages to using 3D printing. For
starters, it improves the overall precision of the printing process.

The 3D printer employs a liquid resin that is solidified at exact points by a
concentrated laser beam. The laser beams' focus point is steered through the resin
by moveable mirrors, leaving behind a hardened line of solid polymer only a few
hundred nanometres broad [20]. Because of the high resolution, highly constructed
sculptures as small as a grain of sand may be created. The structure is created by
employing photoactive polymers that are cemented by the laser on an incredibly
tiny scale. By nanoscale 3D printing standards, this procedure is rapid. The 3D
micro-fabrication technology employed in multiphoton photopolymerization allows
for the creation of ultra-small features [21]. A focused laser is utilized in this
method to trace the required 3D object onto a block of gel. Because of the nonlinear
nature of optical excitation, the gel is only cured to a solid where the laser was
concentrated, and the remaining gel is washed away.

14.3.3 BIOHYBRIDS

The burgeoning subject of bio-hybrid systems blends biological and synthetic
structural components for biomedical or robotic applications [22]. The components
of bio-nanoelectromechanical systems (BioNEMS) are nanoscale in size, such as

DNA, proteins, or nanostructured mechanical parts. Thiol-ene e-beams resist enabling the direct inscription of nanoscale structures, accompanied by the derivatization of the natively reactive resist surface with macromolecules.

14.3.4 SURFACE BOUND SYSTEMS

Several studies have shown that synthetic molecular motors may be attached to surfaces. When restricted to the surface of a macroscopic substance, these rudimentary nanomachines exhibit machine-like behaviours. Surface anchoring motors have the potential to be utilized to move and position nanoscale materials on a surface in the same way that a conveyor belt does.

14.4 NANOMANIPULATION

Nanomanipulation, or positional and/or force control at the nanoscale scale, is a major enabling technique for nanotechnology, bridging the gap between top-down and bottom-up tactics and potentially leading to the advent of recompilation molecular assemblers [23]. At the moment, nanomanipulation may be used to investigate mesoscopic physical processes, biology and the development of prototype nanodevices. It is a basic technique for characterizing the properties of nanomaterials, nanostructures and nanomechanisms, preparing nanoscale building blocks, and assembling nanodevices such as NEMS. Nanomanipulators serve as the positioning device, microscopes serve as the eyes, various end-effectors such as probes and tweezers serve as the fingers, and various types of sensors (force, displacement, tactile, strain and so on) are used to facilitate manipulation and/or determine the properties of the objects [24]. Observation, actuation, measurement, system design and manufacturing, calibration and control, communication and the human–machine interface are all important technologies for nanomanipulation.

14.5 NANOROBOTIC DEVICES

Nanorobotic devices include nanometre-scale tools, sensors and actuators. The ability to operate nanosized items with nanosized tools, measure mass in femtogram ranges, detect force at piconewton scales, and cause GHz motion, among other astounding achievements, is made possible by shrinking device size [25]. A variety of researchers have independently researched top-down and bottom-up ways of producing such nanodevices. Top-down methods are based on nanofabrication and include nanolithography, nanoimprinting and chemical etching technologies. At the moment, they are 2-D manufacturing procedures with poor resolution. Bottom-up tactics are procedures that rely on assembly. At the moment, these strategies include self-assembly, dip-pen lithographic and guided self-assembly. At large sizes, these approaches may yield regular nanopatterns. Nanorobotic manipulation, with the capacity to position and orient tiny objects, is an enabling technology for constructing, characterizing and building a wide range of nanosystems.

Nanotube-based NEMS is still a thriving research field with a plethora of unresolved issues. Nanoscale materials such as nanorods, nanorings and polymers will allow a new family of sensors and actuators for ultrasmall quantity or item detection and actuation with ultrahigh accuracy and frequency [26]. Prototypes have been demonstrated using random spreading, direct growth, optic tweezers and nanorobotic manipulation. Self-assembly methods, on the other hand, will become more critical for integration into NEMS. We expect that dielectrophoretic nanoassembly will play an important role in the large-scale manufacturing of regular 2-D structures.

14.5.1 NANOCOILS ASSEMBLY BY NANOROBOTS

The production of nanocoil-based NEMS requires the assembly of as-grown or as-fabricated nanocoils, which is a considerable problem [27–29]. A variety of innovative techniques have been shown employing a manipulator (MM3A, Kleindiek) mounted in a SEM, focusing on the special characteristics of manipulating nanocoils due to their helical geometry, high elasticity, single-end fixing and strong adherence of the coils to the substrate from wet etching (Zeiss DSM962). Manipulation of fabricated SiGe/Si bilayer nanocoils (thickness 20 nm without Cr layer or 41 nm with Cr layer; diameter D = 3.4 m).

14.6 APPLICATIONS OF NANOROBOTS IN INDUSTRY

Nanorobots enable the production of microscopic electronics and electric devices, such as nanoscale transistors constructed of carbon nanotubes. Because of the extremely small scale, thin and highly flexible items such as plastic solar panels, rechargeable textiles and versatile gas sensors can be printed. Nanorobots are employed in the production of automobiles. Kim, D. et al. proposed a method to use nanobots a wireless power and torque producing device to remotely produce a motion in a motor to achieve a desired motion without providing any power source at the motor. This ability of nanobots can be used to power a motor of a vehicle without using a battery. Polymer nanocomposites are increasingly being used in high-end tyres to boost durability and wear resistance. Furthermore, nanorobots can be used to improve consumer automobile goods such as motor oil. Nanorobots may also be employed to develop more effective and stable lubricants, which can be used in a wide range of industrial applications. Materials at the nanoscale can function similarly to ball bearings in petroleum-based lubricants, keeping things moving smoothly, ensuring uniform distribution and limiting agglomeration. They can keep machine components lubricated even when temperatures or pressures vary rapidly. Nanomachines or nanites – mechanical or robotic devices that function at the nanoscale – are another promising use of nanotechnology. Nanomachines are, for the most part, high-tech and are not currently employed in production. This, however, is anticipated to change shortly – maybe within the next decade. Researchers are already using their inventions in a variety of situations, including medicine, where nanoscopic, self-assembling gadgets may be employed in ways that normal machinery cannot.

Nanorobots are still in their early stages of widespread use. However, the technology has recently gained popularity as fresh research and experimental design have demonstrated how successful these nanotechnologies may be. Popular nanomaterials, such as carbon nanotubes, are already routinely manufactured and used in the production of a wide range of products, including sailboat hulls, bicycle frames and spacecraft components. Nanoscale design in electronics results in very flexible devices and circuit boards. Nanoscale robots, also known as nanomachines or nanites, may soon transform medical device manufacturing.

14.7 DISADVANTAGES OF NANOROBOTS

The original design expense is exorbitant. The nanorobot's design is quite difficult. Electrical devices can generate stray fields, which in biology can trigger bioelectric-based molecular recognition systems. Electrical nanorobots are vulnerable to electrical interference from external sources such as radio frequency (RF) or electric fields, electromagnetic pulses and stray fields from other in vivo electrical devices. Complex interface, customization and design. Nanorobots pose a grave threat in the sphere of terrorism (Table 14.3). Terrorism and anti-groups can employ nanorobots to` torture populations in a new way, as nanotechnology has the power to destroy the human body at the molecular level. Another possible issue associated with nanorobots is privacy. Because nanorobots deal with the building of small and inconspicuous devices, there is potential for much greater eavesdropping than now exists.

14.8 CONCLUSION AND DISCUSSION

Some scientists have aimed to construct a totally mechanical nanorobot; nevertheless, a hybrid device with biological and robotic capabilities may be the most realistic option. To be considered a real nanite, the gadget must be mobile, have the capacity to process information or be programmed, and have a power supply. It should ideally be able to self-replicate. Scientists predict that a working nanite will be available in around 25 years. More advanced units will necessitate the use of specialized nanorobots known as assembler nanobots. In theory, the assemblers would utilize a bottom-up strategy, stacking atoms on top of each other in layers to create the desired nanomachines. These assembler units, however, have yet to be

---

TABLE 14.3
Advantages and disadvantages of nanobot

Advantages of nanobots	Disadvantages of nanobots
Complex operations of affected parts without affecting healthy tissues.	Difficult to design
Faster operation at nanoscale level	Electrical nanobot's electric interference with external sources damage the operation
Clearer picture at atomic level	Terrorists can misuse the technology
Detailed view of microscale defects	
High accuracy	

built. Atoms may now be arranged using atomic force microscopes and scanning tunnelling microscopes, thanks to advances in technology.

The scanning probe family's scanning tunnelling and atomic force microscopes, as well as the electron microscope's transmission electron microscope, have been used to not only resolve objects at the atomic level but also to move atoms and molecules. Researchers made a basic forerunner to working nanobot out of synthetic DNA, and scientists are now attempting to construct DNA strands to independently assemble in a present fashion to build a nanomachines. These developing nanobot featured two arms that rotated in response to a chemical reaction.

The most intimidating challenge from the top-down approach is the power supply. Batteries and solar cells are ineffective solutions to this problem. Nuclear power, which involves painting a small layer of radioactive material onto the surface of the nanorobot, might be one answer. The nanites have the ability to collect energy from decaying atomic particles. Biological nanobots might be programmed to consume small quantities of bodily tissue in order to get energy. Regardless of the design process, a nanorobot should ideally complete its function and then disintegrate, allowing the body to absorb and excrete it.

For a variety of reasons, silicon is an excellent material for the development of nanobots. It is tough, flexible and conducts electricity, yet it will not dissolve in bodily fluids. Because these gadgets are so little, swarms of them will be required to do most activities. Because there are so many foreign particles within the body, biodegradability will be a serious issue for medical applications. Controlled movement is another challenge in performing some jobs, particularly in medical applications. Nanorobots might potentially be used to detect particular substances or poisons, as well as provide early warning of organ failure or tissue rejection. They are also used to capture biometric measures and may be utilized to track an individual's overall health. These devices might be used in a range of industrial applications. They are now being researched for use in the oil sector. Furthermore, recent study is looking at how they might be used in nanophotonics to produce light more effectively. These small gadgets have the potential to manufacture computer circuits. They would be able to produce circuits on a smaller scale than present etching processes, allowing for the production of incredibly tiny CPUs and chips.

REFERENCES

1. Bayda S, Adeel M, Tuccinardi T, Cordani M, Rizzolio F. The history of nanoscience and nanotechnology: From chemical–physical applications to nanomedicine. *Molecules*. 2020 Jan; 25(1): 112.
2. Soto F, Karshalev E, Zhang F, Esteban Fernandez de Avila B, Nourhani A, Wang J. Smart materials for microrobots. *Chemical Reviews*. 2021 Feb 1; 5365–5403.
3. Whitesides GM. Nanoscience, nanotechnology, and chemistry. *Small*. 2005 Feb; 1(2): 172–179.
4. Nasrollahzadeh M, Sajadi SM, Sajjadi M, Issaabadi Z. An introduction to nanotechnology. In Mahmoud Nasrollahzadeh, S. Mohammad Sajadi, Mohaddeseh Sajjadi, Zahra Issaabadi, Monireh Atarod, *Interface Science and Technology* 2019 Jan 1 (Vol. 28, pp. 1–27). Elsevier.

5. Giri G, Maddahi Y, Zareinia K. A brief review on challenges in design and development of nanorobots for medical applications. *Applied Sciences.* 2021 Jan; 11(21): 10385.

6. Sahoo SK, Parveen S, Panda JJ. The present and future of nanotechnology in human health care. *Nanomedicine: Nanotechnology, biology and medicine.* 2007 Mar 1; 3(1): 20–31.

7. Yang T, Xie D, Li Z, Zhu H. Recent advances in wearable tactile sensors: Materials, sensing mechanisms, and device performance. *Materials Science and Engineering: R: Reports.* 2017 May 1; 115: 1–37.

8. Singh AV, Ansari MH, Laux P, Luch A. Micro-nanorobots: Important considerations when developing novel drug delivery platforms. *Expert opinion on drug delivery.* 2019 Nov 2; 16(11): 1259–1275.

9. Cheraghi AR, Shahzad S, Graffi K. Past, present, and future of swarm robotics. arXiv preprint arXiv:2101.00671. 2021 Jan 3.

10. Nayak J, Balas VE, Favorskaya MN, Choudhury BB, Rao SK, Naik B, editors. Applications of Robotics in Industry Using Advanced Mechanisms: Proceedings of International Conference on Robotics and Its Industrial Applications 2019. Springer Nature; 2019 Sep 3.

11. Magnin T, Revol F Life in terms of nano-biotechnologies. In Dirk Evers, Michael Fuller, Antje Jackelén, Knut-Willy Sæther .InIssues in Science and Theology: What is Life? 2015 (pp. 57–65). Springer, Cham.

12. Halder A, Sun Y Biocompatible propulsion for biomedical micro/nano robotics. *Biosensors and Bioelectronics.* 2019 Aug 15; 139: 111334.

13. Blaschke T Object based image analysis for remote sensing. *ISPRS Journal of Photogrammetry and Remote Sensing.* 2010 Jan 1; 65(1): 2–16.

14. Luo M, Feng Y, Wang T, Guan J. Micro-/nanorobots at work in active drug delivery. *Advanced Functional Materials.* 2018 Jun; 28(25): 1706100.

15. Saxena S, Pramod BJ, Dayananda BC, Nagaraju K Design, architecture and application of nanorobotics in oncology. *Indian Journal of Cancer.* 2015 Apr 1; 52(2): 236.

16. Fan J, Njuguna J. An introduction to lightweight composite materials and their use in transport structures. In *Lightweight Composite Structures in Transport* 2016 Jan 1 (pp. 3–34). Woodhead Publishing.

17. Tessier D. Surface modification of biotextiles for medical applications. In *Biotextiles as Medical Implants* 2013 Jan 1 (pp. 137–156). Woodhead Publishing.

18. Poornima KM, Venkatraman S, Andrews J, Varun P Bio-Nanorobotics: The milestone of nanotechnology and medicine. *International Journal of Advanced Research in Computer Science.* 2014 Jul 1; 5(6).

19. Campbell T, Williams C, Ivanova O, Garrett B. Could 3D printing change the world. *Technologies, Potential, and Implications of Additive Manufacturing* 2011 Oct, Atlantic Council, Washington, DC; 3.

20. Prince JD 3D printing: an industrial revolution. *Journal of electronic Resources in Medical Libraries.* 2014 Jan 1;11(1): 39–45.

21. Tripathi R, Kumar A Application of nanorobotics for cancer treatment. *Materials Today: Proceedings.* 2018 Jan 1; 5(3): 9114–9117.

22. Sharma A, Kumar K, Sharma T, Sharma S, Kanwar SS Nanotechnology in molecular targeting, drug delivery, and immobilization of enzyme (s). *Functionalized Nanomaterials I.* 2020 Aug 20: 299–307.

23. Osada M, Sasaki T Two-dimensional dielectric nanosheets: novel nanoelectronics from nanocrystal building blocks. *Advanced Materials.* 2012 Jan 10; 24(2): 210–228.

24. Dong L, Nelson BJ Tutorial-Robotics in the small part II: Nanorobotics. *IEEE Robotics & Automation Magazine*. 2007 Sep; 14(3):111–121.
25. Nelson BJ, Dong L Nanorobotics. In *Springer Handbook of Nanotechnology* 2010 (pp. 1633–1659). Springer, Berlin, Heidelberg.
26. Ke CH, Pugno N, Peng B, Espinosa HD Experiments and modeling of carbon nanotube-based NEMS devices. *Journal of the Mechanics and Physics of Solids*. 2005 Jun 1; 53(6): 1314–1333.
27. Nelson BJ, Dong LX, Subramanian A, Bell DJ Hybrid nanorobotic approaches to NEMS. *InRobotics Research* 2007 (pp. 163–174). Springer, Berlin, Heidelberg.
28. Kim D, Park J, Park B, Shin Y, Kim K, Park HH, Ahn S *IEEE Trans. Magn.* 2020; 56: 1–5. doi: 10.1109/tmag.2019.2948065.
29. General classification of robots. Size criteria Andreea DOBRA Mechatronic Department Politehnica University Timi, Romania andreea.dobra@upt.ro.

15 Nanomanufacturing and Design of High-Performance Piezoelectric Nanogenerator for Energy Harvesting

*Varun Pratap Singh, Ayush Dwivedi, Ashish Karn,
Ashwani Kumar, Subhash Singh,
Shubham Srivastava, and Kashika Srivastava*

CONTENTS

15.1 Introduction..242
15.2 Nanomanufacturing 243
    15.2.1 Applications...245
    15.2.2 Challenges of Nanomanufacturing..245
15.3 Nanogenerator..246
    15.3.1 Maxwell's Equations for Nanogenerators 247
    15.3.2 Polarisation Hypothesis..248
    15.3.3 Current Nanogenerator Transportation Equations.............................249
    15.3.4 Maxwell's Displacing Current Technology Forecasts 249
15.4 Piezoelectric Nanogenerator..250
    15.4.1 Mechanism...252
    15.4.2 Geometrical Configuration Design 252
        15.4.2.1 Single-Wire Generator (SWG)..252
        15.4.2.2 Vertical Nanowires Integrated Nanogenerator
                (VING)...252
        15.4.2.3 Lateral Nanowire Integrated Nanogenerator
                (LING) 254
        15.4.2.4 Nanocomposite Electrical Generators (NEG)..............254
        15.4.2.5 Other Type...255
15.5 High-Performance Piezoelectric Nanogenerator 255
    15.5.1 Variables Influencing the Performance of Piezo-electric
           Nanogenerators 256

DOI: 10.1201/9781003220602-15

                15.5.1.1   Influence of Piezo-Electric Element Matrix on
                           Piezo-Electric Nanogenerator Effectiveness.............257
                15.5.1.2   Influence of Material Micro-Morphology on
                           Piezo-Electric Nanogenerator Effectiveness.............258
                15.5.1.3   Chemical Doping Effect on the Performance of
                           Piezo-Electric Nanogenerators..................................259
                15.5.1.4   The Effect of Device Substrate on the
                           Performance of Piezoelectric Nanogenerators...........259
                15.5.1.5   Composite Thin-Film Material Development
                           to Improve Piezoelectric Nanogenerators'
                           Effectiveness...262
15.6   Piezoelectricity for Energy Harvesting.....................................263
        15.6.1   Piezo Generator: An Approach to Generate Electricity
                  from Vibrations 265
15.7   Energy Harvesting through Piezoelectric Nano Generator......................266
15.8   Applications and Future Scope of Nanomanufacturing in an
        Emerging Technical Field..267
        15.8.1   Self-Powered Nano/Micro Devices 268
        15.8.2   Smart Wearable Systems 268
        15.8.3   Transparent and Flexible Devices 269
        15.8.4   Telemetric Power Transceiver Implant 269
References...269

15.1   INTRODUCTION

Nanotechnology refers broadly to the engineering of functional systems at the atomic or molecular scales in the range of 1–100 nm, and is typically considered a new and emerging technology to create new products or materials with novel functions and characteristics. The modern understanding on the existence of nanomaterials grew after the invention of scanning tunnel microscopy in 1985, although the concept of nanotechnology had gained some momentum right after the publication of Eric Drexler's book *"Engines of creation: The coming era of nanotechnology"* in the 1980s [1]. Norio Taniguchi, a Japanese researcher, coined the term "nano-technology" in 1974 in relation to semiconductor processes such as ion beam milling and thin film deposition, and he defined it as "the processes of separating, consolidating, and modification of materials through one atom or one molecule." [2]. It must be mentioned, however, that although the word "nanotechnology" is new, the phenomenon is not. In fact, there is much evidence to corroborate the numerous applications on nanotechnology in ancient and medieval India such as nano-carbon impregnated metals in weaponry, nanoscale metallurgy, therapeutics and nano-medicines such as nano-sized drugs, highly durable paints, metal ornaments, self-cleaning cloths, carbon nano-forms in cosmeticse [3]. Of course, similar reports on the ancient usage of technology have also surfaced from other parts of the world [4–6]. Thus, the word refers to a technique for regulating the structure of the material down to a few atoms and molecules in order to consolidate nano-sized and ultra-precision accuracy. According to another definition, "nano-technology is the understanding and management of matter at dimensions between 1 and 100 nm where unique phenomena permit novel application" [7].

Nanoscale scientific advancements have resulted in the creation of novel materials and procedures that are altering manufacturing capacities in a variety of industries. Nanotechnology is a large, diverse field of research and invention that is gaining traction around the world as it promises to revolutionize material and product manufacture. Nanotechnology is the manufacturing technique used to achieve the extra precision and super aspects required in items such as circuit boards, optical and electrical devices, mechanical components for compressors, bushings, computer data storage devices and spherical lenses, all of which have an accuracy rate of 1 nm or less. Wearable tech, wireless connectivity, cloud computing, human-machine interface and surgical implants all need dependable energy sources that can run for longer durations with low interference. Nowadays, rechargeable batteries are still used to power electronic devices. On the other hand, individual component battery replacement is unnecessarily complicated and expensive, especially for large-scale remote communication. In addition, the materials used to make batteries harm human health and the environment. Self-powered nanostructures that absorb energy from surroundings or from human body have a lot of promise to fix this problem, and they're conceivable because to nano-devices' low power consumption. Mechanical vibration energy is one of the most prevalent and abundant forms of power acquired from naturally occurring sources for energising the long-term functioning of Micro- and Nano-system in the aforementioned concepts. Various technologies, such as electromagnetism, electrostatic generation and the piezoelectric effect, have been developed to transform mechanical power into electrical power [8].

Because of their direct transformation capabilities and applicability for embedded systems, piezoelectric materials for mechanical energy harvesting are gaining favour. For piezoelectric energy harvesting, Lead zirconate titanate (PZT), is the preferred material. The exceeding brittleness of PZT ceramic and the presence of lead raise concerns about long-term dependability, durability and safety, limiting its use. After their discovery in the 20th century, piezoelectric nanoparticles have demonstrated significant potential for successfully harnessing mechanical energy at the micro and nanoscale. With increased piezoelectric effects, good mechanical properties and extraordinary susceptibility to small-level vibrations, piezoelectric nanostructures outperform their bulk counterparts. Over the last few years, a range of rationally designed piezoelectric nanostructures for mechanical energy harvesting have been included in NGs of diverse shapes [9].

15.2 NANOMANUFACTURING

The use of elements on atomic, molecular and nanostructures levels for industrial applications is known as nanotechnology. The initial and most frequently recognized conception of nanotechnology focused on the particular technological goal of precisely manipulating atoms and molecules to create macroscopic scale objects, known today as molecular nanotechnology [10].

Nanomanufacturing encompasses the creation of nano-scaled material that can be powders or fluids, as well as the fabrication of items "bottom-up" from nano-scaled material like "top-down" in the tiniest increments for higher precision, as seen in laser ablation, etching and other technologies. Nanomanufacturing is distinct from molecular manufacturing, which involves the use of non-biological mechano-synthesis to intricate, nanoscale structures [11].

Nanomanufacturing is the process of creating things or materials with dimensions of one to one hundred nanometres. These techniques produce nano-technology, which includes ultra-compact equipment, frameworks, characteristics and mechanisms with application areas in organic chemistry, molecular genetics, space applications, physics and other fields. A development in generations of nano-manufacturing is shown in Figure 15.1.

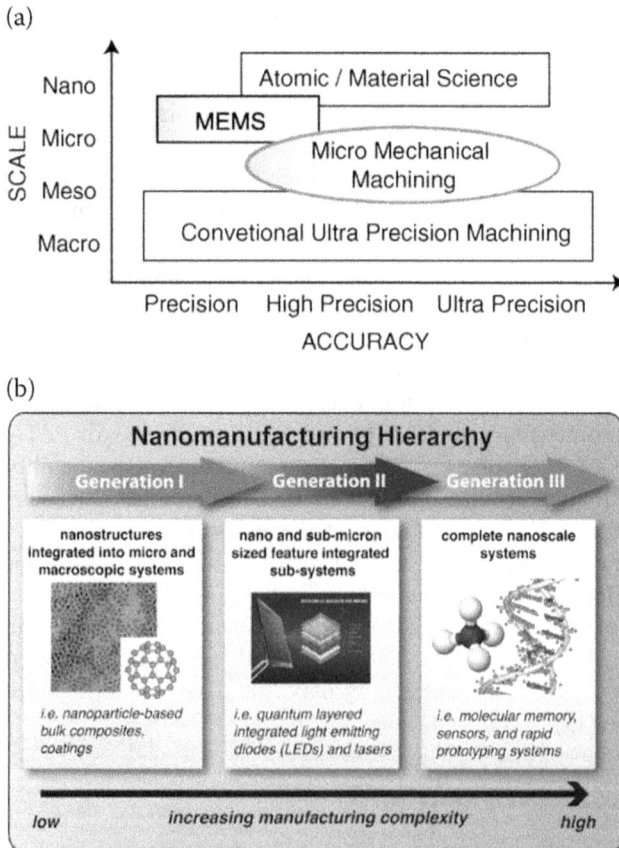

(a)

(b)

FIGURE 15.1 (a) Physical size vs. machining precision [12] and (b) nanomanufacturing generations [11].

15.2.1 Applications

Nanomanufacturing is based on two manufacturing approaches: "top-up" mentioning the slicing of a large matter to obtain nano-sized particles and "bottom-down" referring to the fabrication of different parts from the nano-scaled materials. Nanomanufacturing facilitates the development of new materials and products with applications in removal of material, device subassemblies, diagnostic implants, electrolytic coating and synthetic fibres, and lithography, among many others. Nanomanufacturing is a new field of fabrication that encompasses both a new field of research and a new market. Nanomanufacturing research, unlike traditional manufacturing, necessitates collaboration across traditional technical divides, such as Manufacturers, physicist, biologist, scientists and metallurgists and having application in most of the fields required high precision [13].

The translation of nanotechnology achievements in technology is stymied by a lack of awareness of the constraints to nanoscale production. While reducing dimensionality promises exponential data storage density gains, actual commercial devices will not be viable until the difficulty of combining and interconnecting billions of micro components and avoiding errors and faults in such an assembly is solved. The majority of nanotechnology research is focused on creating a structure or pattern with dozens to hundreds of Nano components (nanoparticles, nanotubes, molecules and so on). Nanoscale elements must be constructed in a quick, vast and directed way at high rates and over large areas. A separate type of basic research concern, such as scaling up assembly processes to capacity utilization, procedure sturdiness and reliability, and integrating nanostructures and gadgets into micro, meso and macroscale products, must be discussed to advance new breakthroughs from the lab to industrial applications. Nanotechnology is a large and diverse topic. This chapter covers nanoscale registration and alignment, dependability and defect control, as well as top-down and bottom-up procedures and mixed top-down and bottom-up approaches to nanoscale registration and alignment, dependability and fault regulation [13].

15.2.2 Challenges of Nanomanufacturing

- Scientists are currently debating the potential implications of nanotechnology. Nanotechnology has the capability to develop a diverse variety of new materials and devices for use in nanomedicine, nanoelectronics, nanomaterials, power generation and consumer goods are all examples of applications for nanotechnology. On the other hand, nanotechnology brings most of the same difficulties that every scientific advance does, such as worries about nanoparticle toxic effects and ecological damage, and also speculations about hypothetical disaster scenarios and their potential ramifications for global economies. Concerns about nanotechnology have created a debate between advocacy organizations and government about whether it warrants special regulation [14]. Challenges of nanomanufacturing and process for advanced materials are shown in Figure 15.2.
- There are a number of challenges in moving nanotechnology from lab demonstration to industrial-scale manufacture.

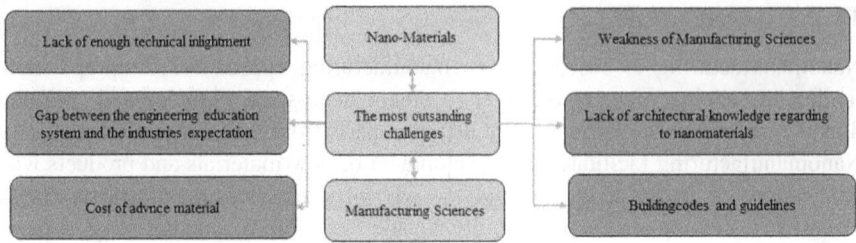

FIGURE 15.2 Challenges of nanomanufacturing and process for advanced materials [14].

- Developing low-cost nanostructured materials and manufacturing methods that produce viable yields.
- Controlling the accuracy of nanostructure assembly in nanoscale manufacturing.
- Testing dependability and developing defect control methods (detection and protection against chemical, biological, radiological and explosive agents).
- Nanoscale instrumentation and metrology skills for nano-electronics, nano-photonics and nano-magnetics.
- Precision in therapeutics, diagnostics and health-care services, as well as energy-related nanoscience research, micro-craft and robotics.

15.3 NANOGENERATOR

Power generation through non-conventional energy sources is always area of attention and covers many areas and form of energy generation and transfer [15–19], methods for generating electricity in the past used electromagnetic generators (EMG), which are predicated on Faraday's electromagnetic theory, which has been originally released in 1831. The energy production technology, on the other hand, could only efficiently extra. EMGs cannot harvest a wider wide variety of environmental forms of energy, such as low magnitude mechanical energy such as wave motion, individual's body movements, and a variety of other kinds of energy such as chemical and thermal energy [20].

A nanogenerator is a device that transforms mechanical energy from small-scale modifications into electric power. Nanogenerators are classified into three categories: piezo-electric, tribo-electric and pyro-electric. Mechanical power may be converted to electrical energy using both piezo-electric and tribo-electric nanogenerators. Pyro-electric nanogenerators may extract excess heat from temperature fluctuations over time.

A range of nanogenerators based on piezoelectric, triboelectric, pyroelectric, thermoelectric, ion streams and other phenomena can transform varied small-scale energy into energy capacity. In a self-powered system, NGs can also utilize energy storage units like batteries and super-capacitors to power operating components. As a result, the NGs' energy can be stored largely in energy storage systems before being used for environmental monitoring, data transfer, data processing and control [20]. The energy levels and respective applications and the power modification mechanism, for harvested power using nanogenerators, for energy storage are shown in Figure 15.3.

(a) (b)

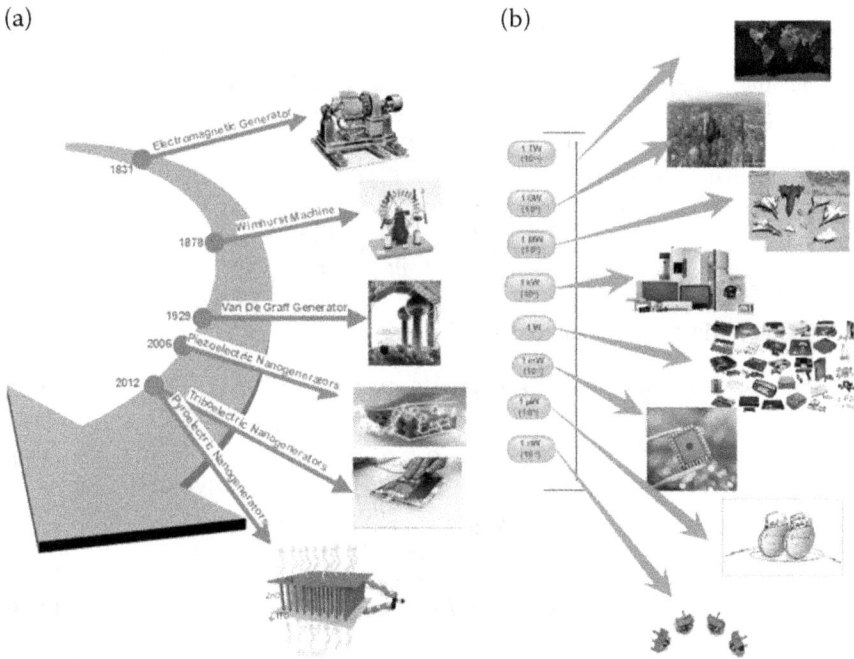

FIGURE 15.3 (a) The energy levels and respective uses (b) the power modification mechanism, for harvested power using nanogenerators, for energy storage and uses [21].

15.3.1 MAXWELL'S EQUATIONS FOR NANOGENERATORS

Here are the fundamental versions of Maxwell's equations:

$$\nabla . D = \rho \tag{15.1}$$

$$\nabla . B = 0 \tag{15.2}$$

$$\nabla \times E = -\frac{\partial B}{\partial t} \tag{15.3}$$

$$\nabla \times H = J + \frac{\partial D}{\partial t} \tag{15.4}$$

Maxwell proposed the displaced current, $\frac{\partial D}{\partial t}$, in 1861 to solve the governing equations for electric currents [22]. $D = \varepsilon_0 E + P$ gives the electric displacement vector D, including an isotropic dielectric material, $P = (\varepsilon - \varepsilon_0)E$, hence $D = \varepsilon E$. The displaced value of current is denoted as follows:

$$J_D = \frac{\partial D}{\partial t} = \varepsilon \frac{\partial E}{\partial t} \tag{15.5}$$

Maxwell's formulas have recently been applied to determine the energy output of nanogenerators. Wang [23] introduced an extra term Ps into D in 2017, wherein Ps indicate the polarisation formed by electrostatic surface-charging susceptible to mechanical stimulation, as opposed to the electric field driven intermediate polarity P. D may be expressed as $D = \varepsilon_0 E + P + P_s$, hence the displacing value of current can be calculated as follows:

$$J_D = \frac{\partial D}{\partial t} = \varepsilon\frac{\partial E}{\partial t} + \frac{\partial P_s}{\partial t} \qquad (15.6)$$

The updated Maxwell's equations can be written as:

$$\varepsilon\nabla.\, E = \rho - \nabla.\, P_s \qquad (15.7)$$

$$\nabla.\, B = 0 \qquad (15.8)$$

$$\nabla \times E = -\frac{\partial B}{\partial t} \qquad (15.9)$$

$$\nabla \times H = J + \varepsilon\frac{\partial E}{\partial t} + \frac{\partial P_s}{\partial t} \qquad (15.10)$$

The above formulations serve as the foundation for calculating nanogenerator output response, wherein the output power, and the accompanying electro-magnetic energy may be computed.

15.3.2 POLARISATION HYPOTHESIS

Whenever the net charge functional $\sigma_s(r, t)$ on the material surface is defined by a basis functions of $f(r, t) = 0$, the polarisation Ps generated by the electrostatic charge density may be described by the following equations [24]:

$$\nabla.\, P_s = -\sigma_s(r, t)\delta(f(r, t)) \qquad (15.11)$$

In this case, the delta function $\delta(f(r, t))$ is used to constrain the medium form. Using the surface energy to solve the scalar electric potential $\varnothing_s(r, t)$

$$\varnothing_s(r, t) = \frac{1}{4\pi}\int\frac{\sigma_s(r', t')}{|r - r'|}ds' \qquad (15.12)$$

the P_s can be obtained by

$$P_s = -\nabla\varnothing_s(r, t) = \frac{1}{4\pi}\int\sigma_s(r', t')\frac{r - r'}{|r - r'|^3}ds' \\ + \frac{1}{4\pi c}\int\frac{\partial\sigma_s(r'', t')}{\partial t'}\frac{r - r'}{|r - r'|^2}ds' \qquad (15.13)$$

This will be the logical continuation for the surface polarization concentration Ps.

15.3.3 Current Nanogenerator Transportation Equations

A surface integration of J_D is used to calculate the displaced current.

$$I_D = \int J_D.\, ds = \int \frac{\partial D}{\partial t}.\, ds = \frac{\partial}{\partial t} \int \nabla.\, D\, dr = \frac{\partial}{\partial t} \int \rho\, dr = \frac{\partial Q}{\partial t} \quad (15.14)$$

when Q represents the maximum amount of available charge on the electrode. The displacing current comprises the inner network of nanogenerators, whereas the capacitance propagation current leads the exterior circuit.

The accompanying general equation may be applied to calculate the current-transport characteristics of different configurations of nanogenerators:

$$\varnothing_{AB} = \int_A^B E.\, dL = \frac{\partial Q}{\partial t} R \quad (15.15)$$

The integral of dL covers a path between A to B point, and $\varnothing_{AB}$ is the potential-drop from A to B electrode.

The current piezoelectric nanogenerator transport equation is

$$RA\frac{d\sigma}{dt} + z\frac{\sigma - \sigma_P}{\varepsilon} = 0 \quad (15.16)$$

where A is the electrodes surface area, z is the thickness of the piezo-electric membrane, and p is the polarisation charge-density.

Throughout the transmission mode, the current transport formula for the piezoelectric material nanogenerator is:

$$AR\frac{\partial\sigma(z, t)}{\partial t} = -\sigma(z, t)[d_1/\varepsilon_1 + d_2/\varepsilon_2] - H(t)[\sigma(z, t) - \sigma_T]/\varepsilon_0 \quad (15.17)$$

H(t) is a function that depends on the interacting rate in-between the respective dielectric materials. The displacing charge, electric potential, output current and power density for four fundamental TENG modes may be determined using the governing equations [25].

15.3.4 Maxwell's Displacing Current Technology Forecasts

A first-term $\epsilon\frac{\partial E}{\partial t}$ of Maxwell's displacing current proposal gives rise to incident electromagnetic hypothesis, and electromagnetism gives rise to antennas, radios, telegraph, television, detection and ranging, microwaves, communication systems and aerospace engineering. The electro–magnetic integration results in modern physics, which serves as the conceptual framework for the discovery of the laser and the advancement of photonic technology. The very first element has fuelled global advancements in communications and laser systems during the previous

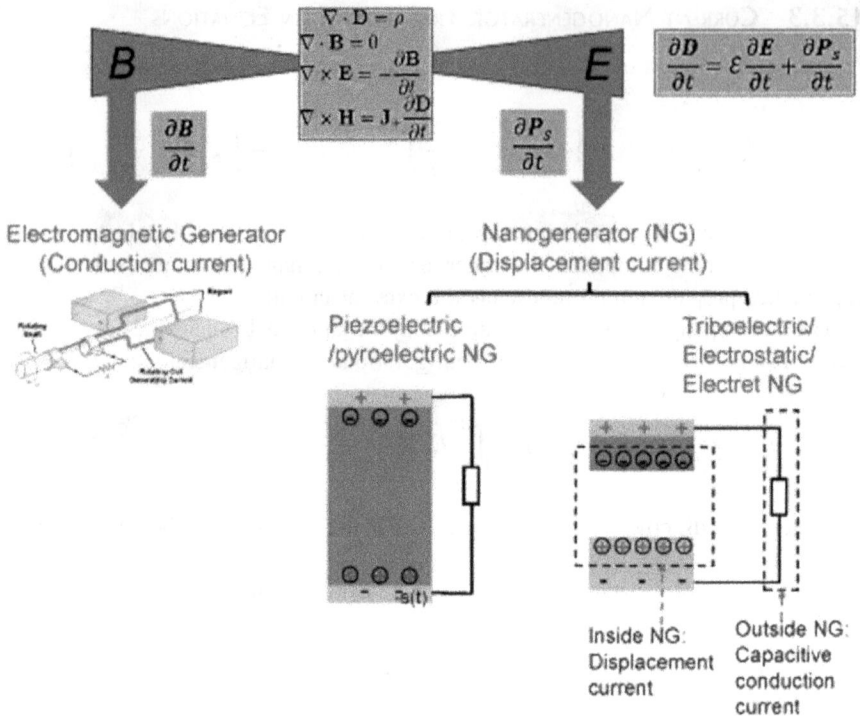

FIGURE 15.4 A diagram of a conducting current dominating electromagnetic generator and a displacing current dominating nanogenerator by leveraging on piezoelectric or pyroelectric and triboelectric or electrostatic or electret phenomena [25].

century. Wang's initial suggested second term $\frac{\partial P_s}{\partial t}$ laid the groundwork for nanogenerators. Including a $\frac{\partial P_s}{\partial t}$ factor in the displacing current, and hence in Maxwell's equations, broadens their applicability to energy [25]. A figure of a conducting current dominating electromagnetic generator and a displacing current dominating nanogenerator by leveraging on piezoelectric or electret phenomena is illustrated in Figure 15.4.

15.4 PIEZOELECTRIC NANOGENERATOR

The word "piezo" comes from the Greek word "piezein," that means "to squeeze or push." The "direct piezoelectric effect" and the "converse piezoelectric effect" are both present in the piezo-material.

A piezoelectric nanogenerator is an energy harvesting instrument that utilizes a nano-structured piezo-electric crystal to transform external kinetic energy into electrical energy. While it can respond to any energy harvesting machine that enables nanostructures to transform different types of ambient energy (including such solar energy and latent heat), are most commonly used in reference to kinetic energy harvesters that use nano-sized piezoelectric crystal, including such thin-film bulk acoustic resonant frequencies [23].

In 1880, while researching the effect of pressure on the formation of electrical charges by crystals, Jacques and Pierre Curie discovered the piezoelectric phenomenon (such as quartz). Whereas the innovation is still very much in initial phases of development, these have been acclaimed as a possibility emergence in the miniaturizing of conventional energy harvesting process, conceivably allowing easy convergence with the other sources of energy harvesting process and the individual implementation of portable electronics without regard for power sources. In Figure 15.5 energy harvesting from solar, thermal, mechanical and biochemical sources is compared to demonstrate their benefits and potential practical constraints.

When crystals are mechanically stressed, the direct piezoelectric effect produces electricity, whereas when an electric potential is supplied, the reverse piezoelectric effect induces stress or strain in crystals. The most widely used crystals are lead zirconate titanate crystals. Sound generation and recognition, high-voltage generation, electronic frequency production, microbalances and ultra-fine focus of optic components are all applications of the Piezo phenomenon. It also powers several devices like scanning-probe microscopes, as well as cigarette lighters and propane barbecues with a push-start feature [11].

Energy source	Solar	Thermal	Mechanical	Biochemical
				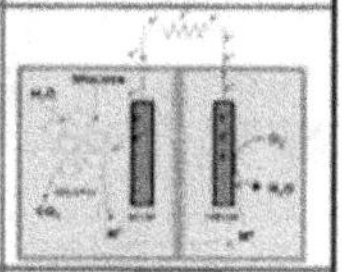
Harvesting Principle	Photovoltaic	Thermoelectric	Electromagnetic/Electro static/ Piezoelectric	Biochemical reactions
Approximate power density	5–30 mWcm^{-2}	0.01–0.1 mWcm^{-2}	10–100 mWcm^{-2}	0.1–1 mWcm^{-2}
Pros	Microfabrication compatible mature technology, long lifetime, DC & high power output	No moving parts required, long lifetime, high reliability, DC output	Ubiquitous and abundant in the ambient, broad frequency and power ranges, high output	Biocompatible/degradable, clean energy, environmentally friendly, abundant in biological entities
Cons	Limited by environmental conditions, not available at night	Low efficiency, large size, a large and sustained thermal gradient is required	AC output, not continuous output	Low power output, poor reliability, limited lifetime
Potential Applications	Remote sensing and environmental Monitoring	Structural-health monitoring for Engines and machines, wearable Biomedical devices	Remote sensing and monitoring, wearable systems, blue energy, internet of things	In vivo applications, environmental Monitoring/sensing, biocompatible application

FIGURE 15.5 Energy harvesting from solar, thermal, mechanical and biochemical sources is compared to demonstrate their benefits and potential practical constraints [23].

15.4.1 Mechanism

The operation of a nanogenerator will be examined for two separate circumstances: pressure exerted perpendicularly of the nanowire and pressure transmitted parallel to the axis of the nano-wire. The operational mechanism is described by a vertically developed nano-wire that's also exposed to a laterally travelling needle. When a movable point applies an external influence to a piezoelectric device, displacement happens. The piezoelectric reaction generates an electrical potential within the nanowire, with a high voltage differential for the extended area and a negative electrical potential for the squashed area having negative strain. This is due to the crystallized material's relative motion of anions and cations [25]:

$$V_{max} = \pm \frac{3}{4(k_0 + k)}[e_{33} - 2(1 + v)e_{15} - 2ve_{31}]\frac{a^3}{l^3}v_{max} \qquad (15.18)$$

where k_0 represents the permittivity in low pressure, is the di-electric constant, e_{33}, e_{15}, and e_{31} represent the piezo-electric correlation coefficient, v represents the Poisson ratio, a represents the radius of the nanorods, l represents the total length of the nanorods, and $vmax$ represents the maximum deflection of the nanorod's edge [20,26].

15.4.2 Geometrical Configuration Design

The majority of nanogenerators can be classified into four classes based on the configuration design of piezoelectric nanostructures: SWG, VING, LING and "NEG." However, as indicated in the other type, there's an arrangement that doesn't fit into any of the above subcategories [22,27].

15.4.2.1 Single-Wire Generator (SWG)

The SWG architecture was frequently employed in the early phases of NG construction. It is made up of two components: a laterally oriented NS and a stretchy substrate. The NS is inserted in the substrates to realize the piezoelectric material's energy production potential. A piezoelectric potential is formed at the end of the NW when the SWG is stretched. The Schottky contact stops electrons from flowing via the external circuit. The NG's charging/discharging processes are regulated with the assistance of the NW by managing the backward and forward passage of electrons through the external circuit [27].

15.4.2.2 Vertical Nanowires Integrated Nanogenerator (VING)

Figure 15.6 shows a VING 3-D design consisting of three layers: a foundation electrode, a vertical grown piezoelectric nanoparticles, and an auxiliary electrode. To construct the piezoelectric microstructures, several synthesis processes are utilized, which is then incorporated with the reference electrode in fully or partially mechanical forces including its end. Following the implementation of a standard formation of VING in 2006 by Dr. Z. L. Wang of the Georgia Institute of Technology [28], in which Wang used the tip of an atomic force microscope (AFM)

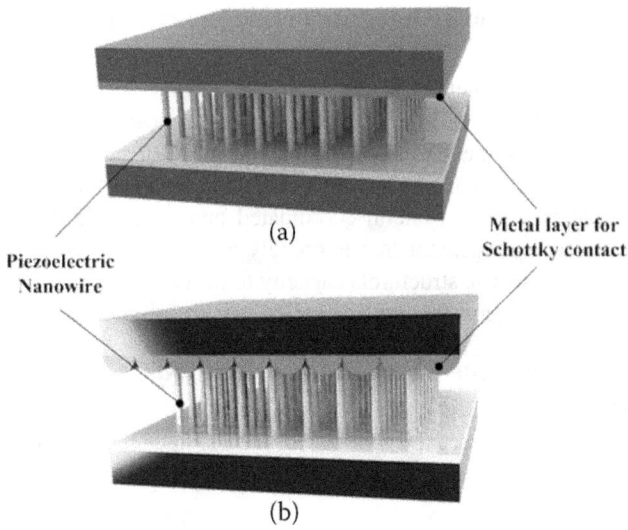

Piezoelectric
Nanowire

Metal layer for
Schottky contact

(a)

(b)

FIGURE 15.6 Schematic view of typical VING nanogenerator [21].

to stimulate the displacements of a long vertical ZnO nanostructure, the initial development of VING took place in 2007. The first VING used a counters electrode with a regular surfaces splitting, comparable to the arrays of AFM pointers, as a movable electrode. Since the reference electrode is not completely in interaction with the pointers of the piezoelectric nanostructures, exterior vibration can cause it to start moving throughout or outside, having caused the piezo-electric nanostructure to denature and generate an electrical potential distribution within each specific nanowire. Inside the instance of an n-type nanowire, its reference electrode is covered with metal to establish a Schottky interface with the nanowire's end, with only the compressed part of the piezoelectric nanowire enabling electrons to pass thru the barrier among its edge and the reference electrode. The toggle features of this setup illustrate its ability to generate dc voltage without the usage of an additional rectifier [20].

The shape of the counter electrode is critical in VING with partial contact. The piezoelectric nanostructures would not be sufficiently deformed by a flat counter electrode, especially when the counter electrode travels in-plane. Following the basic design, which resembled an array of AFM points, a few alternative methods were applied to swiftly manufacture the counter electrode. Using a process similar to that used to construct ZnO nanowire arrays, Doctor Z. L. Wang with his team formed a counter electrode comprised of ZnO nano-rods [29]. Doctor S. W. Kim and his team at Sungkyunkwan University (SKKU) and Doctor J. Y. Choi with his team at Samsung Advanced Institute of Technology (SAIT) in South Korea collaborated to develop a bowl-shaped transparent counter electrode using anodised aluminium with electroplating technologies [26]. They've also created a counter electrode constructed of interconnected single-walled carbon nanotubes (SWNTs) on a substrate material that is both efficient and transparent at energy conversion [30].

15.4.2.3   Lateral Nanowire Integrated Nanogenerator (LING)

Figure 15.7 shows a schematic structure of LING having base-electrode, the vertically developed piezoelectric nano-structure, and the metalic electrode for Schottky-contact make up LING, a two-dimensional design. In most cases, the width of the substrate layer exceeds the radius of the piezo-electric nano-structure, leading in purely tensile force on the specific nanowire. LING is a single-wire generator (SWG) development that combines a lateral coordinated nanowire on a flexible substrate. SWG is a technical arrangement that is broadly applied in the initial phases of research to test a piezoelectric structure's capacity to generate electric power [8]. LING generates AC electrical signals, similar to VINGs with full mechanical contact [27].

15.4.2.4   Nanocomposite Electrical Generators (NEG)

Figure 15.8 shows a schematic diagram of NEG, which is made up of metal backplate electrode, a vertical produced piezo-electric nano-structure, and polymer matrices that filled in the spaces inside the piezo-electric nano-structure. Momeni et al. were

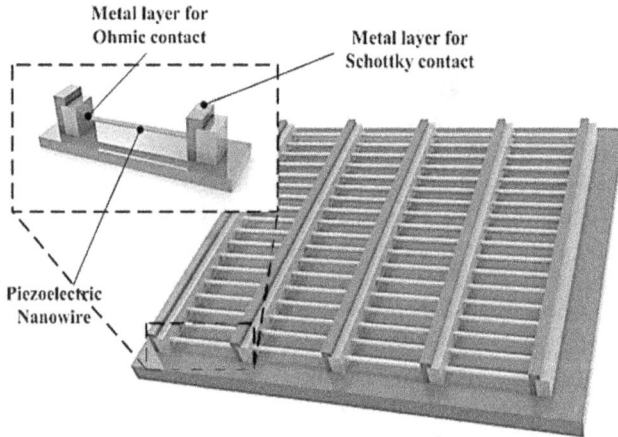

FIGURE 15.7   Schematic structure of LING [21].

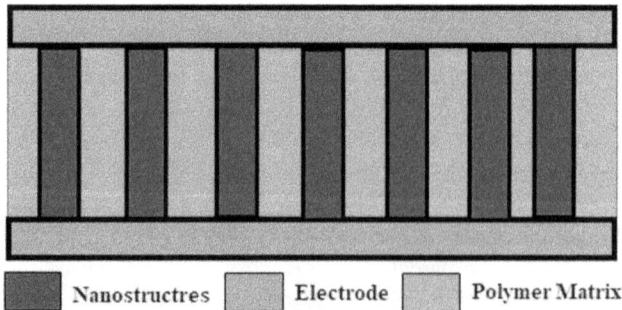

FIGURE 15.8   Schematic diagram of NEG.

the first to introduce NEG. The efficiency of the NEG was demonstrated to be higher than that of the basic nanogenerator design having a ZnO nanowire bent by an AFM tip. NEG also can be considered as a more sustainable energy producer [27].

15.4.2.5 Other Type

Professor Zhong Lin Wang proposed the fabric-like geometrical structure in 2008. VINGs' counter electrode is formed by coating any one microfiber with metal to make a Schottky-contact. The deformation of the nanostructure on the stationary microfiber occurs while the mobile microfiber is stretched and voltage generated. Its operation is analogous to that of VINGs with real physical contact, resulting in the generation of a DC electrical signal [30].

Because the piezoelectric constant is so important to a piezoelectric nanogenerator's overall performance, a further area of research to enhance sensor performance is to produce alternative materials with a considerable piezoelectric response. Lead-magnesium-niobate-lead-titanate (PMN-PT) is a future piezo-electric material having an extraordinarily high piezo-electric characteristic whenever the optimal compositions and orientations are obtained. A hydro-thermal approach was utilized in 2012 to create PMN-PT nano-wires having significant piezoelectric characteristic and can be used in energy-harvesting devices. The development of a single crystal PMN-PT nano-belt, which had been subsequently employed as a key component in the construction of a piezo-electric nanogenerator, increased the hitherto unrivalled piezo-electric characteristic even further [31].

15.5   HIGH-PERFORMANCE PIEZOELECTRIC NANOGENERATOR

The evolution of piezo-electric nanogenerators and related development towards significant energy production. Many forms of piezo-electric nanogenerators, as well as their characteristics and applicability scope, are also investigated. A number of suggestions for boosting the effectiveness of piezo-electric nanogenerators are also presented. Piezo-electric nanogenerators had come a long way over the years. Therefore, in contrast to typical engines, they can transform the atmosphere's low and irregular mechanical energy into electricity. According to published data, mechanical power density in the surroundings may achieve $1-10$ mW/cm^2, which is only significantly below solar energy density ($10-100$ mW/cm^2). Piezo-electric nanogenerators offer a high power density and thus are completely compatible with future microelectronic devices since they are dependent on the piezo-electric action of piezo-electric materials [31].

With the widespread availability of various piezoelectric nanogenerators, researchers are concentrating their efforts on figuring out how to transform physical energy into electricity effectively and fast. Several piezoelectric nanostructures have been employed to build nanogenerators in recent years (NGs). The growth of high piezo-electric nanogenerators since the beginning of 2006 is shown in Figure 15.9. ZnO was first contributed in NG development during these piezoelectric NGs and has been extensively investigated. Perovskite piezoelectric materials also have a high piezoelectric coefficient, making them ideal for piezoelectric NGs. New materials with high piezoelectric coefficients are employed, such as ZnO, GaN, CdS, InN,

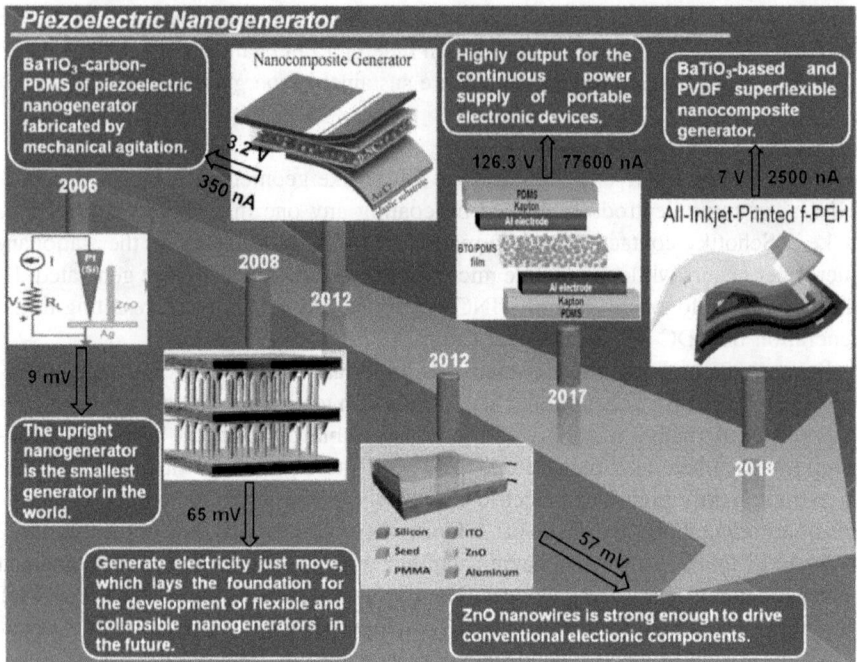

FIGURE 15.9   Growth of high piezo-electric nanogenerators since the beginnings in 2006 [35].

PbZr0.52Ti0.48O$_3$, NaNbO$_3$, PVDF and BaTiO$_3$. The most well-studied semi-conductors piezo-electric materials, including PZT and lead-free BaTiO$_3$ nano-particles, were utilized to build several forms of piezo-electric nanogenerators with different purposes [32,33]. Piezo-electric nanogenerator topologies have progressed from individual nano-wires to an interconnected lateral and vertical nano-wire arrays. The incorporation of semiconductor materials like multiwall carbon nano-tubes, graphene-oxide, as well as other chemical agents enhanced performance level and generator operational life, resulting in piezo-electric nanogenerators to greater output, flexibility, consistency and longevity than earlier generations [34].

The voltage output of piezo-electric nanogenerators has enhanced from the millivolt to 100 V, and the value of current also has significantly increased, enabling each other to keep driving compact micro-electronic hardware. Enhancing the performance level of piezo-electric nanogenerators through increasing the structural elements of piezo-electric nanogenerators and creating a range of novel piezo-electric materials were significant research subjects [35].

15.5.1   VARIABLES INFLUENCING THE PERFORMANCE OF PIEZO-ELECTRIC NANOGENERATORS

The major purpose of piezo-electric nanogenerators is to enhance system adaptability and wattage generation. commonly used methods for boosting the effectiveness of piezo-electric nanogenerators. Employing different piezo-electric materials, chemical

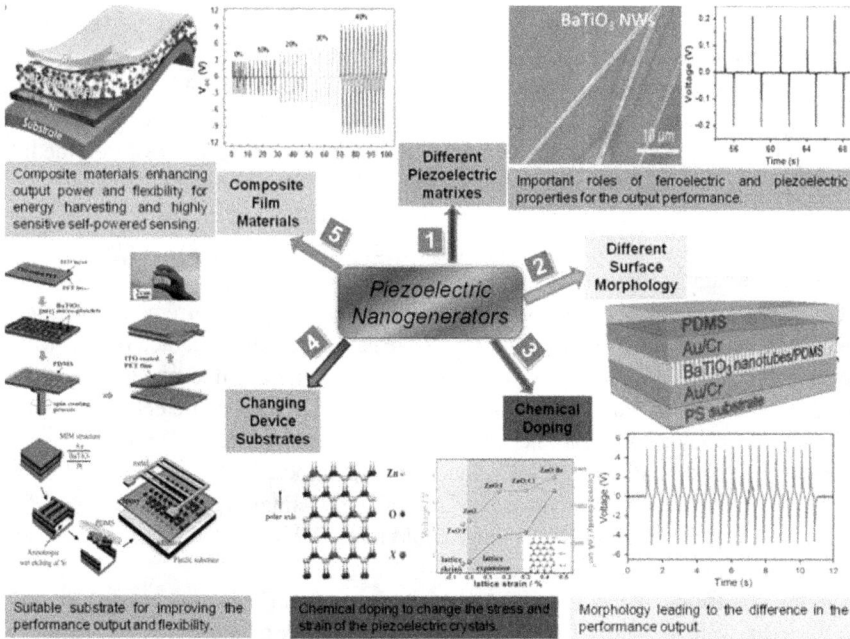

FIGURE 15.10 Factors influencing piezo-electric nanogenerator output performance [35].

dosing of piezo-electric materials, changing the microscopic morphology of conventional piezo-electric crystals, selecting suitable interfaces and developing polymer thin sheet components are a few instances [25] (as shown in Figure 15.10).

15.5.1.1 Influence of Piezo-Electric Element Matrix on Piezo-Electric Nanogenerator Effectiveness

The performance of piezo-electric nanogenerators is heavily influenced by the matrices used in their construction. The first semi-conductive piezo-electric material crystal studied was ZnO. ZnO is a significant functional substance for generating piezo-electric nanogenerators due to its outstanding optoelectronic, semiconducting and piezo-electric capabilities. It has a simple chemical structure and crystalline structure, and it's simple to control purity, size and orientation. Additionally, the high surface area of ZnO leads the nano-wires to stretch and bend under stress, leading in a voltage differential. Lateral Zno nano-wires were twisted employing atomic force microscopes (AFM) devices to generate mechanical work in prior study. Moreover, the ZnO nanowires' piezo-electric activity triggered spontaneously polarization of the charge, transforming mechanical energy into electricity and momentarily conserving it in the nanowires [36].

It is essential to improve the electrical output of a piezo-electric nanogenerator for practical uses by merging a significant number of system nanowires on a stretchable framework. As a consequence, integrated vertical/horizontal Zno nanowire grids were employed to produce a multi-layer alternator-based foldable nanogenerator [34,37].

As a consequence, nanowires of varying densities and orientation may be created for a variety of applications. To generate constant alternating current, a new AC piezo-electric nanogenerator founded on laterally developed ZnO nanowire clusters was suggested. The most significant distinction between this and earlier generators is that the upper layer of ZnO has been coated with an insulating layer of polymethyl-methacrylate, followed by a layer of metal electrodes. This generator produced 10 V and 0.6 A of voltage and current by piezoelectric potential and electrostatic coupling, respectively, with a maximum energy density of 10 mW/cm^2 [34].

While certain voltages and currents may be generated using ZnO-based piezo-electric materials, they are inadequate to fulfil the energy requirements of micro-electronic systems. This is owing to ZnO's weak piezoelectric characteristics, which necessitated the search for other compounds. PZT-based ceramic materials with superior crystallization, large dielectric properties and strong piezo-electricity became the favoured choice for piezo-electric nanogenerators. As a response, piezo-electric nanogenerators depending on PZT materials have attracted a lot of interest. For example, hydrothermal synthesis was used to make PZT single-crystal nanowires. At a frequency of 25.2 Hz, the output voltage could reach 2.7 V and the power density was 51.8 W/cm^2 [38].

Finally, various piezo-electric materials possess different piezo-electric capabilities, as well as faults that restrict the effectiveness of piezo-electric nanogenerators. While ZnO piezo-electric materials are easy to manufacture, these possess low piezo-electric properties, and although BaTiO$_3$-based piezo-electric materials possess significant piezo-electricity, the piezo-electric nanogenerators produce possess poor elasticity. Generally, the performance-intensive of BaTiO$_3$-based piezo-electric generator is satisfactory, and they have a lot of future possibilities. Other strategies, such as carefully manipulating the micro-morphologies of piezo-electric materials, can be used to improve piezoelectric nanogenerators in connection to generating high duty piezoelectric materials [39].

15.5.1.2 Influence of Material Micro-Morphology on Piezo-Electric Nanogenerator Effectiveness

By varying the process temperature with other variables throughout the material development, several micro morphologies of the similar materials, such as nanowire, nanoribbon, nanosheet, nanotube, nanorod and nanopillar, may be produced. The piezoelectric characteristic of similar piezoelectric-material with differing micro morphologies impact the effectiveness of generated piezoelectric nanogenerators. Generally, the piezoelectric nanogenerator composed of ZnO nanowire beats the piezoelectric nanogenerator constructed of ZnO nanorods. BaTiO$_3$ nanowires outperform BaTiO$_3$ nanotubes in piezoelectric properties, and the outputs of produced piezoelectric nanogenerators have also improved. The mean output voltage produced by ZnO nanorods was around 9.5 mV using contact mode Atomic Force Microscopy [40].

The same may be said of the various micro-morphologies of PZT-based piezoelectric compositions. The overall performance of piezoelectric nanogenerators constructed using crystalline PZT nanowires, vertical aligned PZT nanorod configurations and lateral spreading PZT nanofibers varied significantly. As previously stated, different micro-morphologies of the same piezoelectric material results in considerable variances in piezoelectric properties. Nanowires show greater piezo-electricity than some other micro-morphologies in ZnO and PZT-based piezo-electric substances. 2-D ZnO nano-sheets, on the other side, outperform 1-D ZnO nanowires in terms of piezoelectricity. The piezoelectric characteristics of nano-wires and nanopillars in $BaTiO_3$-based piezoelectric materials are comparatively superior, and this strong piezoelectricity ensures the good performance of the piezoelectric nanogenerators created. As a result, by altering the micro-morphology of piezoelectric materials, high-output piezoelectric nanogenerators can be created [41].

15.5.1.3 Chemical Doping Effect on the Performance of Piezo-Electric Nanogenerators

The characteristics of piezoelectric nanogenerators composed of ZnO, PVDF, PZT and $BaTiO_3$ piezo-electric materials are all distinct, but not ideal. Although the enhancement is small, modifying the micro morphology of the piezoelectric material enhances the effectiveness of manufactured piezoelectric nanogenerators. In order to enhance the effectiveness of piezoelectric nanogenerators, scholars devised a method of chemically-doping for materials in order to boost the piezoelectric coefficient, which improves the materials' piezoelectricity and enhances the effectiveness of piezoelectric nanogenerators [42,43]. The coefficient of piezoelectricity and dielectric-constant of piezoelectric material can be changed by doping with different chemical components, allowing for high-efficiency energy collection processes. There are two types of ZnO doping techniques: n-type and p-type doping. ZnO is a high-performance semiconductor material. Using n-type doping, the dopant could induce crystal framework deformation along the polar c-axis of the Zno lattice, improving the piezoelectric sensitivity and enhancing the effectiveness level of piezo-electric nanogenerators. If the dopant's ionic radius becomes too large, more lattice defects may occur, prohibiting energy from passing through the external circuit and decreasing effectiveness. P-type doping in ZnO nanowires can reduce the insulating effect of the dopant's free electrons. Chemical doping improves the effectiveness of piezoelectric nanogenerators made of ZnO and $BaTiO_3$, as well as the efficiency with which mechanical power is transformed into electrical energy in PVDF and PZT piezoelectric nanogenerators [33].

Dopants with chemical constituents may enhance the piezoelectric effect, fer-romagnetism, dielectric properties and interfacial conductance of nanocomposite, leading to an increase in piezoelectric nanogenerator energy capacity. Table 15.1 summarizes many of the electrospinning application's mentioned parameters including their impacts on the PVDF synthesized. Whenever it comes to gaining high piezoelectric effectiveness PVDF film, the table helps to grasp the significance of each parameter.

15.5.1.4 The Effect of Device Substrate on the Performance of Piezoelectric Nanogenerators

The piezoelectric materials employed in piezoelectric nanogenerators (PEN) have a big impact on their performance. The performance of a piezoelectric material with

TABLE 15.1

The Effect of Electrospinning Process Parameters [26]

Parameter	Parameter consideration	Effect
Solution	Solution-concentration	*As the percentage of PVDF rises, the following results are obtained:* • Increased solution-viscosity • Increased interfacial-tension • Increased nanofiber-diameter *At low-concentrations:* • Electro-spinning can be transformed into electro-spraying. • Have a detrimental effect on β-phase concentration
	Solvent-systems	*Solvent-systems are used to evaluate:* • The spin ability, Viscus nature • A PVDF solution's interfacial-tension • Nano-fibres uniformity • Improved β-phase characteristics • PVDF dissolve in solvents with a high boiling point. • Adding volatile solvent causes morphological alterations.
	Molecular weight	*An increased PVDF molecular mass produces the following outcomes:* • Viscosity and surface tension increases • Changes in nanofibre surface morphology • The content of β-phase increases • Expansion of fibre diameter • Restricts jet interruption
Processing variables	Voltage	*Voltage adjustments have an impact on:* • Nanofibre production • Minimal contributions to the morphology and β-phase concentration *An extremely high-voltage causes the following:* • Jet instabilities • Decrease in the rate of evaporation of a solvent • Non-uniform, beaded fibres with tiny diameter are produced.
	Feed rate	*Feed rate parameters (high/low) have an impact on:* • A PVDF solution's properties (viscosity) • It is hard to form Taylor cones at a lower rate. • A higher feed rate causes an unsteady jet and electro-spinning disturbance. • Increasing the feed rate causes the creation of the β-phase, which results in more homogeneous nanofibres.

TABLE 15.1 (Continued)
The Effect of Electrospinning Process Parameters [26]

Parameter	Parameter consideration	Effect
	Tip-to-collector distance	*Increase the gap, which allows more chance for the jet to traverse, resulting in nanofibres that are thinner and bead-free.* • The spacing between the tip and the collectors is normally between 10 and 20 cm. • A PVDF membrane was reported to have the highest β-phase concentration at ambient temperature (25°C). • As the atmospheric temperature rises, the diameter of the fibres reduces.
Environment-al conditions	Temperature	• As the atmospheric temperature rises, the diameter of the fibres reduces. • A PVDF membrane was reported to have the highest β-phase concentration at ambient temperature (25°C).
	Humidity	*The variation of humidity has an impact on:* • Pores are created on the surfaces of PVDF nano-fibres. • The diameter of the fibre • Nanofibers' smoothness • A volatile solution dissipates quickly in a dry environment, resulting in syringe obstruction.

excellent piezoelectric characteristics is unquestionably superior to that of a piezoelectric material with weak piezoelectric characteristics. For example, the PEN layer might have a substantial influence on the product's effectiveness [8,36].

PEN constructed on PET have greater output effectiveness over PEN depending on other materials, as shown in Table 15.2. PET is extremely versatile and may be used as a flexible framework to relieve the displacement of framework for effectiveness under inferior mechanical surrounding loads, enabling effective mechanical to electrical energy conversion [44].

TABLE 15.2
The Yield of ZnO Piezoelectric Generators with Various Substrates [35]

Substrates	Morphologies	Voltage output: (V)	Current output: (nA)
PET based	Nano-wires	2.40	380.00
Si based	Nano-wires	0.65	49.70
Polyster based	Nano-wires	0.50	200.00

Furthermore, its insulating properties can significantly reduce electrical energy loss during charge flow, enhancing the performance parameters of piezoelectric nanogenerators. Because of its superior insulating qualities, PDMS can decrease electric power waste throughout transmission and boost the performance of piezoelectric nanogenerators. This can shield piezoelectric components against degradation while also avoiding mechanical stress as a substrate material. As a consequence, PET and PDMS are now the best substrates for manufacturing foldable and piezoelectric nanogenerators with superior efficiency [44].

15.5.1.5   Composite Thin-Film Material Development to Improve Piezoelectric Nanogenerators' Effectiveness

The creation of alternative piezoelectric material, varied micro morphology of the material, chemical-doping, as well as the adoption of suitable substances have all been considered thus far as tactics for improving the effectiveness of piezoelectric nanogenerators.

These efforts, even so, fall short of meeting the dual objectives of piezoelectric nanogenerators: adaptability and effectiveness. A large percentage of legendary piezoelectric materials were also inorganic ceramic materials with low fracture toughness and adaptivity. Furthermore, their piezoelectric properties are poor. Studies have suggested combining biodegradable materials with inorganic nanoparticles to create nanostructured thin layer material to increase piezoelectric nanogenerators' adaptability and output efficiency. As a result, another advancement in piezoelectric nanogenerators is the creation and production of deposited thin film material with a high elasticity and piezoelectric properties.

When compared to conventional piezoelectric nanogenerators made purely of synthetic nanostructures, the use of nanocomposites for piezoelectric nanogenerators has several benefits, along with price, ease of manufacture, developing sustainability, adaptability, relatively low power dissipation, responsiveness for vast mechanical stresses, easiness of large-scale manufacturing and the capabilities to provide a significantly greater power output. The endurance and improved performance of composites broaden the spectrum of conceivable implementations in electronics products, environmental control and sensor technologies. Many BaTiO3-based composite substance piezoelectric nanogenerators can produce nanocomposite thin films with PDMS, PVDF, P(VDF-HFP), P(VDF-TrFE), P(VDF-HFP) and PVC, giving rise in nanostructured thin film materials which can be used to improve piezoelectric nanogenerator performance level. However, future challenges for composite thin film materials include the development of greater response thin film piezoelectrics, improved surface roughness control and the investigation of nanoscale and microscale piezoelectric devices [21,45].

ZnO, PZT, BaTiO3 and PVDF are now the most common piezoelectric materials used in piezoelectric nanogenerator research and development. Existing techniques and approaches for enhancing the efficiency of piezoelectric nanogenerators include the use of numerous piezoelectric materials, chemical doping of piezoelectric crystals, different sources of piezoelectric materials, the availability of appropriate surfaces and the advancement of composite thin films [34].

As a consequence of continuous advancements in piezoelectric materials, development procedures and building system, the effectiveness of piezoelectric nanogenerators has increased by many times greater. The voltage output has achieved 100 V in specific, however, the reduced output current remains the most significant barrier to the widespread use of piezoelectric nanogenerators. As a result, increasing current output has become a significant issue in this industry. Scientists are exploring materials with improved piezoelectric properties to enhance the effectiveness of piezoelectric nanogenerators by altering the nano microscopic morphology of piezoelectric material, doped them with synthetic chemical components, and making preparations composite thin-film materials. Piezoelectric nanogenerators will be employed to deliver high performance and adaptability in the future.

The output of a piezoelectric nanogenerator is largely determined by the piezoelectric matrix and surface topography. A piezoelectric nanogenerator's power output is presently fairly modest (W). By chemically doping piezoelectric crystals and selecting appropriate substrate, the voltage output of piezoelectric nanogenerators will be increased by many times greater [22]. A performance comparison of different piezoelectric materials geometries is mentioned in Table 15.3.

15.6   PIEZOELECTRICITY FOR ENERGY HARVESTING

PEHs are based on the direct piezoelectric effect. The following are the parametric formulations that link the mechanical (stress T and strain S) and electrical domain (electrical charge E and charge density D):

$$\begin{bmatrix} Converse \\ Direct \end{bmatrix} = \begin{bmatrix} S \\ D \end{bmatrix} = \begin{bmatrix} s^E & d^t \\ d & T \end{bmatrix} \begin{bmatrix} T \\ E \end{bmatrix} \qquad 15.19$$

where s^E is the compliant below a continuous electric fields, T is the dielectric permittivity underneath a regular stress, and d and d^t are the matrices for directly and indirect piezoelectric effect, with the superscript t denoting transpose. Major portion of harvesters has a frequency range that is significantly lower than the resonance frequency of the piezoelectric devices utilized. As a result, the piezoelectric elements are analogous to parallel plate capacitors. We can approximately compute the accumulating charge Q on the electrode, the voltage V over the material, and the overall transformed electricity generation U for a piezoelectric material (surface area S and thickness t) applied with a stress s as:

$$U = \frac{1}{2}QV = \frac{1}{2}(d \times \sigma \times S) \cdot (g \times \sigma \times t) = \frac{1}{2}d \times g \times \sigma^2 \times Volume \quad (15.20)$$

where d is the current constant and g is voltages constant corresponds to the operating mode's individual coefficients When a piezoelectric material is effectively strained, the modified Equation (15.20) shows that a material or operating mode with a high value will exhibit a high energy capacity.

TABLE 15.3
Performance Comparison of Different Piezoelectric Materials Geometries [34]

Materials	Types	Geometries	Voltage output	Power output	Synthesis	Institute
ZnO(n-type)	Wurtzite	D~100 nm, L200~500 nm	V_P = ~9 mV, R = 500 MΩ	~0.5 pW per cycle (estimated)	CVD, hydrothermal process	Georgia Tech.
ZnO(p-type)	Wurtzite	D~50 nm, L~600 nm	V_P = 50~90 mV, R = 500 MΩ	5~16.2 pW per cycle (calculated)	CVD	Georgia Tech.
ZnOZnS	Wurtzite (Heterostructure)	Not stated	V_P = ~6 mV, R = 500 MΩ	~0.1 pW per cycle (calculated)	Thermal-evaporation and etching	Georgia Tech.
GaN	Wurtzite	D25~70 nm, L10~20 μm	V_{avg} = ~20 mV, V_{max} = ~0.35, R = 500 MΩ	~0.8 pW per cycle (average, calculated)	CVD	Georgia Tech.
CdS	Wurtzite	D~100 nm, L1 μm	V_P = ~3 mV	Not-stated	PVD, Hydro-thermal Process	Georgia Tech.
BaTiO$_3$	Perovskite	D~280 nm, L~15 μm	V_P = ~25 mV, R = 100 MΩ	~0.3 aJ per cycle (stated)	High-temperature chemical reaction	UIU
PVDF	Polymer	D0.5~6.5 μm, L0.1~0.6 mm	V_P = 5~30 mV	2.5 pW~90 pW per cycle (calculated)	Electro-spinning	UC Berkeley
KNbO$_3$	Perovskite	D~100 nm, L~few cm	V_P = ~16 V, R = 100 MΩ	–	Electro-spinning	SUTD/MIT

15.6.1 PIEZO GENERATOR: AN APPROACH TO GENERATE ELECTRICITY FROM VIBRATIONS

When compressive or tensile load is applied mechanically on an element made of poled-piezoelectric-ceramic, a voltage is detected at output. A voltage with the same polarity as the poling voltage is produced by compression parallel to the polarisation direction or tension orthogonal to the polarisation direction (Figure 15.11).

Piezoelectric motors, sound or ultrasonic generators and a range of other objects can all benefit from this idea. Piezoelectric motors, sound or ultrasonic producing gadgets, and a variety of other products have had their motor action modified. Generator activity is used in fuel-igniting gadgets, solid-state batteries and other goods. The crystal has now reached a state of electrical neutrality. All of these sides make an electric dipole, and when dipoles are near together, they form "Weiss domains." Piezo crystals are most commonly used to generate the voltage in electric cigarette lighters. A spring-loaded striker impacts a piezoelectric crystal as the user pushes the lighter's switch, providing a high enough voltage to cause an electric current to pass across a small spark gap, heating and lighting the gas. Some substances, such as quartz, can produce potential differences of thousands of volts by their direct piezoelectric effect.

Piezoelectric Flexible: Because of their capacity to bear huge quantities of strain, materials are appealing for power harvesting applications. When the stresses are higher, more mechanical power is available for conversion into electric power. Using a more effective coupling mode is another method for increasing the quantity of energy received from a piezoelectric. The following are the several types of mechanical vibrations:

Longitudinal Vibrations: This term refers to vibrations that occur in rods or more stretched objects in a direction that is parallel to the wave's propagation direction, i.e., normal to the wave-front. Without referring to the longitudinal effect, the term "longitudinal" is used here. The terms "compressional" and "extensional" are used to describe this type of vibration. Longitudinal vibrations can be produced by both fluids and solids.

Transverse Vibrations: In the same way that "longitudinal vibrations" are used to describe longitudinal waves, "transverse vibrations" are used to describe transverse (distortional) waves. The pulsating particles travel in a direction that is parallel to the

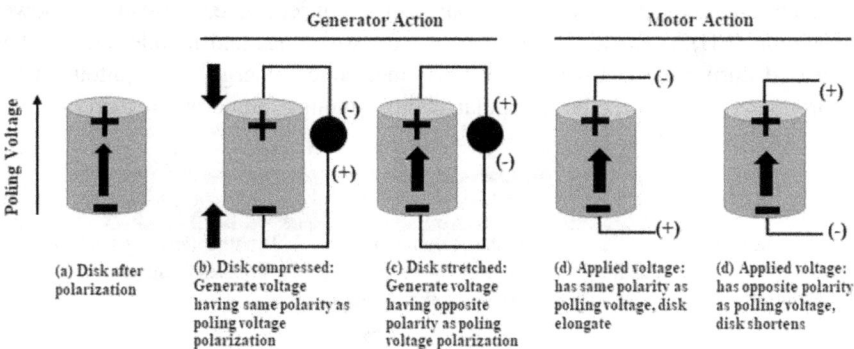

FIGURE 15.11 Working mechanism of simple piezo transducer.

wave front and normal to the propagation direction. Examples include electromagnetic radiation, oscillating strings, membranes and thin slabs. Transverse vibration in piezoelectric crystals can develop when the electric field is directed in such a way that it creates shear stresses along some plane. Wave propagation is more likely to occur in a direction parallel to another dimension if this plane is parallel to one of the parallelepiped's fundamental dimensions. Such vibrations are also known as shear vibrations.

Because flexural vibrations occur in stretched plates or bars, they are usually referred to as "transverse" or "lateral" vibrations. To differentiate these from the transverse oscillations explained above, the word "flexural" is preferable. They have to do with the sample flexing in a specific plane. "Flexural vibrations in the YZ plane" is a good example.

When adjacent cross-sections encounter a corresponding deflection (shear strain) along the figure's axis, which is usually cylindrical or prismatic, torsion vibrations develop. For example, we talk about torsional vibrations "around the X-axis." From what has been discussed, it is evident that referring to the "direction of vibration in a crystal" without also specifying the sort of vibration is ambiguous.

15.7 ENERGY HARVESTING THROUGH PIEZOELECTRIC NANO GENERATOR

Sunlight, electromagnetic waves, environment mechanical power, human physical heating and human physical power are the most frequent considerable energy sources. Human body power harvesters, with exception of solar energy, electromagnetic radiation and environmental vibrational energy, that seem to be strongly dependent on the environment, could be incorporated with daily human activities to energize a range of devices [17].

Mechanical power can be converted into electricity in three different ways: electro-magnetic, electrostatic or triboelectric, and piezoelectric, as shown in Figure 15.12. The application dictates which of the three ways is used, although piezoelectricity is the most extensively researched [22].

Low-power microelectronics have become a reality in recent decades as a result of technological advances, most notably improved circuit and power storage system performance. Electrical power ranging from 10 to 100 mW is currently available for many microelectronics applications. As a consequence, curiosity in power generation (EH), a viable source of power for wearables and mobile devices, has increased during the last decade. Because mechanical energy is ubiquitous in the environment, it has become a popular choice amongst numerous resources (solar

Piezoelectric Energy Harvesting	Electromagnetic Energy Harvesting	Triboelectric Energy Harvesting
Non linearity	Sprung eccentric rotor	Ultrathin flexible single-electrode
Double pendulum system	Frequency up conversion	Core-shell structure mechanism
Frequency up conversion	Spring clockwork mechanism	Air-cushion mechanism
Circuit management	Spring-less system	Liquid metal electrode

FIGURE 15.12   Techniques used for conversion of mechanical power into electric power.

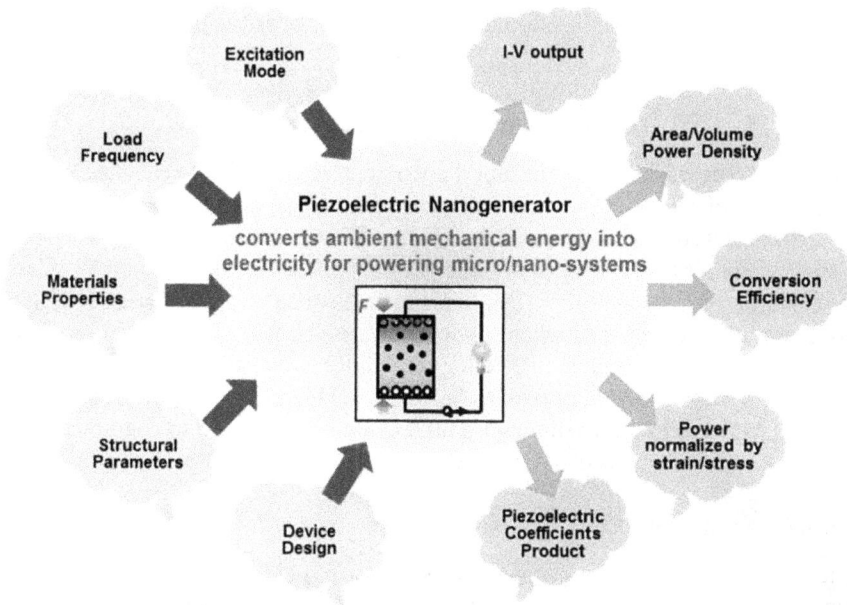

FIGURE 15.13 Schematic of measurement and assessment of performance for a piezoelectric nanogenerator [31].

energy, thermal energy, mechanical energy) (human movement, ocean waves oscillation and mechanical vibrations). As a result, a significant effort has gone into developing sophisticated vibration-based extractors that convert atmospheric vibration to electric energy.

Piezoelectric energy harvesting is the solution if the demand requires high energy concentration for long time duration with low mechanical damping and with the caveat that piezoelectric materials can be brittle or inflexible, as well as poisonous (Figure 15.13). Electromagnetic devices can be used instead of piezoelectric transducers in situations that do not require external sources. They feature a high current output at low impedance with no connections, however, coil losses and low efficiency at low frequencies are an area of concern. Because of the power demand, speed growth gears are used to achieve the desired rotational velocity. It still has reliability and durability issues, and its working mechanism is unknown. Each energy harvesting approach has advantages and disadvantages, and the literature has offered various methods for efficiently collecting energy from human body motion [34].

15.8 APPLICATIONS AND FUTURE SCOPE OF NANOMANUFACTURING IN AN EMERGING TECHNICAL FIELD

The nanogenerator will be employed in a range of cyclic kinetic energy systems, working with large winds and wave energy to tiny muscle movement generated by a heartbeat or lungs inhalation. Among the other possible uses are those listed hereunder [13].

15.8.1   SELF-POWERED NANO/MICRO DEVICES

These are some of the nanogenerator's possible uses is as a self-contained or additional source of energy for nano/micro equipment that uses a little source of power but has a consistent stream of kinetic energy (Figure 15.14). In 2010, Professor Zhong Lin Wang's team, for instance, created a self-powered pH or UV sensing with VING and an output power of 20–40 mV [29]. Since the generated electric power is still relatively low for functioning nano/micro equipment, its use as a backup power supply for the batteries is still restricted. Integrating the nanogenerator with some other energy-collecting systems, including photovoltaic cells or biological photovoltaic devices, has been explored as a breakthrough. This technique is expected to contribute to the creation of a self-contained source of energy [46].

15.8.2   SMART WEARABLE SYSTEMS

One of the nanogenerator's potential uses is a clothing that is incorporated or comprised of fabrics incorporating piezoelectric fabric. Piezoelectric threads transform the kinetic energy of the human body into electric power, which may then be used to energize compact electronic equipment such as health-monitoring systems linked to Intelligent Wearable Technologies [47]. The nanogenerator, for instance, may be simply put into a footwear that simulates human movement patterns. A power-generating prosthetics would be another use. The potential was

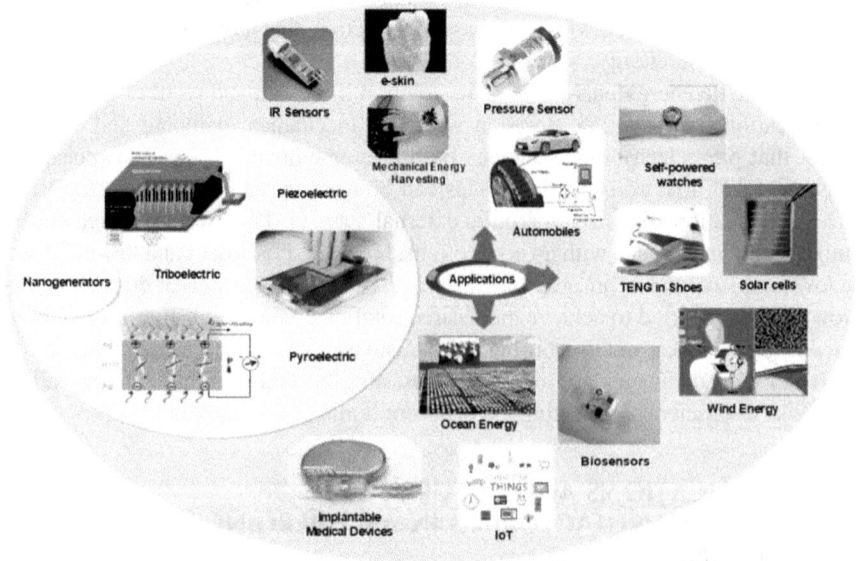

FIGURE 15.14 An overview of the main application sectors for nanogenerators as micro/nano-energy sources [21].

proven by Doctor Zhong Lin Wang's group employing a foldable SWG coupled to a running hamster to create a Voltage output of up to 100 mV [48].

15.8.3 TRANSPARENT AND FLEXIBLE DEVICES

To create piezoelectric nanostructures, a range of substrates, particularly flexible and transparent organic substances, can be employed. The SKKU laboratory of Doctor Sang-Woo Kim and the SAIT team of Doctor Jae-Young Choi successfully created a transparent and extensible nanogenerator, which can be used as a self-powered tactile sensing device, with the objective of adapting the innovation to an energy-efficient touchscreen monitor. They intend to improve the product's transparency as well as price through substituting the Indium-Tin-Oxide (ITO) electrodes with a graphene sheet [31].

15.8.4 TELEMETRIC POWER TRANSCEIVER IMPLANT

The nanogenerator depending on ZnO nanoparticles can be employed in implantable medical devices since ZnO is biocompatible and it can be manufactured on an organic substance. The implanted device with the nanogenerator receives exogenous ultrasonic vibrations, that are transformed to electricity generated by the piezoelectric nano-structure outside the human body [34].

Despite the fact that energy-harvesting technology has been extensively researched for years, there is still a big discrepancy among realized and expected performance. The majority of energy harvesters are designed for general use and examined using simple harmonic vibrational modes. This method delivers technology that is far from ready for deployment in possible applications. Despite the fact that various application-oriented harvesters have been tested in situ, their dependability, durability and interoperability have still not been thoroughly investigated. Further investigation is planned to be conducted to address these issues. Similarly, further system-level studies are planned, in which energy harvesters are combined with electronic power circuits, energy storage components, sensor and electronic control. Such study will assist in translating decades of energy harvesting investigation into concrete improvements in our daily lives.

Overall, we have seen considerable advancements in energy-harvesting technologies during the previous decade. It is getting closer to self-powered autonomous functioning of wearable electronics, healthcare gadgets, automobile sensors and wireless sensor surveillance systems.

REFERENCES

1. K. E. Drexler, "Molecular engineering: An approach to the development of general capabilities for molecular manipulation," *Proc. Natl. Acad. Sci. U.S.A.*, vol. 78, no. 9 I, pp. 5275–5278, 1981, doi: 10.1073/pnas.78.9.5275
2. N. Taniguchi, "Proceedings of the international conference on production engineering, Tokyo," *Soc. Precis. Eng. Part II, Japan*, vol. 2, 1974, [Online]. Available at: https://www.worldcat.org/title/proceedings-of-the-international-conference-on-production-engineering-tokyo-1974/oclc/4192092

3. V. A. Gargade, "Nanotechnology in ancient India," *Hist. Nanotechnol.*, no. March, pp. 37–55, 2019, doi: 10.1002/9781119460534.ch3

4. M. F. Hochella, M. G. Spencer, and K. L. Jones, "Nanotechnology: Nature's gift or scientists' brainchild?," *Environ. Sci. Nano*, vol. 2, no. 2, pp. 114–119, 2015, doi: 10.1039/c4en00145a

5. M. F. Hochella *et al.*, "Nanominerals, mineral nanoparticles, and earth systems," *Science (80)*, vol. 319, no. 5870, pp. 1631–1635, 2008, doi: 10.1126/science.1141134

6. A. E. Seago and V. Saranathan, "Nature's nanostructures," *Nat. Publ. Gr.*, vol. Ch.12, no. March, pp. 313–327, 2012, [Online]. Available: 10.1038/scientificamerican0512-74

7. S. Bayda, M. Adeel, T. Tuccinardi, M. Cordani, and F. Rizzolio, "The history of nanoscience and nanotechnology: From chemical-physical applications to nanomedicine," *Molecules*, vol. 25, no. 1, 2020, doi: 10.3390/molecules25010112

8. Z. L. Wang and W. Wu, "Nanotechnology-enabled energy harvesting for self-powered micro-/nanosystems," *Angew. Chemie - Int. Ed.*, vol. 51, no. 47, pp. 11700–11721, 2012, doi: 10.1002/anie.201201656

9. J. G. Monroe, M. Bhandari, J. Fairley, O. J. Myers, N. Shamsaei, and S. M. Thompson, "Energy harvesting via thermo-piezoelectric transduction within a heated capillary," *Appl. Phys. Lett.*, vol. 111, no. 4, 2017, doi: 10.1063/1.4996235

10. M. J. Jackson, *Micro and nanomanufacturing*. Springer Science Business Media, LLC, 2007.

11. A. P. Malshe *et al.*, "Tip-based nanomanufacturing by electrical, chemical, mechanical and thermal processes," *CIRP Ann. –Manuf. Technol.*, vol. 59, no. 2, pp. 628–651, 2010, doi: 10.1016/j.cirp.2010.05.006

12. J. Chae, S. S. Park, and T. Freiheit, "Investigation of micro-cutting operations," *Int. J. Mach. Tools Manuf.*, vol. 46, no. 3–4, pp. 313–332, 2006, doi: 10.1016/j.ijmachtools. 2005.05.015

13. A. K. Harpreet Sharda, "Applications and future scope of nanomanufacturing in an emerging technical field," *Int. J. Sci. Res. Dev.*, vol. 07, no. 06, pp. 75–78, 2019, doi: 10.6084/m9.figshare.9484163

14. M. Mahdavinejad, L. H. Rafsanjani, and M. Karimi, "Mechanism of manufacturing and adoption of nano materials in contemporary architectural project of developing countries," *Adv. Mater. Res.*, vol. 748, no. August, pp. 1150–1154, 2013, doi: 10.4028/ www.scientific.net/AMR.748.1150

15. V. P. Singh, S. Jain, and J. Gupta, "Performance assessment of double-pass parallel flow solar air heater with perforated multi-V ribs roughness—Part B," *Exp. Heat Transf.*, pp. 1–18, 2022, doi: 10.1080/08916152.2021.2019147

16. V. P. Singh, S. Jain, and A. Kumar, "Establishment of correlations for the Thermo-Hydraulic parameters due to perforation in a multi-V rib roughened single pass solar air heater," *Exp. Heat Transf.*, no. April, pp. 1–20, 2022, doi: 10.1080/08916152.2022. 2064940.

17. V. P. Singh, M. Saini, A. Sharma, S. Jain, and G. Dwivedi, "Solar thermal receivers— A review," *Lect. Notes Mech. Eng.*, vol. II, p. 1_25, 2022, doi: 10.1007/978-981-16-8341.

18. V. P. Singh, S. Jain, and J. M. L. Gupta, "Analysis of the effect of variation in open area ratio in perforated multi-V rib roughened single pass solar air heater – Part A," *Energy Sources, Part A Recover. Util. Environ. Eff.*, pp. 1–20, 2022, doi: 10.1080/ 15567036.2022.2029976

19. V. P. Singh, S. Jain, and J. Gupta, "Analysis of the effect of perforation in multi-v rib artificial roughened single pass solar air heater: Part A," *Exp. Heat Transf.*, no. Oct., pp. 1–20, Oct. 2021, doi: 10.1080/08916152.2021.1988761

20. Y. Zi and Z. L. Wang, "Nanogenerators: An emerging technology towards nanoenergy," *APL Mater.*, vol. 5, no. 7, pp. 1–13, 2017, doi: 10.1063/1.4977208

21. S. S. Indira, C. A. Vaithilingam, K. S. P. Oruganti, F. Mohd, and S. Rahman, *Nanogenerators as a Sustainable Power Source: State of Art, Applications, and Challenges*, vol. 9, no. 5. 2019.

22. Q. Xu, J. Wen, and Y. Qin, "Development and outlook of high output piezoelectric nanogenerators," *Nano materials*, vol. 86, pp. 1–15, April, 2021, doi: 10.1016/j.nanoen.2021.106080

23. Z. L. Wang, "On Maxwell's displacement current for energy and sensors: The origin of nanogenerators," *Mater. Today*, vol. 20, no. 2, pp. 74–82, 2017, doi: 10.1016/j.mattod.2016.12.001

24. Z. L. Wang, T. Jiang, and L. Xu, "Toward the blue energy dream by triboelectric nanogenerator networks," *Nano Energy*, vol. 39, no. May, pp. 9–23, 2017, doi: 10.1016/j.nanoen.2017.06.035

25. Z. L. Wang, "On the first principle theory of nanogenerators from Maxwell's equations," *Nano Energy*, vol. 68, no. November 2019, p. 104272, 2020, doi: 10.1016/j.nanoen.2019.104272

26. K. Y. Lee, B. Kumar, J. Seo, K. Kim, J. I. Sohn, and S. N. Cha, "P-type polymer – Hybridized high-performance piezoelectric nanogenerators."*Nano Lett.*, vol. 12, no. 4, pp. 1959–1964, 2012, 10.1021/nl204440g.

27. A. M. M. Roji, G. Jiji, and A. B. T. Raj, "A retrospect on the role of piezoelectric nanogenerators in the development of the green world," *RSC Adv.*, vol. 7, no. 53, pp. 33642–33670, 2017, doi: 10.1039/c7ra05256a

28. X. Wang, J. Song, J. Liu, and L. W. Zhong, "Direct-current nanogenerator driven by ultrasonic waves," *Science (80)*, vol. 316, no. 5821, pp. 102–105, 2007, doi: 10.1126/science.1139366

29. Y. Zhang, M. K. Ram, E. K. Stefanakos, and D. Y. Goswami, "Synthesis, characterization, and applications of ZnO nanowires," *J. Nanomater.*, vol. 2012, pp. 1–22, 2012, doi: 10.1155/2012/624520

30. Y. Qin, X. Wang, and Z. L. Wang, "Microfibre-nanowire hybrid structure for energy scavenging," *Nature*, vol. 451, no. 7180, pp. 809–813, 2008, doi: 10.1038/nature06601

31. W. Wu, "High-performance piezoelectric nanogenerators for self-powered nanosystems: Quantitative standards and figures of merit," *Nanotechnology*, vol. 27, no. 11, pp. 1–6, 2016, doi: 10.1088/0957-4484/27/11/112503

32. Y. Huan et al., "High-performance piezoelectric composite nanogenerator based on Ag/(K,Na)NbO3 heterostructure," *Nano Energy*, vol. 50, no. March, pp. 62–69, 2018, doi: 10.1016/j.nanoen.2018.05.012

33. Y. Zhao et al., "High output piezoelectric nanocomposite generators composed of oriented BaTiO3 NPs at PVDF," *Nano Energy*, vol. 11, pp. 719–727, 2015, doi: 10.1016/j.nanoen.2014.11.061

34. Z. Yang, S. Zhou, J. Zu, and D. Inman, "High-performance piezoelectric energy harvesters and their applications," *Joule*, vol. 2, no. 4, pp. 642–697, 2018, doi: 10.1016/j.joule.2018.03.011

35. D. Hu, M. Yao, Y. Fan, C. Ma, M. Fan, and M. Liu, "Strategies to achieve high performance piezoelectric nanogenerators," *Nano Energy*, vol. 55, no. September 2018, pp. 288–304, 2019, doi: 10.1016/j.nanoen.2018.10.053

36. H. Ghayour, H. R. Rezaie, S. Mirdamadi, and A. A. Nourbakhsh, "The effect of seed layer thickness on alignment and morphology of ZnO nanorods," *Vacuum*, vol. 86, no. 1, pp. 101–105, Jul. 2011, doi: 10.1016/J.VACUUM.2011.04.025

37. A. Karn, N. Kumar, and S. Aravindan, "Chemical vapor deposition synthesis of novel indium oxide nanostructures in strongly reducing growth ambient," *J. Nanostructures*, vol. 7, no. 1, pp. 64–76, 2017, doi: 10.22052/jns.2017.01.008

38. T. Morita, "Piezoelectric materials synthesized by the hydrothermal method and their applications," *Materials (Basel).*, vol. 3, no. 12, pp. 5236–5245, 2010, doi: 10.3390/ma3125236

39. M. Acosta *et al.*, "BaTiO3-based piezoelectrics: Fundamentals, current status, and perspectives," *Appl. Phys. Rev.*, vol. 4, no. 4, pp. 1–53, 2017, doi: 10.1063/1.4990046

40. N. Sezer and M. Koç, "A comprehensive review on the state-of-the-art of piezoelectric energy harvesting," *Nano Energy*, vol. 80, no. August 2020, p. 105567, 2021, doi: 10.1016/j.nanoen.2020.105567

41. S. H. Shin, Y. H. Kim, M. H. Lee, J. Y. Jung, J. H. Seol, and J. Nah, "Lithium-doped zinc oxide nanowires-polymer composite for high performance flexible piezoelectric nanogenerator," *ACS Nano*, vol. 8, no. 10, pp. 10844–10850, 2014, doi: 10.1021/nn5046568

42. A. Karn, M. Kumar, V. N. Singh, B. R. Mehta, S. Aravindan, and J. P. Singh, "Growth of indium oxide and zinc-doped indium oxide nanostructures," *Chem. Vap. Depos.*, vol. 18, no. 10–12, pp. 295–301, 2012, doi: 10.1002/cvde.201207001

43. S. K. Si *et al.*, "A strategy to develop an efficient piezoelectric nanogenerator through ZTO assisted γ-phase nucleation of PVDF in ZTO/PVDF nanocomposite for harvesting bio-mechanical energy and energy storage application," *Mater. Chem. Phys.*, vol. 213, pp. 525–537, 2018, doi: 10.1016/j.matchemphys.2018.04.013

44. C. Yoon, B. Jeon, and G. Yoon, "Enhanced output performance of sandwich-type ZnO piezoelectric nanogenerator with adhesive carbon tape," *Sensors Actuators, A Phys.*, vol. 318, p. 112499, 2021, doi: 10.1016/j.sna.2020.112499

45. M. G. Kang, W. S. Jung, C. Y. Kang, and S. J. Yoon, "Recent progress on PZT based piezoelectric energy harvesting technologies," *Actuators*, vol. 5, no. 1, pp. 1–17, 2016, doi:-10.3390/act5010005

46. S. Das Mahapatra *et al.*, "Piezoelectric materials for energy harvesting and sensing applications: roadmap for future smart materials," *Adv. Sci.*, vol. 8, no. 17, pp. 1–73, 2021, doi: 10.1002/advs.202100864

47. B. P. Bharathan, R. Rajagopal, A. Alfarhan, M. V. Arasu, and N. A. Al-Dhabi, "Advancement in nanomaterial synthesis and its biomedical applications," *Emerg. Nanomater. Adv. Technol.*, no. February, pp. 419–462, 2022, doi: 10.1007/978-3-030-80371-1_14.

48. T. Zhang, T. Yang, M. Zhang, C. R. Bowen, and Y. Yang, "Recent progress in hybridized nanogenerators for energy scavenging," *iScience*, vol. 23, no. 11, p. 101689, 2020, doi: 10.1016/j.isci.2020.101689

Index

2D materials, 2, 189
3D bioprinting, 75
3D nanosystems, 4
3D printer, 74
3D printing, 18, 73, 79, 86, 232
3D systems, 7
3rd-generation products, 4
4D printing, 88

additive manufacturing, 18
alginate, 210
atomic force microscopy (AFM), 26
atomic–molecular levels, 1
ANOVA, 51, 52

biocompatible, 208
biodegradable, 109, 205
biomass, 169
biomedical, 25
biomedicines, 95, 115
biomaterial, 83, 92

carbon nanofillers, 62
chemical vapour deposition (CVD), 18
chemical etching, 203
coating, 181
computer-aided design (CAD), 75
computed axial lithography (CAL), 85, 91
cytocompatibility test, 210

dopants, 9, 67, 68
doping, 67

ECDM, 137
economy, 17
energy levels, 32
energy conversion, 165
energy beam machining, 39
extraordinary precision, 1

FEM, 155
film deposition techniques, 64
fluidics, 4
FLSI, 22
F-statistical test, 57
fused deposition modelling (FDM), 74

good manufacturing practice (GMP), 98
global environment, 59

high precision, 22
hydrogels, 81, 213
hydroxyapatite, 220

inkjet 3D printing, 76, 182
integrated tissue organ printer (ITOP), 85

layer-by-layer, 75
lifecycle, 18, 95
lithium ion batteries, 171
limit of detection (LOD), 67
lower explosion limit (LEL), 60

macroscopic systems, 25
MEMS technology, 23, 139
microfluidization, 108
micro-hardness, 53, 55
molecular electronics, 4
MOGA, 46

nano-actuation, 21
nanodevices, 21
nanotechnology, 1, 17
nanofabrication, 137, 175
National Nanomanufacturing Network
      (NNN), 100
Nanotechnology Characterization Laboratory
      (NCL), 98
nanomanufacturing, 2, 18, 97, 189, 243
nanoscales, 2, 17, 40, 97, 228
nanostructures, 3, 184
nano cutting, 14
nanoproduct, 19, 27, 95
nanoelectronics, 20
nanoimprint lithography, 35, 38
noble metals, 62, 167
nanomedicine, 102
nanonization processes, 105
nano-emulsion processes, 107
nanocarriers, 110
nanorobots, 227
nucleic acid robot (NUBOT), 232
nanogenerator, 241, 246

optimization techniques, 46
optic-electronic components, 1
optoelectronics, 4
oncological, 13
oxide semiconductor gas sensors, 62

parallel processing, 19
photovoltaic, 175
physical vapour deposition (PVD), 18
piezoelectric, 258
prerequisite parameters, 80

quantum dots (QDs), 33
quantum rods (QRs), 34

regenerative biomedicine, 79
RSM, 46

saw devices, 154, 158
selectivity, 66
SEM, 47
semiconductors, 1, 17
sensor, 60, 149
sensor response, 65, 66
single-layered graphene, 4

single-wire generator (SWG), 252
spindle speed, 50
standardization, 22
stereolithography (SLA), 74
storage devices, 165
supercritical fluids (SCF), 106
surface modification, 68
surface roughness, 46

tissue, 205, 213
TONIC (The Translation of Nanotechnology in Cancer), 99
top-down method, 6, 35, 96, 195
transition metal dichalcogenides (TMD), 63

ultra-precision, 23
UV wavelength, 40

volatile organic compounds (VOCs), 60

For Product Safety Concerns and Information please contact our EU
representative GPSR@taylorandfrancis.com
Taylor & Francis Verlag GmbH, Kaufingerstraße 24, 80331 München, Germany

www.ingramcontent.com/pod-product-compliance
Lightning Source LLC
Chambersburg PA
CBHW060343220326
41598CB00023B/2788

9 7 8 1 0 3 2 1 1 5 8 6 3